THE OTHER SABER-TOOTHS

THE OTHER SABER-TOOTHS

SCIMITAR-TOOTH CATS OF THE WESTERN HEMISPHERE

Edited by
VIRGINIA L. NAPLES
LARRY D. MARTIN
JOHN P. BABIARZ
Graphics Editor, H. TODD WHEELER

THE JOHNS HOPKINS UNIVERSITY PRESS
Baltimore

The Johns Hopkins University Press
2715 North Charles Street
Baltimore, Maryland 21218-4363
www.press.jhu.edu

Library of Congress Cataloging-in-Publication Data

The other saber-tooths : scimitar-tooth cats of the Western
Hemisphere / edited by Virginia L. Naples, Larry D. Martin,
and John P. Babiarz ; graphics editor, H. Todd Wheeler.
        p. cm.
   Includes bibliographical references and index.
   ISBN-13: 978-0-8018-9664-4 (hardcover : alk. paper)
   ISBN-10: 0-8018-9664-9 (hardcover : alk. paper)
   1. Scimitar cat—Western Hemisphere.   2. Animals, Fossil—
Western Hemisphere.   3. Paleontology—Pleistocene.   I. Naples,
Virginia L.   II. Martin, Larry D.   III. Babiarz, John P., 1947–
   QE882.C15O84   2010
   569'.75—dc22   2010023461

A catalog record for this book is available from the British Library.

Special discounts are available for bulk purchases of this book.
For more information, please contact Special Sales at 410-516-6936
or specialsales@press.jhu.edu.

The Johns Hopkins University Press uses environmentally friendly
book materials, including recycled text paper that is composed of
at least 30 percent post-consumer waste, whenever possible.

To our good friend and colleague
George Chuck Sim Lee Jr.
July 23, 1945–November 16, 2004

And to all the professional and amateur
paleontologists, paleopathologists,
paleoengineers, paleoartists, hobbyists,
and rock hounds of the world

CONTENTS

List of Contributors   ix
Preface   xi
Acknowledgments   xv

1 Introduction   3
LARRY D. MARTIN, JOHN P. BABIARZ, VIRGINIA L. NAPLES

2 Experimental Paleontology of the Scimitar-tooth and
Dirk-tooth Killing Bites   19
H. TODD WHEELER

3 Pathology in Saber-tooth Cats   35
BRUCE M. ROTHSCHILD, LARRY D. MARTIN

4 The Osteology of a Cookie-cutter Cat, *Xenosmilus hodsonae*   43
LARRY D. MARTIN, JOHN P. BABIARZ, VIRGINIA L. NAPLES

5 The Musculature of *Xenosmilus*, and the Reconstruction of
Its Appearance   99
VIRGINIA L. NAPLES

6 Osteology and Myology of *Homotherium ischyrus* from Idaho   123
JONENA M. HEARST, LARRY D. MARTIN, JOHN P. BABIARZ,
VIRGINIA L. NAPLES

7 Revision of the New World Homotheriini   185
LARRY D. MARTIN, VIRGINIA L. NAPLES, JOHN P. BABIARZ

8 A Saber-tooth Cat Skull from Tajikistan, Central Asia,
   and the Relationships between Eurasian and North American
   Homotheres   195
   PETER E. KONDRASHOV, LARRY D. MARTIN

9 A Framework for the North American Homotheriini   201
   LARRY D. MARTIN, JOHN P. BABIARZ, VIRGINIA L. NAPLES

Appendix A   211
Appendix B   212
Glossary   219
Literature Cited   225
Index   231

CONTRIBUTORS

John P. Babiarz
Babiarz Institute of Paleontological Studies

Jonena M. Hearst
Guadalupe Mountains National Park

Peter E. Kondrashov
Kirksville College of Osteopathic Medicine

Larry D. Martin
University of Kansas Museum of Natural History

Virginia L. Naples
Northern Illinois University

Bruce M. Rothschild
Northeastern Ohio Universities College of Medicine
University of Kansas Museum of Natural History

H. Todd Wheeler
George C. Page Museum
John Day Fossil Beds

Xenosmilus hodsonae
"Cookie - Cutter" Cat

Three views of *Xenosmilus hodsonae*: skull and mandible *(top)*;
reconstruction of the animal's head in lateral view *(middle)*;
reconstruction of the entire animal *(bottom)*. (drawing Mary
Tanner)

PREFACE

When someone says "saber-tooth cat," the first image that usually comes to mind is the fierce-looking, well-known genus *Smilodon*, from the tar seeps of Rancho La Brea. It is the purpose of this book to broaden our understanding beyond that genus to include a rare group of sabercats, which are not only contemporary with *Smilodon* but had a unique appearance and behavior. Until the past couple of decades, our knowledge of scimitar-tooth cats, the other saber-tooths, was very limited. Indeed, the genus *Xenosmilus* was only recently discovered. Virtually complete skeletons of these animals, detailed in this book, demonstrate how interesting and spectacular the scimitars were. We are able to reveal as much about their anatomy, appearance, and behavior as is known about *Smilodon*, including the history of their diseases.

Our book was inspired by the 1932 classic, *The Felidae of Rancho La Brea,* by Merriam and Stock. We provide comparable illustrations of skeletal anatomy by artist Mary Tanner. Life-like restorations were done by paleoartist Mark Hallett. The text was mostly written by the three principle authors, Larry D. Martin, Virginia L. Naples, and John P. Babiarz. Four other authors contribute to the chapters, bringing additional breadth and completeness to the subject. Jonena M. Hearst discovered a remarkable skeleton of *Homotherium* in the Pliocene of Idaho and included the first description of that specimen in her Ph.D. dissertation. H. Todd Wheeler, a mechanical engineer, was recruited by Larry Martin and Virginia Naples in 1996 to build a mechanical device to test the various hypothetical bite models of the saber-tooth *Smilodon*. Four years later, this device, called "Robocat," was filmed by the BBC and featured in the Discovery Channel series. Bruce M. Rothschild, a practicing rheumatologist and a leading figure in the study of paleopathology, summarizes the diseases and resultant pathologies found in fossil sabercats. Peter E. Kondrashov, a paleontologist formerly attached to the Russian Academy of Sciences in Moscow and an expert on the Cenozoic fauna of Central Asia, describes a *Homotherium* skull discovered along the route taken by these cats when they settled in North America.

Smilodon fatalis defending its kill, a *Bison* calf, from a trio of Dire wolves. (drawing Mark Hallett)

The initial chapter contains an introduction to sabercats, with discussion of the three principal morphotypes: dirk-tooth, scimitar-tooth, and cookie-cutter. We offer a model to explain how these cats may have partitioned their environment, including differences in their modes of prey capture and killing behavior, and how their origins and evolutionary trends are related to the broad spectrum of climate change during the past seven million years. Included are sidebars on dating the fossil record and a brief history of taxonomy.

The second chapter looks at the saber-tooth bite from the viewpoint of simple mechanics. We describe the workings of a mechanical bite simulator that was tested on animal cadavers. These experiments illustrate the limitations of the dirk-tooth and scimitar-tooth ecomorphs in prey killing, and why the two morphotypes may have evolved.

The third chapter is a survey of the evidence for pathology on the skeletal remains of scimitar-tooth cats and a compari-

son with *Smilodon*. Once again, we have insight into the behavior of long-extinct predators, including the discovery of evidence for tuberculosis in the scimitar-tooth *Homotherium serum*. This disease was probably universal in the North American bison population, which suggests that bison were a prey item for that species.

Chapter four provides the first detailed description of the skeletal anatomy of the cookie-cutter cat *Xenosmilus hodsonae*. The short-legged, bear-like proportions of *Xenosmilus* are compared to the long, slender, almost cheetah-like legs of a Pliocene *Homotherium*, showing how differently these related forms dealt with problems of capturing large prey. This chapter, along with chapter six, establishes a basis for comparisons with other predators.

Chapter five explains the relationship of the musculature to the skeleton and how this can be used to restore the appearance of extinct animals and gain insight into their behavior. The anatomical structure of *Xenosmilus* is described and related to the biomechanical requirements for the lifestyle of a robustly built, short-limbed ambush predator; it is then compared with the anatomy of *Smilodon*.

Chapter six describes in detail one of the most complete skeletons of *Homotherium ischyrus*. The proportions of the North American Pliocene scimitar-tooth cat are compared to those of the stout, short-legged cat, *Xenosmilus hodsonae*. This chapter also includes a sidebar on the discovery and site stratigraphy of *Homotherium ischyrus* from Idaho (Box 6.1).

Chapter seven is a summary of the taxonomy of *Homotherium* in North America. It concludes that the North American Pliocene sample consists of two long-legged species, *H. ischyrus* and *H. crusafonti*. Included is the skull description of *H. crusafonti* from Delmont, South Dakota. These species can be separated from each other on size and cranial characters, with *H. crusafonti* being smaller, and more like the later species *H. serum*.

The eighth chapter describes a homothere skull from the Pliocene of Central Asia. This sabercat samples the Beringian *Homotherium* taxa found along the high latitude pathway connecting the exchange between Eurasia and North America. When compared with the contemporary North American species, it seems evolutionarily precocious, suggesting a northern center for advancement in the *Homotherium* lineage.

Chapter nine pulls together the common threads of homothere evolution into an understandable pattern of climatic change and distributional events. The record of scimitar-tooth cats in South America is also discussed.

The arrangement of these chapters progresses from a general discussion of scimitar-tooth sabercats in the Plio-Pleistocene, including how we determine their appearance and behavior, to a detailed description of the anatomy of homotheres. Their species diversity and biogeographic distribution is related to their way of life and the overall imprint of climatic change. Their success and ultimate extinction must, in some way, relate to these factors.

As in all basic science, truth is elusive. Some of what we have to say is speculation, supported by bits of evidence from diverse and often unrelated sources. As such, the study of scimitar-tooth cats is incomplete. With that caveat, we now submit our conclusions for your review. We believe this book contributes to the understanding of the repeated cycle of evolution and subsequent extinction of saber-tooth cats. We also hope that it will inspire more investigations into these fascinating animals.

ACKNOWLEDGMENTS

We would like to thank the following for assistance and perseverance: Mary Tanner, who prepared the shaded illustrations and graphics for *Xenosmilus hodsonae, Homotherium ischyrus, H. crusafonti,* and *H. crenatidens*; Mark Hallett, who provided art for the book jacket and additional artwork; Jerry Stark, who took many of the photographs; Don Hurlbutt, who provided the digital image of the *Xenosmilus hodsonae* cranial muscle reconstruction; H. Todd Wheeler, who did computer graphics and image editing; Paula M. Ott, who did the postcranial reconstruction of *X. hodsonae*; Bill Akersten, who generated the first testable model of saber-tooth biomechanics; David Kronen, who did the molding and casting; Mike Karash, who engineered the *X. hodsonae* skeletal mount; the late Phillip V. Wells, who generously provided the North American Late Pleistocene vegetation data, and Mary Brooks for the colored map; Brad Archer, for critical reading of the manuscript and for preparing tables, images, and the glossary; Tia Eaton, for technical review, word processing, and formatting of all chapters and tables.

For useful comments, discussions, photography, images, access to collections, and casting we would also like to thank: Mauricio Antón; Kenny Bader; Remie Bakker; R. Ballesio; Ross Barnett; Japheth Boyce; David Burnham; Matt Christopher; G. Corner; Shelley Cox; Douglas Due; Elizabeth Ebert; Burkhart Engasser; A. Falk; Aisling Farrell; M. Frank; Ted J. Fremd; Henry Galiano; John M. Harris; Kees van Hooijdonk; Gordon Hubbell; R. C. Hulbert Jr.; Cris Jass; George T. Jefferson; Cliff Jeremiah; the late Bjorn Kurtén; W. Langston; Wann Lanston Jr.; Wilre van Logchem; E. Lundelius; Bruce MacFadden; R. McCarty; M. McKenna; Guy Marwick; T. J. Meehan; Dick Mol; Lyndon Murray; Francisco Prevosti; Asconio Rincón; Holliston L. Riviere; Timothy Rowe; the late C. B. Schultz; Kevin Seymour; Chris Shaw; Ron Stebler; George Stevenson; Craig Sundell; Kent Sundell; R. Tedford; Blaire Van Valkenburgh; M. Voohies; D. Webb; Lars Werdelin; P. Whistler; and Mike Wigger, who assisted the study in a variety of ways.

We appreciate and thank the following institutions for allowing us access to, or loans from, their collections: American Museum of Natural History; Bone Clones, Inc.; Florida Paleontological Society; Idaho Museum of Natural History; John Day Fossil Beds National Monument; Los Angeles County, George C. Page Museum; Mark Hallett Paleoart; Museum of Natural History, Lyon; Natural History Museum, London; Naturhistorisches Museum, Basel; Russian Academy of Sciences, Moscow; School of Dentistry, Oregon Health and Sciences University; Silver River Springs Museum; Texas Memorial Museum; United States National Museum, Smithsonian Institution; University of Florida Museum of Natural History; University of Kansas Museum of Natural History, Vertebrae Paleo Lab; University of Nebraska State Museum; and a special mention to BIOPSI for financial support.

We especially thank Deb L. Hodson, Jean Martin, and Vicki L. Wheeler for their forbearance throughout this endeavor.

THE OTHER SABER-TOOTHS

Homotherium serum, Friesenhahn Cave. (drawing Mark Hallett)

LARRY D. MARTIN

JOHN P. BABIARZ

VIRGINIA L. NAPLES

Introduction

MUCH CAN BE learned about cats, not only from the size of their bodies but from the size and shape of their canine teeth (fig. 1.1). The canines of all felid-like carnivores fall into two categories: conical-tooth and saber-tooth. The conical tooth's rounded cross section punctures and grips the flesh of the prey. The saber tooth's flattened cross section forms an edge that can slice as well as puncture. The conical-tooth cats rely (as do most carnivores) on the crushing and slicing ability of the carnassial teeth in conjunction with the canines to kill their game, whereas the saber-tooth cats use their canines as the primary weapon.

As there are no living saber-tooth predators, it is difficult to speculate on their prey killing behavior. In order to sort through the various hypotheses, careful observation of living predators is imperative. All predators must be able to capture and kill consistently the species that constitute their diet. Risk of injury must remain low, because they hunt many times over their lifespan. Game that is killed may be slower or more careless animals, thus driving the evolution of faster, more attentive prey. Prey species primarily respond to predation by reproducing at a rate commensurate with their losses. Animals that are larger or more difficult to capture and kill reproduce at a slower rate than smaller prey species. The larger species can become vulnerable to attack, or even extinction, when exposed to a new, more effective predator. This is the central thesis of the overkill model for Pleistocene megafaunal extinction (P. S. Martin, 1967). Early humans may have refined hunting techniques to a point where they could kill large prey more successfully than other predators. If this kind of change happened rapidly, the reproductive rates of large animals may have been unable to adjust quickly enough to compensate for higher losses.

PREDATION

Body size is a major controlling factor in many aspects of vertebrate evolution. Herbivores gain advantages from scaling as they increase in size. One advantage is increased stride length, which permits them to cover larger areas in search of food and to run faster from predators. Also, as herbivores become larger, they can process less nutritious but more abundant foods than can their smaller relatives (Janis, 1976). Larger herbivores are stronger and therefore more difficult to kill. Thus predation on members of larger-bodied species is limited to larger or more specialized preda-

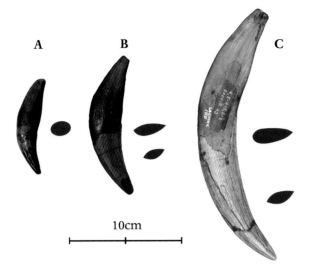

Figure 1.1. Labial and cross-sectional views of upper canines: A, conical-tooth cat *Panthera spelaea* (Alaska) BIOPSI 0201; B, scimitar-tooth cat *Homotherium serum* (California) LACMHC 70001; C, dirk-tooth cat *Smilodon fatalis* (California) LACMHC 631. (*Smilodon* photo Aisling Farrell, courtesy The George C. Page Museum)

tors. A narrowed range of predators helps to compensate for the longer periods required for gestation and maturation of offspring of larger-bodied species and the resultant lower reproductive potential (Hudson, 1985). Any increases in this range of predators would certainly have had a negative impact on reproduction.

All lineages of saber-tooth carnivores began with animals not much larger than a bobcat. As an example, consider the primitive early Oligocene nimravids, *Dinictis felina* at 20 kg and *Hoplophoneus primaevus* at 18 kg, compared to mid to late Oligocene species such as *Dinictis eileenae* at 40 kg and *Hoplophoneus occidentalis* at 74 kg (Van Valkenburgh and Ruff, 1987; Meehan, 1998). The earliest known Smilodontins, such as *Megantereon,* were roughly the size of a mountain lion at 50 kg (Christiansen and Aldolfssen, 2007), while the latest examples of *Smilodon fatalis* may have exceeded the size of a lion at 200 kg. This means that, as the body size of the predators increased, the body size of their prey species also shifted. However, no saber-tooth cat became bigger than the largest extinct conical-tooth felids (Van Valkenburgh and Ruff, 1987). Modern lions and tigers are comfortable bringing down prey twice their body size and are capable of bringing down and dispatching prey over four times their size on occasion (Schaller, 1972; Seidensticker and McDougal, 1993; Biknevicius and Van Valkenburgh, 1996). Unless the saber-tooth adaptation presents an advantage beyond that available to ordinary lions and tigers, it is hard to explain why it should have evolved at all. For this reason, most researchers studying saber-tooth carnivores have suggested that they were specialized predators on prey of such large body size that they could not ordinarily be taken by lions or tigers. Their

entire cranial and postcranial osteology was carried to mechanical extremes rarely, if ever, seen in other predators. Such extraordinary specialization implies unusual rewards and opportunities beyond those available to more ordinary carnivores.

Saber-tooth cats include some of the largest mammalian carnivores known. Their thin, elongated, flattened, and serrated upper canine teeth represent the extreme adaptation of teeth for cutting, and are only suitable for a meat diet that provides little or no contact with bones. The most ubiquitous, super-large prey species, the proboscideans, are thought by many as the preferred prey of saber-tooth carnivores. But there can be no one-to-one correlation. Saber-tooth carnivores were at the apex of diversity and population density in the Oligocene (33.6–24 M.Y.A.) and early Miocene ages of North America and Eurasia (24–5 M.Y.A.) when no Proboscidea were available in those regions as prey species (Savage and Russell, 1983; Emry et al., 1987). The largest prey species available to these carnivores were rhinoceroses and entelodonts (large, extinct pig-like artiodactyls). The early nimravid saber-tooth carnivores did not reach the body size of their later cousins, rarely exceeding that of a jaguar, and probably weighing no more than 100 kg. It was only when large animals such as proboscideans became available as prey species that saber-tooth carnivores reached the size of lions (Merriam and Stock, 1932; Meade, 1961; Rawn-Schatzinger, 1992).

The Ice Age fauna of North America included two types of proboscideans, browsing forest-dwellers (mastodons) and longer legged grass-eating relatives of the modern elephants (mammoths; Agenbroad, 1984; King and Saunders, 1984). Only the largest and most sophisticated killers are successful in taking elephants (fig. 1.2). The North American lion, *Panthera atrox*, may have occasionally taken a proboscidean calf, but like its modern African relatives, *P. leo,* it probably specialized on large bovids and equids.

Cats, more often than dogs, individually overpower their prey when making an attack. As a result, cat-like carnivores tend to be larger and stronger compared to canids from the same fauna. They are fundamentally solitary hunters, and members of a lion pride, even when hunting together, hunt as individuals, with cooperation as much due to coincidence as to intention (Schaller, 1972). Saber-tooth killing behavior reconstructions have ranged from solitary ambush to organized packs (Bohlin, 1940; Simpson, 1941; Kurtén, 1952; Martin, 1980; Miller, 1983), but even the most optimistic reconstruction of their social behavior would probably not exceed the social behavior documented in modern lions.

The fundamental method of prey capture and dispatch used by modern conical-tooth cats is to stalk from concealment until close enough to overtake prey in a short burst of speed. The ability to accelerate quickly, but not to maintain this rate of speed, is characteristic of all living cats and presumably of their extinct, cat-like relatives. When overtaken, the prey is grappled using strongly muscled forelimbs with

Figure 1.2. *Homotherium serum* stalking a juvenile *Mammuthus columbi* playgroup. (drawing Mark Hallett)

paws bearing retractable claws, permitting the cat to immobilize and position the prey temporarily for a lethal bite delivered by a short, broad mouth. This killing bite is directed to the throat whenever possible, crushing and piercing the blood vessels that supply the brain (Leyhausen, 1965; Schaller, 1972; Ewer, 1973; Biknevicius and Van Valkenburgh, 1996). An attempt may be made to crush or pierce the spinal column or to occlude the nose causing suffocation, but the neck is the preferred target. Because of the differences in their canine-tooth shapes, saber-tooth cats killed their prey in a different manner than do modern conical-tooth cats.

Saber-tooths fall into two groups (Martin, 1980): those with relatively short, broad, coarsely serrated sabers (scimitar-tooths), and those with relatively long, narrow, and unserrated or finely serrated canines (dirk-tooths; Kurtén, 1952). Based on his estimation of the forces these animals could generate during mandibular closure, Therrien (2005) suggested that cats belonging to these distinct ecomorphs subdued prey differently. In addition to differences between their crania and dentitions, these animals also reflect divergent morphology of other skeletal areas. Because of their shorter legs, dirk-tooth carnivores extend the cat gestalt beyond modern felids in postcranial morphologies. They became bear-like, suggesting that they were ambush predators, physically stronger than lions and tigers of equivalent size. Bjorn Kurtén (1952) proposed a killing behavior for saber-tooth cats consistent with their known anatomy and especially the pronounced bear-like body form of dirk-tooth predators. According to Kurtén, bears in Scandinavia sometimes prey on domestic cattle. To do this, the bear surprises the prey from ambush, running alongside it until it can grapple with, and then immobilize, the animal. During the final attack, the bear is actually standing on its hind feet with one paw thrown over the shoulder and the other over the nose until the head of the prey animal is bent around in a bulldogging fashion. The bear then bites the exposed arch of the throat. There are a number of advantages to this type of attack. Large animals are more easily and safely immobilized by controlling the front end rather than the hind end (Leyhausen, 1965; Kleiman and Eisenberg, 1973). Bending the neck creates an arc with an apex that easily fits into the mouth. The ventral blood vessels are brought upward and outward during this rotation, and because the neck is stretched, they become taut under the skin and therefore easier to pierce. A saber-tooth cat biting in this manner would find it easier to get a "corner" of the neck into the mouth. This kind of bite would cut an arc, effectively slicing through that corner. Such a bite placement would sever the carotid artery and the jugular vein, bringing rapid death. Because the entire cutting-stroke would be the result of the downward drive of the saber, the canines would be subjected to very little lateral stress during the stroke. Most of the force for the bite would come from the head depressor muscles. The tip of the saber-tooth canine moves in an arc, cutting

itself out during the bite (Wheeler, 2000; 2004; chapter 2, this volume), eliminating the need to dislodge the canine using cranial elevation. It is hard to imagine a quicker or more efficient way to kill any large prey animal. Bending the neck in this manner results in repositioning of the carotid artery and jugular vein to within six to eight cm (2–3") of the surface of the neck, which is within easy reach of the sabers. It is presumed that this is the secret behind the saber-tooth adaptation. Such a highly specialized bite requires good control of the prey, and saber-tooth cats needed to have strong forelimbs with sharp, retractile claws to grip and wrestle with the victim. The dirk-tooth's short hind limbs provided a stable base so that the attack could be made with the predator standing up. The development of a plantigrade (flat-footed) stance by many dirk-tooth predators provides strong morphological evidence for this behavior.

Scimitar-tooth cats, with their shorter, thicker, more coarsely serrated canines, could be more opportunistic in their choice of prey (fig. 1.2). The canines of scimitar-tooth cats have greater strength and more resistance to twisting forces. Normally, scimitar cats have more elongated limbs for pursuit of prey. Unlike the short-legged dirk-tooths, they would more likely be grappling with the rear of the animal. In such a situation, biting at the hindquarters or abdomen is possible but would provide less predictability of the outcome for each attack. It seems likely that given a choice, both scimitar-tooth and dirk-tooth cats would prefer the throat, where a quicker and more certain kill was assured (fig. 1.3).

There were limitations to the size of prey that might be killed by either type of cat-like carnivore. One limitation is the size of the "cylinder" that can be introduced into the mouth. To be bitten, an object has to fit between the upper and lower canines. While the amount of jaw opening of saber-tooth cats may be relatively enormous, most of this rotation is simply used to clear the upper canines. The effective jaw opening (distance between the tips of upper and lower canines) is usually no more than that of living predators, about 60° (Emerson and Radinsky, 1980). A saber-tooth cat bite would create a long wound, but it would be superficial. There are really only two plausible areas where a saber-tooth could attack prey effectively. Both have been suggested by a variety of authors. One is the throat (Kurtén, 1952; Antón and Galobart, 1999; Martin et al., 2001; Antón et al., 2004), and the other is the abdomen (Akersten, 1985). The lining of the abdominal cavity is richly supplied with blood vessels, and a long ripping wound could certainly cause death. However, the abdomens of large prey have such a large curvature that it would be nearly impossible to get enough flesh into the mouth to reach the lining successfully (Wheeler, 2000). Because of this problem, it is often thought that dirk-tooth cats must have attacked the throat of large-bodied prey, rather than the abdomen. For scimitar-tooth cats, the greater importance of pursuit in the prey-capture strategy makes it more likely that they attacked the animal rear end first, making attacks on

Figure 1.3. *Homotherium serum* hunting a *Mammuthus columbi* juvenile. *A*, The cats leaving cover and initiating a short burst of speed to establish contact with the prey. *B*, Grappling with and taking down the prey. *C*, A killing bite to the throat and cooperation with another cat. (drawing Mark Hallett)

the abdomen more likely. Their relatively shorter canines and more procumbent incisors would also be more effective in creating a deep wound in a large curvature. In either case, it is possible that a particular scimitar-tooth cat might get the broad abdomen into the mouth and produce a wound at least six cm deep at its apex. At this depth the wound should eventually have a fatal result.

In *Homotherium*, all permanent and deciduous incisors are serrated (Rawn-Schatzinger, 1992), with I^1 and I^2 almost as large as I^3 (Merriam and Stock, 1932; Meade, 1961). Biknevicius et al. (1996), in an analysis of carnivoran incisors, found *Smilodon* to have a more felid-like incisor arcade, with shape ratios similar to those found in extant felids, in contrast to the more semicircular arcade arrangement found in the Oligocene nimravids and felid scimitar-tooths. It was proposed that the incisors appeared to be used more for killing and feeding than are those of modern big cats, and that the positive correlation between canine and incisor development in the nimravids supported Martin's (1980) proposal of increased reliance on incisor use. Heavy incisor use is confirmed for *Homotherium* by Marean and Erhardt (1995), tabulating a breakage rate of 29.6% for upper incisors and 60% for the lowers, compared to 7.6% and 4.3% respectively for the extant lion. Marean mentions a breakage rate of 9.6% for *Smilodon*, which is not absolutely comparable to his figures because of different criteria for wear and breakage (Van Valkenburgh, 1988; Van Valkenburgh and Hertel, 1993). The canines of scimitar-tooths are shorter, broader, more coarsely serrated, and nearly continuous with the incisor arcade than in either the conical- or dirk-tooth cats. These features are most prominent in *Xenosmilus hodsonae* (Martin et al., 2000; fig. 1.4), a newly described early Pleistocene form that broke Martin's paradigm by combining scimitar-sabers with the short limbs of an ambush predator. In *Xenosmilus,* the sharp, coarsely serrated, procumbent incisor arcade, coupled with the canines, could have cut out a lump of flesh about as big as the cat could swallow. Such terrible bites would have quickly put the prey into shock. Death might have resulted from multiple wounds rather than through a directed killing bite, although these cats could make such a bite if the opportunity presented itself. This bite would resemble the action of a cookie cutter taking out a unit of dough, and we have applied the name "cookie-cutter cat" to those saber-tooths that have a large procumbent incisor arcade that incorporates the canine in this arch without a distinct diastema. A diastema is present posteriorly between the canine and the first premolar. In addition, these cats have an elongated temporal fossa and a shortened postcranial skeleton. This contrasts with the more typical scimitar-tooth cats, with shortened temporal regions and elongated postcranial skeletons.

In the late Pleistocene of North America, we have three potential predators on mastodons and mammoths: the dirk-tooth cat *Smilodon,* the scimitar-tooth cat *Homotherium,* and the conical-tooth cat *Panthera atrox.* The discovery of a *Ho-*

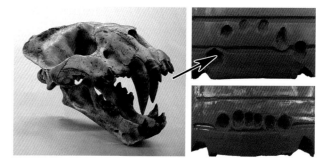

Figure 1.4. Anterolateral view of the skull and mandible of *Xenosmilus hodsonae* showing the massive incisor arcade not separated from the canines by a diastema *(left).* (courtesy John Babiarz) The characteristic impression in clay of the rounded arch, including incisors and canines, of the bite of a cookie-cutter cat, *Xenosmilus hodsonae (upper right),* compared to the bite impression of the dirk-tooth saber-tooth, *Smilodon fatalis,* where the incisors are in a row separated by a diastema, as shown in the impression made by the leading edges of the canines *(lower right).* (courtesy Virginia Naples)

motherium den in Texas along with the remains of 300 to 400 baby mammoths suggests that these cats, at least, were predators of the young of elephants (Meade, 1961; Rawn-Schatzinger, 1992; fig. 1.5). The Friesenhahn Cave site is from the Late Wisconsin (Kurtén and Anderson, 1980), but Rawn-Schatzinger (1992) lists 16 other sites with similar associations, ranging from Villafranchian to (late) Rancholabrean. No similar associations have been found for *Smilodon* or *P. atrox.*

A constraint can be established for available prey size for *Smilodon* and possibly *Homotherium*. Getting the prey into the mouth of a predator is the most important criterion for making a bite (chapter 2, fig. 2.12A). Body areas of prey that have too large a circumference prevent the cat from being able to encompass enough of the circle to give the canine tips sufficient purchase to grasp them, even at maximum gape.

Second, when attacking a prey animal, the predator must be as tall as, or taller than, the shoulder height of the prey (fig. 1.6). This enables the predator to manipulate the prey by reaching over its shoulder in a bulldogging fashion, resulting in making the kill. Thus, if the shoulder height of the prey exceeded that possible for the predator to grasp, this feature would certainly be a good defense against any attack.

Prey size directly correlates with predator size and physical parameters, not the numbers of animals hunting. Thus, formation of packs by saber-tooth cats using the technique of biting the throat would have had no influence on the size of the prey killed. However, large-bodied prey would most likely have provided a meat resource in excess of what a single saber-tooth cat, or even a hunting couple accompanied by cubs, could consume. It may be that this surplus could have been shared with other members of their own species, and certainly would be a bounty for scavengers such as dire wolves (*Canis dirus*) and birds that might occur contemporaneously with dirk-tooth and scimitar-tooth cats. Potentially,

Figure 1.5. Skeleton of four-month-old cub, *Homotherium serum* (TMM 933-3235), found in the late Pleistocene of Texas in the Friesenhahn Cave locality. (courtesy Bruce M. Rothschild)

the greatest value in pack structure would be the sharing of surplus meat, thereby reducing the number of hunting ventures required for any member of the group. Social behavior might also be of value in protecting a carcass from other social carnivores, including grey wolves (*Canis lupus*), possibly dire wolves (*C. dirus*), and predatory birds. Even extant lions in Africa have problems protecting their kills from packs of hyenas, and dire wolves are hyena-like enough to imply some shared behavior (Kruuk, 1972). The ability of the social group to protect and consume a kill for an extended period is a characteristic of predators that kill large prey.

Feranec (2008) studied the rate of canine tooth growth, comparing modern lions to the scimitar-tooth *Homotherium* and the dirk-tooth cats *Smilodon gracilis* and *Smilodon fatalis*. He found the rate of canine tooth growth in *Homotherium* to be similar to that of modern lions, while in the *Smilodon* species the canines grew at a faster pace. He suggested that these differences reflected ecological differences between scimitar-tooth and dirk-tooth saber-tooths. The canine tooth growth rate in young of the *Smilodon* species suggests that the animals had to learn rapidly how to make the precise killing bites necessitated by their long and relatively fragile canines. These observations further suggest that young dirk-tooths

had elongate canines in their mouths from early in their learning process, but do not allow the inference of group behavior beyond that of a family unit with their mother or parents and cubs.

EXTINCTION

Saber-tooth cats belonging to the genus *Homotherium* were most likely capable of short-distance rapid pursuits, while the short-legged *Smilodon* probably relied entirely on ambush. *Homotherium* may have included more agile and wary prey in its diet, but in either case the predator would have required sufficient cover to approach the prey closely enough that it could be overtaken in the initial rush. Obviously, *Smilodon* would have to get closer to the intended victim and would have required significantly more cover than would *Homotherium,* a cat that was probably at least as agile as a modern lion. However, even lions require vegetation as cover for prey capture. The extinction of all large-bodied prey species in North America, except for the bison, left large predators, including lions and saber-tooth cats, with a single prey species of appropriate body size. Unfortunately for the cats, the American bison became an abundant inhabitant of short grass steppes where the stalking behavior of cats was inappropriate, and the occasional meandering of bison into woodlands may not have provided sufficient opportunities

Figure 1.6. An adult male mastodon, *Mammut americanum,* can attain a height exceeding three meters at the shoulder (Osborn, 1936, Vol. 1). *Top,* A female of over two meters, a juvenile mastodon of over one and one-half meters, along with the dirk-tooth cat *Smilodon fatalis,* of approximately one meter. The dashed line shows the approximate maximal reach of *S. fatalis* and the near impossibility of an attack on individuals taller than the juvenile (modified from Martin, Naples, and Wheeler, 2001). An adult male mammoth, *Mammuthus columbi,* can attain a shoulder height approaching four meters (Osborn, 1936, Vol. 2). *Bottom,* A female at over three meters, a juvenile approaching two meters, a smaller juvenile at about one and one-half meters, along with the scimitar-tooth cat *Homotherium serum,* approximately a little over one meter. The dashed line shows the approximate maximum reach of *H. serum* and the near impossibility that this cat could make a throat attack on individuals much larger than the small juvenile, although anterior wounding bites (fig. 1.3B) may have permitted a larger juvenile (approaching two meters) to be taken through attrition. All measurements indicate shoulder height. (drawing Mark Hallett)

to support giant felid predators. Consequentially, we feel that, as a result of sudden habitat change and prey-species degradation, saber-tooth cats became extinct.

Saber-tooth cats have hardly ever missed the opportunity to go extinct. Why do they have this special vulnerability? We might claim that they died out because everyone else was doing it, as 40% to 55% of all the genera in the same fauna were also going extinct at the end of the Pleistocene; the loss of so many of their usual prey species would certainly have been a detriment to saber-tooth survival. However, this theory is still not informative enough to be scientific fact. It would be more interesting to know if there were specific attributes of large carnivores, or saber-tooth cats in particular, that made them especially vulnerable to extinction. What were the conditions that characterized periods of extinction? If we look at those periods of time, and carefully survey the sedimentary record, we see evidence that periods of extinction generally fall within times of marked global cooling (fig. 1.7 top; modified from Martin and Meehan, 2005, fig. 5). It may be that these periods of extinction were at the

ends of episodes of cooling of as much as six degrees Celsius, and this cooling interval would have extended over a few hundred thousand years, with the more severe episodes taking place in perhaps less than three thousand years. Global cooling had two overall climatological effects. One is that it reduced evaporation on a worldwide level, and this reduction of moisture in the air dried everything out. The other is that cooling and aridity increased the impact of seasonal change. While the overall differences in seasonality, relative cooling, and precipitation may not have been any greater than previously, the system was already closer to biological limits, and therefore the effect of the seasonal change was greater, or we could say the effective seasonality was increased. An increase in effective seasonality selected against trees and favored understory plants, especially ones that died or went dormant during part of the year. The loss of adequate young foliage for grazers or browsers during an extended portion of the year would have had a severe, perhaps catastrophic, effect on the herbivore population, reducing population sizes and perhaps causing extinctions. Naturally this would have also

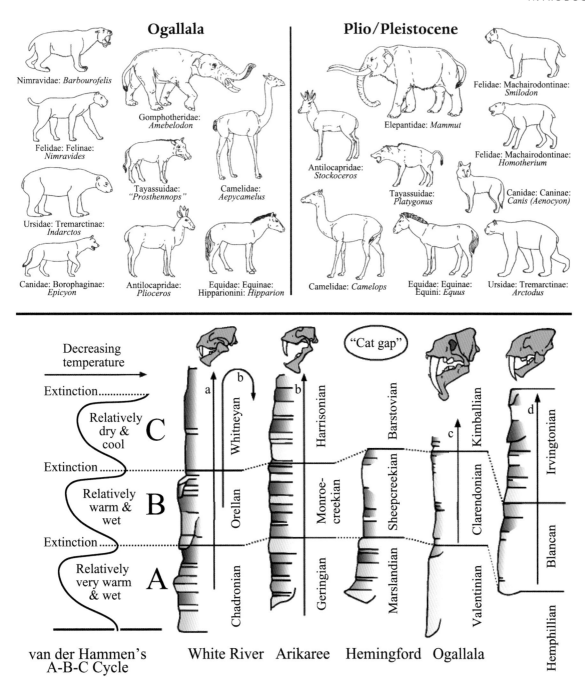

Figure 1.7. Representative mammals that became extinct at the end of two chronofaunas, showing the extinction of similar adaptive types, including the dirk-tooth cats *Barbourofelis* and *Smilodon*, and the scimitar-tooth cats *Nimravides* and *Homotherium* (modified from Martin and Meehan, 2005, fig. 5) *(top)*. Van der Hamman's cyclic model (Meehan, 2003) compared to Stout's (1978) sedimentary/climatic cycles in Nebraska *(bottom)*. The distribution of stratigraphic units is related to the North American Land Mammal Ages (Webb, 1977). The distributions through time of four dirk-tooth cat lineages from left to right, *Hoplophoneus, Eusmilus, Barbourofelis,* and *Smilodon,* are related to this model (modified from Martin and Meehan, 2005).

affected the predator population. Some of the herbivores that would have been most affected were browsers of large body size, a group that may have been favored as prey by saber-tooth predators. As pointed out by Janis (1976), the remaining herbivores would have been more widely distributed and less likely to intersect the home range of any given predator at a given time. To be successful, predators would have required greatly expanded home ranges, and this would have favored predators with efficient locomotor systems, in particular, long strides compared to their body mass. Nothing in this scenario would have been good for cats in general, especially any short-legged dirk-tooth predators.

Dating the Fossil Record

A recent paper by Van Dam et al. (2006) extended the Milankovitch orbital variations to include a 2.4 million-year climatic cycle that terminates in periods of global cooling and drying. Previously, astronomical parameters were only shown to have climatic impact at about 100,000 year intervals. The new model provides a mechanism for a pattern of climatically forced biologic change, first noted by Van der Hammen in the 1950s. Martin (1985) described a similar pattern from the North American mammal record, drawing many examples from the history of saber-tooth cats. Van der Hammen concluded that repeating climatic cycles of various scales are the ultimate basis for global biostratigraphic correlation, and Martin and Meehan (2005) concluded that these cycles influenced the patterns of evolution found in the fossil record (fig. 1.7 bottom). These models show great promise for understanding evolutionary history but rely on precise dating of the geological record.

One of the basic principles behind geological thought is the Law of Superposition. This concept applies to sedimentary rocks composed of detritus that has settled out of water or air. Simply stated, it means that particles that settle to the bottom first reside there longer than particles that settle out later and overlie them. Lower rock strata are thereby older than progressively higher ones. Fossils found in rock layers can be placed in a comparative sequence of different ages (relative dating). This does not reveal how old a fossil is in years. To determine that, you must go to an absolute dating technique, and these are usually based on radioactivity. Radioactive materials transform at a statistically constant rate into new isotopes and elements. If the ratio of the original radioactive material (the parent) can be compared to the amount of new element or isotope (the daughter product), the amount of time required for that degree of transformation can be estimated, yielding a radiometric age. Such measurements rely on a variety of assumptions, and if any are violated, the age assessment will be in error.

The most commonly used radioactive dating system follows the transformation of Carbon 14 (C^{14}). Carbon 14 is a radioactive isotope of carbon produced in the upper atmosphere by cosmic rays. It is transported as carbon dioxide by atmospheric circulation to the earth's surface and incorporated into plants. As long as the cosmic ray bombardment is constant, the concentration of Carbon 14 available should also be constant, and comparable to the overall concentration of naturally occurring Carbon 13 (C^{13}). After an organism dies, no further radioactive Carbon 14 is incorporated into its tissues, and what is there is reduced by radioactive decay. The ratio of Carbon 14 to Carbon 13 in a sample is a measure of how long this radioactive decay process has progressed. Accuracy is largely dependent on the ability to measure differences between these ratios. Eventually, the amount of Carbon 14 remaining is too small to measure accurately, putting a limit on ages that can be estimated using Carbon 14. For practical purposes, this limit is between 40 and 50 thousand Y. B. P. (years before present).

Other radioactive elements with different rates of decay (half-lives) must be used for older deposits and geologically older saber-tooth cats. A variety of radioactive isotopes is available, but the most common ones used for older deposits are potassium-argon (K-Ar) and argon-argon (Ar-Ar). We can use these isotopes to date the whole history of saber-tooth cats. Unfortunately, proper rocks for dating are not uniformly present, and radioactive dating is expensive and time consuming. There are enough radioactive dates to form a general framework (the North American Land Mammal Ages, NALMA), but not enough to delineate closely most evolutionary processes. We still rely on superpositional relationships extended geographically through the use of time-restricted fossils. In the marine section, distinct intervals, characterized by unique fossil assemblages, are called zones. In the terrestrial record, plants and animals have also been used to characterize discrete geological units. One of the most successful applications is the NALMA, set up by the Wood Committee and later refined by Woodburne (1987). We are using a slightly modified version of their arrangement (fig. 1.7 bottom).

The NALMA, as the name implies, were applicable only to North America. To extend our understanding of age relationships to other continents, we must use a more global framework, provided by the epochs and periods of the Age of Mammals (Cenozoic). Unfortunately, the boundaries of these epochs vary from author to author, and are subject to controversy whenever they are invoked. Mostly these are arguments over semantics. The important conclusions concerning the sequence of rocks and relative ages of fossils are usually not affected.

Taxonomy

To place our saber-tooth cats within the previously defined geological context we use scientific taxonomy, the language scientists use to describe relationships among organisms. This is like telling some friend that we share a mutual cousin or distant ancestor. Because different branches of life share complex relationships, taxonomists developed their own shorthand. Taxonomy, on the one hand, is mostly the practice of a special kind of linguistics. Systematics, on the other, is an attempt to reconstruct the history of genealogies. Any reconstructed genealogy will necessarily have a unique taxonomy. The rejection of spontaneous generation by Pasteur, and the acceptance of evolution as an explanation for biodiversity, results in the generalization that there is only one genealogy (phylogeny) of life, beginning with a single common ancestor. This generalization may not hold at its highest taxonomic levels, but no evolutionary biologist doubts that modern plants and animals are derived from such a common ancestor.

Taxonomy did not begin from this hypothesis but was simply a system for grouping organisms on the basis of shared characters. There was no need to seek an explanation of why these characters were shared, beyond the mind of the Creator. An unexpected result of these groupings was that they had predictive power. From knowledge of a few shared features, other characters might be predicted, searched for, and found. Cuvier, a creationist, recognized the significance of this observation as the necessity of features being linked to a common function. For example, long legs in a grazing animal should also result in a long neck, so that the mouth could reach the height of the food. Cuvier called this the Principle of Correlation, and it is one of the fundamental axioms of comparative anatomy. It was only through the study of comparative anatomy that scientists were able to make sense out of the fossil bones they discovered. The Principle of Correlation gave rise to a mostly incorrect but widely believed myth that scientists could restore an extinct animal from a single bone. However, some features are so closely associated with a special place in the ecosystem that a large number of other correlated structures can be inferred. More commonly, a trained observer may identify an animal from a very small fragment. If a complete specimen of the same animal has previously been discovered, the scientist might describe that specimen to an astonished layman who would then think that the description was from the fragment at hand.

Linnaeus, who invented the system that we use in all modern taxonomy, was a creationist, and therefore only looked for groupings based on shared characters. He did not concern himself too deeply with why these characters were shared. The Theory of Evolution provided an explanation that subsequent workers generally use as a rationale for their taxonomy. There has been an increasing interest in phylogeny, the historical side of this endeavor. Phylogeny may predict morphological evolution and behavior, and to some extent develop a more profound understanding of evolutionary processes as well as preferred paleo-ranges. Unfortunately, phylogenies and their resulting taxonomies are, and must always be, based on imperfect knowledge. While there may be only one family tree, there may be, and usually are, differing hypotheses claiming to describe this tree. This is a direct result of the independent origin of characters not present in the common ancestor; when these are used to define groups they create a false branching pattern.

In the 1960s, an attempt was made to get around this problem by using newly developed computer technologies and multivariate statistical analyses (numerical taxonomy); these approaches created taxonomies based solely on the numerical superiority of their defining characters versus those defining competing groups. Numerical taxonomy did not claim to reconstruct phylogeny, although it was thought that characters representing phylogenetic history should overwhelm characters that developed independently. Numerical taxonomy was widely supported because it did not require special knowledge of any taxonomic group beyond the ability to recognize and describe morphological features, thereby greatly increasing the population of scientists who could address the relationships of any particular group. It soon became clear that shared but independently acquired features, resulting from shared behavior, were often organized into correlated sets that had the power to degrade this type of analysis.

In the late 1960s, a taxonomic school superseding numerical taxonomy appeared based on the tenets laid down by German entomologist, Willi Hennig (1966). He claimed that taxonomy should mirror phylogenetic history, thus, the phylogenetic or cladistic school. Its basic axiom is that only derived characters carry phylogenetic information. At first glance, this may seem to be too specific a generalization, as derived and primitive characters must necessarily form a yin and yang relationship. Individual characters are only primitive (earlier) and derived (later) in terms of their temporal relationship to each other and phylogeny. The basic feature of a cladistic taxonomy is that it is a three-taxon grouping of organisms, in which two share a feature because it was present in their common ancestor and is not shared by the third taxon (the outgroup).

(continued)

(continued)

An outgroup is often referred to as a sister group, with the implication that this group appeared one step before the common ancestor. In reality, such a statement is untestable in cladistics, and the nearness of the sister group to the common ancestor is a comparative statement based on a second outgroup whose nearness must likewise be equivocal. Cladistic taxonomies essentially float free until they are anchored by some widely accepted relationship in conventional taxonomy.

Conventional taxonomies were based on morphological distance, not on the number of branching steps. The relative worth of morphological distance was set by the historical pattern of developing taxonomy in each group. This procedure may seem to lack rigor, but it has been remarkably successful in organizing and conveying information about the several million known fossil and living organisms. We would agree that shared derived characters (synapomorphies) carry the signature of special relationship, and if all are uniquely derived, such a system is adequate to work out phylogenetic (genealogical) relationships. Unfortunately, as was learned by the numerical taxonomists, many independent acquisitions of shared structures have occurred that lead to false hypotheses. To minimize this concern, cladists have turned more and more to the tenets and computer applications (pattern cladistics) of the numerical taxonomists they sought to replace. If all the features of an organism could be analyzed, the assumption that phylogeny will swamp the record of independent acquisitions of derived characters (phylogenetic noise) may hold. Taxonomists presently use infinitesimally small subsets of the total character base, and the most sophisticated computers and computer programs probably could not analyze much larger ones successfully. We think that small numbers of rigorously analyzed features would give more meaningful phylogenies, and that most independently acquired, shared, derived characters can be recognized if they are adequately investigated. We suggest that there is less harm in throwing out a questionable character that might still be valid than risking the introduction of features that were not present in the common ancestor.

How do we establish the relationships of the various sabercats? This is done in the same way that we might seek to establish the parentage of our own offspring. We include characters that are shared with the parent, or in this case, a common ancestor. These characters are only informative if they are unique to the group they are used to establish. Shared characters that are independent of the common ancestor give false evidence of relationship. It is the distinction between such characters and genuine evidences of relationship that is the greatest problem in the development of any family tree (fig. 1.8A). There are many shared characters between the extinct nimravids and later cats that might argue for a close relationship, or even ancestry, of one to the other. Certainly, this was the opinion of most of the early workers studying these animals, including one of the most famous of all the mammalian paleontologists, William Diller Matthew. He proposed that all modern cats were derived from *Dinictis,* and the later saber-tooths such as *Smilodon* were derived from an animal somewhat like *Hoplophoneus.* Both *Dinictis* and *Hoplophoneus* are nimravids from the American Oligocene. He divided the cat family, Felidae, into a subfamily of saber-tooth cats, the Machairodontinae, and the cats with conical canines, the Felinae (Matthew, 1910). Alternatively, cats can be divided into two families, the Machairodontidae and the Felidae, which are united into a single group called the Feloidea. No less a scholar than George Gaylord Simpson used the latter scheme in his monumental work, *The Principles of Classification and a Classification of the Mammals* (1945), and it persisted to some extent in the work of Schultz et al. (1970). Most of the earlier saber-tooths are now separated from cats entirely, and the term Machairodontinae refers just to a later radiation of felids that possibly includes the Tribes Homotheriini and Smilodontini.

Figure 1.8 A, A chart modified from Bell et al. (2004), summarizing the temporal boundaries of the Hemphillian, Blancan, Irvingtonian, and Rancholabrean. The subages proposed by Schultz et al. (1978), the arvicoline rodent zones of L. D. Martin (1979), and the arvicoline divisions proposed by Repenning, Weasma, and Scott (1994) are identified. B, The phylogeny of scimitar-tooth cats as compared to conical-tooth felids. Derived Characters:

1. Bilaminar septum in auditory bulla; cruciate sulcus on brain.
2. Shortened muzzle and elongated cranium.
3. Elongation of the muzzle in relation to a shortened cranium.
4. Elongation of the limbs and a bladelike upper canine.
5. Reduction of the carnassial metaconid and reduction of the lower canine.
6. Rounded cranium and slightly compressed upper canine.
7. Lower canine reduced and incisor-like; upper canine compressed and blade-like.
8. Elongation of the muzzle without much elongation of the cranium.
9. Elongation of the upper canine and increase in size of the dependant flange on the mandibular ramus.
10. Short tail.
11. Elongation of an unserrated or finely serrated upper canine; shortening of the distal leg segments.
12. Shortening of the muzzle; coarsely serrated upper canines; serration of the incisors.
13. Shortening of the distal limbs and elongation of the temporalis region.

A

		SUBAGES (Schultz et al, 1978)	ARVICOLINE ZONES (Martin, 1979)	ARVICOLINE DIVISIONS (Repenning, 1987; Repenning et al., 1990)
PLEISTOCENE	.1 Ma Rancholabrean	- - - - - - - - - -	Zone VII	Rancholabrean
	.6 Ma Irvingtonian	Sheridanian	Zone VI	Irvingtonian III
		- - - - - - - - - - Cudahyan	Zone V	Irvingtonian II
	2.2 Ma	Sappan	Zone IV	Irvingtonian I
PLIOCENE	2.5 Ma Blancan	- - - - - - - - - - Senecan	Zone III	Blancan V
		Rexroadian	Zone II	Blancan IV
				- - - - - - - - - - Blancan III
MIOCENE	5.3 Ma Hemphillian (Restricted)	- - - - - - - - - -	Zone I	Blancan II
				- - - - - - - - - - Blancan I
	7.2 Ma Early Hemphillian Kimballian			Hemphillian

B

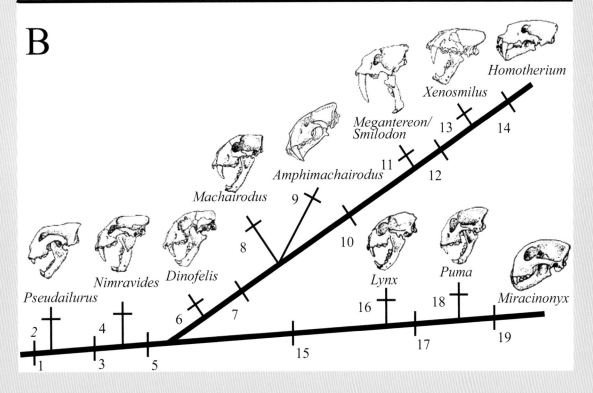

14. Elongation of the limbs and shortening of the temporalis region.
15. Felid grooves on the canines.
16. Short tail.

17. Small, domed cranium.
18. Small lower carnassial.
19. Elongated limbs; enlarged ectotympanic contribution to the auditory bulla area.

Figure 1.9. The distributions of *Xenosmilus hodsonae* and *Homotherium serum* in the late Pleistocene of North America are compared to the faunal provinces of Martin and Neuner (1978). Nearby localities have been grouped, and many represent more than one site. (Alaskan specimens may represent *H. latidens*.)

DISPERSAL AND DISTRIBUTION OF SABER-TOOTH CATS

Saber-tooth carnivores evolved in biogeographic circumstances quite different from the present time. In the modern world, the high northern latitudes produce a severe climatic filter. This filter is not simply one of cold, but a filter of food, including vegetation. Easily crossed by pursuit predators accustomed to hunting without cover, it is an insurmountable barrier to the type of predation practiced by most cats, in which cover is required to stalk prey and prey capture

relies on surprise. The separation of North America and Asia presently is a water gap, the Bering Strait. In contrast, the Panamanian Isthmus provides a forested land connection between North and South America that is suitable for cat-like predators. We do not have to go far into the geological past to change all these connections (fig. 1.8B). During much of the Tertiary, North America was connected to Asia, and hence to Europe, by a contiguous extent of land across Beringia, and for much of the time, this connection was forested. Cats, in particular saber-tooth cats, required this cover to extend their ranges. However, until a few million years ago, the presence of a water barrier prevented the dispersal of placental (eutherian) carnivores to South America, where endemic marsupial carnivores thrived. Large carnivores are good dispersers because they generally have large home ranges and young are usually forced from their mother's territory before the next litter arrives. A home range may include some unfavorable habitat, but because the animals travel widely, patches of unfavorable habitat do not constitute a barrier. As a result, very large terrestrial carnivores, such as saber-tooth cats, may have enormous species ranges. For instance, the members of the non-saber-tooth cat genus *Panthera* eventually occupied Africa, Asia, Europe, and North and South America. The only saber-tooth cat that succeeded in achieving such a vast range was *Homotherium,* but it was common for saber-tooth cats to have a Holarctic distribution extending from Eurasia through North America (fig. 1.9). Only continued research into the fossil record will provide us with a better understanding of these dispersals and subsequent extinctions.

The Robocat killing bite test fixture *(top)*. A cast of *Smilodon*
LACMHC 2001-249 skull and mandible with a composite mounted
skeleton from the George C. Page Museum *(bottom)*. (courtesy
Vicki Wheeler)

2

Experimental Paleontology of the Scimitar-tooth and Dirk-tooth Killing Bites

H. TODD WHEELER

CATS AND CAT-LIKE carnivorans are the ultimate hypercarnivores, and since Martin (1980) classified them into different ecomorphic groups, independent of phylogeny, their functional morphology has been defined by the shape of their canines. Each of the three basic types of upper canine morphologies—conical-tooth, scimitar-tooth, and dirk-tooth—are associated with unique killing bites. The saber-tooth cats (both felid cats and cat-like mammals) are aptly named. In addition to the visual similarity between their curved upper canines (sabers) and the traditional cavalry weapon, they share a basic functional similarity, the "draw-cut." As an essential part of the saber-tooth killing bite, this action is associated with the curved arc of the saber blade, in which the blade movement parallel to the long axis facilitates a slicing movement perpendicular to the blade (Burton, 1884). Machairodont (felid or "true cat" saber-tooth) upper canines are not sharp enough to cut animal hide or other connective and muscular tissues unless there is some associated slicing movement along the tooth axis. Both canine morphologies are defined by how this is achieved.

Scimitar-tooth cats bit into their prey with short, broad, serrated upper canines. It is suggested that they used an inflected arc trajectory, not unlike that of conical-tooth cats and other carnivores. Once the teeth stop moving, they must change direction and either be withdrawn by opening the mouth, as extant cats do, or be pulled free perpendicular to the tooth axis, as piranhas do.

Dirk-tooth cats, with longer and relatively slender, finely serrated upper canines, apparently used another method to cut through hide and tissue. They carved through prey in a continuous arc, keeping the incisors free as they cut their way out. Force was applied perpendicular to the tooth axis as the bite progressed, pulling the tooth into a slicing arc posterior to the tooth axis. This produces a wound several times the width of the tooth, and the trajectory results in the upper canines cutting their way out of the prey without any need to open the mouth. This bite becomes a fluid, even movement, without apparent change in the direction of tooth or head movement.

Kurtén (1963) first applied the terms scimitar-tooth and dirk-tooth when distinguishing between the Machairodont Tribes Homotheriini and Smilodontini. Martin (1980) expanded the use of the scimitar- and dirk-tooth terms to all varieties of cat-like saber-tooth ecomorphs, using these designations to characterize animals that share cranial and postcranial as well as dental features. Prior to the identification of *Xenosmilus hodsonae* (Martin et al., 2000), scimitar-tooth forms were exclusively associated with a more cursorial postcranial morphology as compared to the shorter-legged dirk-tooth forms. However, in this chapter I address only cranial and dental morphologies and the killing bite models they facilitate, so that the terms are used exclusively in the cranio-dental context. From this perspective, the distinction between scimitar- and dirk-tooth is based only on the type of killing bite the morphology supports. In addition to the upper canine proportions and distinctive characters on the cutting blade margins, the shape of the cemento-enamel junction (CEJ) is a key diagnostic feature in this analysis (Riviere and Wheeler, 2005).

THE CONICAL-TOOTHS

All extant pantherine cats are conical-tooths, with oval upper canines shorter than those of the saber-tooth. Typical crown height to anterior-posterior diameter (APD) ratios in pantherines are around 2.4:1, and (APD) to labial-lingual diameter (LLD) are around 1.3:1 (Van Valkenburgh and Ruff, 1987). These values conform well to recent models of optimum tooth aspect ratio (Freeman and Lemen, 2007a). The upper canines are separated by a diastema from a row of small, unserrated, somewhat spatulate incisors. Extant pantherine cats use one of two killing bites, depending on the relative size of the prey and the skill and experience of the predator (McDougal, 1977; Sunquist and Sunquist, 2002). For a quick kill, if the cat is confident of being successful, a bite to the nape of the neck, with the intent of severing the spine, will be used. If the cat is uncertain that it can do this successfully, the default killing bite will be a throat or nose clamp, with the cat holding onto the throat or muzzle, asphyxiating the prey over a period of several minutes.

THE DIRK-TOOTHS

The dirk-tooths are the most spectacular saber-tooths, with upper canine crown heights reaching 180 mm. Dirk-tooth machairodonts have elongate sabers approaching an APD to crown height ratio of 4:1, with typical APD to LLD of 2:1 (fig. 2.1). The upper canines have a sharp margin (by dental standards, approximately 0.2–0.5 mm radius) on both the posterior and incisal half of the anterior margin, frequently with fine serrations (although serrations may be lost early in life through wear). The incisors are larger and more robust when compared to conical-tooth standards (Biknevicius and Van Valkenburgh, 1996). The dirk-tooths also have a somewhat procumbant incisor row, separated from the upper canines by a pronounced diastema. In conical-tooth cats, the upper and lower incisors directly occlude I^1 on I$_1$, I^2 on I$_2$, and so on. In *Smilodon,* the arch may approach the straight line of the conical-tooths, but the incisors interlock (Akersten, 1985).

The I^{1}s fit between the I$_1$s, so that the incisors form a zigzag pattern between the uppers and lowers, fitting between each other straight up and down and in a straight row. They are posteriorly recurved, with much larger roots than a conical-tooth. Akersten (1985) was convinced that this implied a functional role in the killing bite. An alternative explanation is that the incisors of a saber-tooth must serve all of the pulling, tugging, grasping, accessory functions in prey capture, feeding, cub transport, and so on, that are normally supported by the canines of extant cats.

Typically, the deciduous upper canines approach scimitar-shape, with a crown height about half that of the secondary (permanent) upper canine. Machairodont dentary-maxillary gapes are about 90° (Akersten, 1985), requiring a shortened bite radius with substantial skull adaptations. The lower canine is dramatically reduced, particularly in taxa with longer upper canines such as *Smilodon.*

The highly derived nimravids, *Eusmilus, Hoplophoneus* (*Eusmilus*) *sicarius,* and also *Barbourofelis,* approach an APD to LLD ratio of 4:1 and may achieve a gape of 120°. Gapes exceeding 90° are associated with hystricomorphic features (Naples and Martin, 2000). The extant cat with the longest upper canines in proportion to skull size is *Neofelis nebulosa* (clouded leopard), which has a crown height to APD ratio of 3.28:1. It has been argued that *Neofelis* is more similar to the machairodonts than are any of the other extant "big" cats (Christiansen, 2006; 2008). Because the lower canines of *Neofelis* are also proportionally longer, the dental morphology is more like a small cat puncture or skewer bite and the antithesis of the cutting function of saber-tooth upper canines. The strong cranial similarities of *Neofelis* and the machairodonts are gape adaptations (Slater and Van Valkenburgh, 2008).

A distinctive characteristic of dirk-tooth machairodont sabers is an inverted bell-shaped CEJ. Starting at the anterior margin of the upper canine, the CEJ on both the labial and lingual sides curves toward the crown for a distance about equal to the APD of the tooth. At about the midpoint it curves apically (toward the root) and returns to the posterior margin, at a point a few millimeters away from the alveolar bone on a fully erupted upper canine (Riviere and Wheeler, 2005). The lingual side of the upper canine CEJ of nimravids is similar, but not necessarily the labial.

Interestingly, the more derived nimravids and *Barbourofelis* (Bryant, 1988) have deciduous sabers similar in profile to the permanent upper canines and that also approach them in crown height. Prior to the eruption of the permanent sabers, substantial wear on the rest of the permanent dentition has already occurred (Bryant, 1988). This delayed eruption could be beneficial. Should a subadult cat break a deciduous saber in the course of learning the killing bite technique, the cat will get a fresh start with the eruption of the permanent canines.

Figure 2.1. The profile of a *Smilodon* upper canine showing characteristic features of the CEJ, LACMHC 2000 R-43; George C. Page Museum collection. (photo Aisling Farrell, courtesy The George C. Page Museum)

ROBOCAT AND DIRK-TOOTH BITE MODELS

Not surprisingly, dirk-tooth cats have always received the most attention from researchers and the public alike. Their charismatic nature notwithstanding, the fact remains that the absence of any living proxy precludes realistic insight on how the dentition functioned from direct observation. This has led to countless theories, some disputed, more often controversial, but all largely theoretical as to the method or methods of dispatching prey.

There is an extant mammal with "teeth" that come close in external appearance to the dirk-tooth: the Chinese water deer *Hydropotes inermis* (fig. 2.2A). These teeth are better described as tusks, because they have no opposing lower teeth and only a vestigial enamel coating. They are only well developed in the males and serve no useful function apart from mating and display. These tusks are used with a closed mouth to lacerate the withers of other males during struggles to establish dominance. The tusks have a very short root, with an extremely thick periodontal ligament (fig. 2.2B), allowing them to fold back in their sockets (fig. 2.2C) and not restrict normal behavior such as grazing. Because they provide a modern analogue of what a breeding display saber is like, these tusks refute the hypothesis that saber-tooth cat sabers had a primary breeding display purpose, as they have low sexual dimorphism (Van Valkenburgh and Sacco, 2002), and the massive roots demonstrate that they apply substantial force in biting.

Dirk-tooths have been the focus of most preceding killing bite studies (Matthew, 1901; Bohlin, 1940, 1947; Simpson, 1941; Kurtén, 1952; Martin, 1980; Akersten, 1985; Bryant, 1996). Akersten based his model on a specific specimen (LACMHC 2001-2, George C. Page Museum collection from Rancho La Brea, RLB), considering it to be typical of *Smilodon,* and stipulated measurable parameters, thereby creating a testable hypothesis that could be verified experimentally. The initial focus of this research was to test the Akersten killing bite model by replicating it and comparing it to generalized throat killing bites using a mechanical device (Wheeler, 2000; fig. 2.3).

The practice of physical experimentation is more visible in other fields such as archaeology (Coles, 1973), with a published category of "combat archaeology" involving tests of replica weapons on various replicated armor or proxies (Molloy, 2008). There have been some excellent quantitative laboratory studies in the paleontology field yielding *Tyrannosaur* bite force (Erickson et al., 1996) and effect of serration size, shape, and sharpness on cutting ability in prey proxy material (Abler, 1992). Abler constructed a fixture capable of applying measured force perpendicular to a test blade, while applying an independent measured force parallel to the blade, adding the draw-cut to the other quantitative variables. The intent

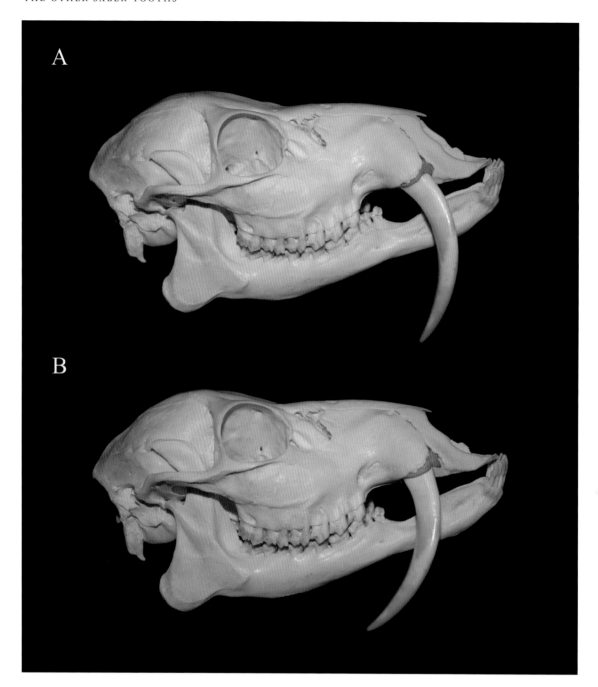

Figure 2.2. A specimen of the Chinese water deer, *Hydropotes inermis*: *A*, The tusks in the forward combat position. *B*, The tusks in the hinged back, grazing position. *C (opposite)*, The thick short periodontal ligament permits hinge-like movement and is shown by a foam replica of the same dimensions as the root socket on the right tusk. (courtesy Vicki Wheeler)

of my design was primarily qualitative, to establish which targets and sequence of movements were workable and which were not. The capability to measure the magnitude and direction of forces was a means of assessing the relative effectiveness of the various proposed bite models in the existing literature. My assumption was that an existing bite model would work and the rest would not prove feasible.

This mechanical saber-tooth cat acquired the nickname "Robocat." It used both steel and resin casts as interchangeable *Smilodon* teeth (LACMHC 2001-249) from RLB. This specimen was chosen because of the availability of high-quality casts, the presence of an associated mandible with good fit and occlusion, and the representative condition of the dentition apart from the sabers (LACMHC 2001-249, like

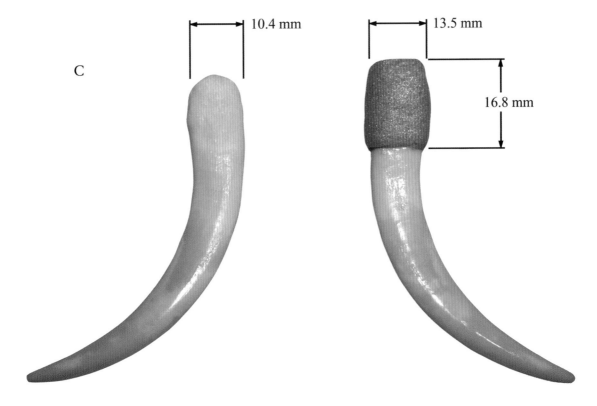

most of the collection, was recovered without sabers). The sabers used for the cast specimens fit the sockets (suggesting they are probably a bit small) and result in the tips being close to the representative condition and that of LACMHC 2001-2, although the CEJ falls too far from the alveolus by about 2 cm. The geometry of our resulting model is identical to that of the original skull (LACMHC 2001-2) and mandible (Akersten, 1985), reduced by 3% to correspond to the difference in overall size of LACMHC 2001-249 (see figure on page 18). Interchangeable steel, steel/resin composite, and cast sabers were modeled from other RLB specimens, sized to LACMHC 2001-2, reduced by 3%. The construction of multiple sets of different interchangeable teeth was done so that we could start with the strongest all-steel sets and, as our skill improved, progress to increasingly more accurately detailed but weaker sets. Ultimately, we intend to determine experimentally and quantify the effect of differences in serrations and wear on the forces involved. The forces and observations in this chapter reflect results with unserrated, steel-edged composite sabers, with 0.1 to 0.2 mm radius sharpness. Particular care was taken to match the erupted upper canine height and angle, determining the tip location relative to cranial and muscular morphology. Robocat is mounted on a mini track hoe that employs hydraulic force to move the model of the head and mandible independently through the full range of movement that a dirk-tooth cat could achieve while attempting to bite prey. It is designed to exert measurable forces substantially in excess of those a living cat could achieve. Rather than attempting to limit the device to an estimated strength of an extinct cat, the design could simulate various bite models, and subsequent analyses could show if the forces and model are practical. The first test session used a 559 kg female *Bison bison* cadaver as a prey proxy for *B. antiquus,* an identified prey species for *Smilodon* at RLB (Jefferson and Goldin, 1989; Coltrain et al., 2004). I replicated a "canine shear-bite" as described by Akersten (1985) on the animal's abdomen (fig. 2.4), and, for comparison, a similar bite on the throat. Observers present included John Harris, Larry D. Martin, Virginia Naples, and Chris Shaw.

The first observation was that *Smilodon* sabers are not sharp enough to cut through bison hide unless there is relative motion between the cutting edge of the upper canine and the hide. As long as this motion included a component parallel to the tooth axis, a draw-cut, the sabers sliced the hide. No reasonable amount of force (generated by a living cat), applied perpendicular to the cutting edge of the tooth, will cut in that plane without associated up or down movement. If the proper trajectory is followed, the sabers cut through the prey in one continuous arc without the cat having to reopen its mouth (Akersten, 1985). However, the draw-cut stalls if the incisors hook into the hide or if for any reason the tooth stops moving in the vertical plane. With a mouthful of tissues, the incisors haven't reached occlusion, and the shear part of the upper canine shear-bite does not occur (Akersten, 1985). The bite stalled as the incisors engaged the hide in the abdomen attempt (fig. 2.5A); we applied a force of 5,400 Newtons perpendicular to the upper canine axis without pulling the sabers through the hide (fig. 2.5B)

Figure 2.3. *A,* The Robocat killing bite test fixture. *B,* Robocat mounted onto a Bobcat X-331 model hydraulic track hoe. (courtesy Vicki Wheeler)

Figure 2.5. *A,* Video sequence from a canine shear-bite attempt on the abdomen of a bison specimen. *B,* The bite stalls as the incisors engage. The bison starts to lift off the ground as the force is increased to 5,400 Newtons. (courtesy Vicki Wheeler)

Figure 2.4. The Robocat test fixture being positioned for making a canine shear-bite on the abdomen of a bison specimen. (courtesy Vicki Wheeler)

and discontinued the attempt after the carcass started to lift off the ground.

Examination of the resulting wound (fig. 2.6) indicated that, although the sabers did pierce the hide and fatty tissue, they did not penetrate the abdominal cavity deeply enough to produce an imminently fatal injury. The unanimous consensus of those present was the bite as tested was not viable.

Subsequently, Akersten (2005) pointed out that, because the forelimbs of *Smilodon* are extremely robust, the resultant force on the sabers was not limited to a force equal to the prey weight. The cat could apply force in excess of the weight of the prey by pushing away from its prey. During the 2008 Saber-tooth Symposium organized by Akersten, we replicated the shear-bite sequence on the abdomen of a bull *Cervus* (elk) carcass with a body mass of about 180 kg (fig. 2.7).

On the initial attempt (fig. 2.8), we broke a composite saber. The next attempt was too superficial to sever the intestine, but we subsequently succeeded in executing an abdominal bite. Because of the smaller diameter of the elk abdomen, the bite penetrated the abdominal cavity and severed the intestines. This wound might not be immediately disabling but would certainly be fatal. The force required to pull out of the wound was over 11,000 Newtons, with the hide tearing rather than being cut by the sabers. As the mass (1,148 kg) required to keep the elk on the ground was over six times that of the prey, we used the track hoe to hold it down. Freeman and Lemen (2007b) experimentally determined the

Figure 2.6. An abdominal wound on a bison specimen after being cut open for examination, demonstrating that the wound in the hide and fatty tissue was superficial and did not extend into the abdominal cavity. (courtesy Virginia Naples)

Figure 2.7. The optimum position for making a canine shear-bite being marked on the abdomen of an elk specimen. (courtesy Vicki Wheeler)

Figure 2.8. Hide of the abdomen of the experimental elk specimen under tension from the stalled sabers at the point of composite saber failure. (courtesy Vicki Wheeler)

anterior-posterior ultimate strength of *Smilodon* upper canines to be 7,000 Newtons.

After study of the photographic sequences (1 frame/second) of the various bite attempts, it was apparent that the canines had stopped slicing and stalled (fig. 2.9A) when the upper incisors reached the elk's hide. When the slicing stopped, the relative movement of the fixtures also stopped, lifting the carcass off the ground. Our next attempt, with beams to hold the elk down, stretched the hide with the canines but ran out of travel before reaching the elastic limit (fig. 2.9B).

We then repositioned the carcass with larger beams closer to the wound. At a peak force over 11,000 Newtons, the hide tore at the canines. When the hide failed, it was still pinched between the upper and lower incisors and can be seen to pull free without cutting (fig. 2.9C). In the next frame (fig. 2.9D), the intestine is severed as the canines tear free.

Subsequent wound examination showed that the incisors did not cut the hide (fig. 2.10). This indicates that the force required to stall the bite was relatively low and probably occurred when the incisors started pulling the hide along in the same relative motion as the upper canines. Our tests revealed that cutting by spatulate incisors does not appear to be possible, necessary, or helpful to free the upper canines, and would appear to have no more role in the dirk-tooth killing bite than it does in the bites of extant cats. Because of the large forces involved, steel canines and incisors were used. Although resin cast incisors are sharper, the steel replicas were no more rounded than many worn incisor specimens in the RLB collection. When we retest with sharper incisors, I doubt it will change the results. The steel saber margins used were sharper than typical specimens in the RLB collection.

Therefore, the canine shear-bite model as proposed by Akersten (1985, 2005), in which the incisors are occluded and shear through the hide and tissue until they cut free of the body of the prey, is not a workable hypothesis, not on a bison or even thinner-skinned prey such as *Cervus*. Additionally, because the bite force of 10,350 Newtons needed to occlude the incisors failed to produce cutting, it is also clear that the incisor row is not capable of shearing through the hide of a large prey animal.

A dirk-tooth throat bite model similar to that of Martin

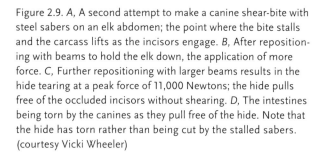

Figure 2.9. *A,* A second attempt to make a canine shear-bite with steel sabers on an elk abdomen; the point where the bite stalls and the carcass lifts as the incisors engage. *B,* After repositioning with beams to hold the elk down, the application of more force. *C,* Further repositioning with larger beams results in the hide tearing at a peak force of 11,000 Newtons; the hide pulls free of the occluded incisors without shearing. *D,* The intestines being torn by the canines as they pull free of the hide. Note that the hide has torn rather than being cut by the stalled sabers. (courtesy Vicki Wheeler)

(1980) and Kurtén (1952) was replicated successfully, with the proviso that the incisors did not engage the hide. Done well, the resulting wounds (fig. 2.11) are deceptively small clean cuts but result in the severing of the jugular or carotid arteries and would be almost instantly fatal.

This draw-cut bite on the throat can be performed experimentally, but requires extreme precision, using no more force than a live dirk-tooth cat could reasonably generate (fig. 2.12A). As our technique improves I anticipate the force required will substantially decrease. The precise vector must be applied if the saber is to slice out of the prey (fig. 2.12B). If the bite trajectory is too shallow, the wound will be superficial. If the trajectory is too deep and follows the natural arc of the tooth axis, the bite will stall, and the cat will be lucky to escape with the sabers intact (fig. 2.13C). As long as the proper trajectory is maintained, the sabers will arc out the bottom of the neck (fig. 2.13D).

Real-world application of all the bite models is difficult due to the amount of distortion of the prey's body as the bite

progresses. Large-scale movement of the prey relative to the predator can be constrained by powerful forearms, but fine-scale adjustments in upper canine placement require tactile input from whiskers, lips, and nerves in the periodontal ligament and pulp cavity. Tactile input from the tooth cervix and the question whether gingiva could exist on it (Riviere and Wheeler, 2005) is a subject still under study (Werdelin and Sardella, 2006). This issue is not central to the proposed bite model but merely to our understanding of the source(s) of the tactile input to achieve the bite.

At the initiation of the dirk-tooth killing bite, substantial jaw adduction force is involved at high gape angle when the upper canines pierce the hide and underlying tissue. As the bite progresses, and the arced path of the slicing saber diverges posterior to the projected arc of the saber axis, the jaw adduction component of the vector sum acting on the upper canines diminishes. As the head depresses, neck and forelimb musculature become the predominant source of force. Therefore, for dirk-tooths, jaw adduction musculature and mandible beam strength are very poor indicators of "strong" versus "weak" saber-tooth killing bite strength. Sophisticated computer techniques, such as finite element analysis (FEA), no matter how well executed, are no better than the mathematical model on which they are based. If the model involves a series of movements, such as a saber-tooth killing bite, the results can be no better than the assumed sequence and given conditions. As Cook (Cook et al., 2002) states in the introduction to his text, "It is possible to use FEA pro-

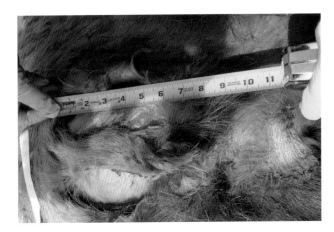

Figure 2.10. The wound produced by the sabers, made while the preceding photos were taken. Note the incisor print at the far right (located by the anatomist's thumb) and the evidence of tearing of the hide at portions of the wound. The tape at the left was a target mark for the track hoe operator. (courtesy Vicki Wheeler)

Figure 2.11. A wound from a successful throat bite on a bison specimen. The deceptively clean cuts reached the jugular vein and the carotid artery. (courtesy Vicki Wheeler)

grams while having little knowledge of the analysis method or the problem to which it is applied, inviting consequences that may range from embarrassing to disastrous." Even in a virtual world, there is a need for experimental paleontology involving real force measurement of real biological tissue (Erickson and Olson, 1996; Erickson et al., 1996; Thomason et al., 2001; Manning et al., 2005; Freeman and Lemen, 2006; 2007a; 2007b). Details of the machairodont dirk-tooth killing bite are still emerging, and the implications for those nimravids and *Barbourofelis,* with gapes exceeding 90°, and hystricomorphic morphology, are largely unstudied. Further experimental work should help in the development of improved bite models.

If our bite modeling reflects reality, the absence of a practical default killing bite was a serious problem for dirk-tooth

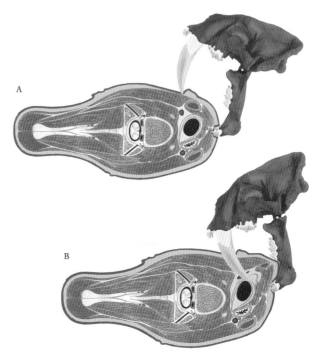

Figure 2.12. A cross section of a bison neck: *A,* The start of the dirk-tooth killing bite on the bison throat. *B,* The bite progressing, with the sabers on a suitable trajectory to bite through the prey. (drawing H. Todd Wheeler)

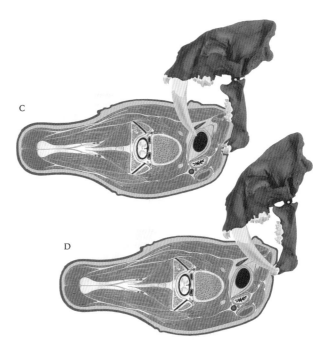

Figure 2.13. A cross section of a bison neck: *C,* Bite trajectory that is too deep because the cat is allowing the saber to follow its natural axis into the prey instead of biting through the prey. *D,* A bite progressing on a suitable trajectory to bite through and out of the prey in a continuous arc. (drawing H. Todd Wheeler)

cat populations. If for some reason the cat could not perform the specialized throat bite that was modeled successfully, or there were no suitable large prey available (assuming smaller prey species were agile and difficult to capture), scavenging would be the only remaining option. The required precision of the dirk-tooth bite is underscored by the presence of dirk-tooth specimens that have broken one or both upper canine teeth. Clearly, these individuals survived and continued to use the canines, because specimens often show wear on the broken surfaces. Therefore, it is reasonable to assume that dirk-tooth sabercats, such as *Smilodon,* could survive on an individual basis as scavengers, as do injured, ill, or aged extant felids. As a taxon, however, the dirk-tooth killing bite is apparently so prey-specific that even relatively small changes or disruption of the suitable prey supply would result in extinction for the cat.

THE SCIMITAR-TOOTHS

In contrast to dirk-tooths, there is very little published literature on the scimitar-tooth killing bite. The issue of prey selection and pursuit has been addressed in terms of the more cursorial ecomorph, but not the actual bite, despite the considerable cranio-dental differences between dirk- and scimitar-tooth morphologies. It seems logical that the evolution of a dirk-tooth would involve starting with a scimitar-tooth, but to date, acknowledged transitional forms in the fossil record have not been recognized. A possible candidate might be the *Pogonodon* specimen (JODA 5841) from the Arikareean of the John Day Basin (Bryant and Fremd, 1998). Sabers with enamel and a clear CEJ are not commonly described on nimravid saber-tooth cats. More often than not, the sabers of nimravid skulls are either missing, found broken, worn, or damaged to such an extent that the CEJ is not available for study. However, JODA 5841 is remarkably well preserved. It shows a level of detail not commonly present for nimravids. The specimen's sabers have a coarsely serrated anterior margin over the full length of the enameled crown, but the lingual CEJ shows remarkable convergence with *Smilodon* (Wheeler et al., 2004). *Pogonodon* is a remarkably long-lived saber-tooth genus, but it has not been extensively studied and may contain surprising interspecific differences, possibly including both scimitar- and dirk-tooth forms.

Be that as it may, the scimitar-tooth is a viable morphology in its own right and not just a precursor to the dirk-tooth. The tacit assumption that the scimitar-tooth is a junior version of the dirk-tooth, and that similar killing bite theories apply to the scimitar-tooth, is seriously flawed, as examination of the largely non-overlapping differences in tooth and craniodental morphologies indicate. Scimitar-tooths have shorter upper canines, usually with coarsely serrated cutting edges, an APD to crown height ratio rang-

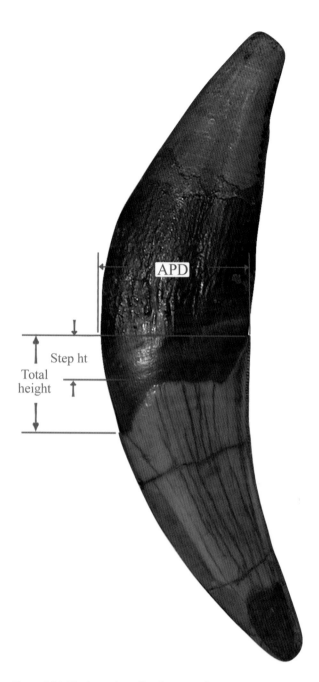

Figure 2.14. The lingual profile of a *Homotherium* upper canine shows the characteristic features of the CEJ, including the "step height" proportions. LACMHC 70001, George C. Page Museum. (courtesy John M. Harris)

ing from 2:1 to 2.6:1, and typical APD to LLD ratios ranging from 2.1:1 to 2.8:1 (fig. 2.14). The sharp, serrated margin on the anterior and posterior edge of the upper canines extends over the full length of the enameled crown. The incisors are larger and more robust than those of dirk-tooth machairodonts and are separated by a diastema from the upper canines (Biknevicius et al., 1996). The procumbant incisor arcade and short mandible result in an even shorter effective bite radius for the upper canines. As is the case for the dirk-tooths, the lower canine is dramatically reduced, and gape angles are about 90°, so the shorter upper canines have more clearance for bite placement and the required withdrawal.

The center of the arc of the tooth axis (center of upper canine curvature), which is several centimeters ventral to the temporomandibular joint (TMJ) in *Smilodon* (Akersten, 1985) and other dirk-tooths, is close to the TMJ in the scimitar-tooth. Unlike the dirk-tooth morphology, an arc drawn from the tip of the upper canine and centered at the TMJ falls within the upper canine alveolus. This, in conjunction with the full-length, sharp, serrated anterior margin allows simple rotation of the head and mandible to cut as the mouth closes or until the incisors can penetrate no farther.

The CEJ of the scimitar-tooth upper canine is distinctly different in shape from the dirk-tooth canine, with the distance from the alveolus greatest at the anterior margin and sloping apically up to a somewhat horizontal "step" and finally sharply sloping apically up at the posterior margin. Both the total and "step" heights in proportion to the APD may vary between scimitar-tooth cats and are quantitatively distinguishable between *Machairodus* and the more derived *Homotherium* (Wheeler, 2004).

While the functional basis for these CEJ profile variations is still being studied for dirk-tooths, CEJ patterns are a useful generic-level diagnostic feature (Werdelin and Sardella, 2006). For scimitar-tooths, the cementum-covered region at the cervix (neck) of the canine, between the CEJ and the alveolar bone, is conventional enough in shape and relative size that we can reasonably expect it was covered with gingiva, as is the case with all extant cats and most similar carnivorous thecodont mammals. Tactile feedback from this region would be a great asset during the scimitar-tooth killing bite. The flat step in the CEJ profile permits the upper canine to be inserted to the full crown height before the gingiva would contact the prey, triggering a reflexive flinch and mouth opening. The predominance of gingiva at the anterior margin of the tooth would ensure that the reflex would involve a substantial posterior component of force, producing a wound-widening motion by the serrated posterior margin, as the canine was withdrawn. Even the least-derived of the scimitar-tooths have larger and more procumbant incisor arcades, and in the less derived forms, the gingiva between the upper canines and the incisor arcade is positioned to trigger the reflexive releasing flinch as soon as the prey is pinched between the upper canine and incisors (fig. 2.15A). This suggests that the incisors do not play a role in the killing bite. In the more derived Homotheriini, this is not the case, and the incisors become progressively more involved in making the bite in some forms.

This progression culminates with *Xenosmilus hodsonae* (Martin et al., 2000); the incisors are larger and the arcade

more procumbant than in any other scimitar-tooth described. The upper and lower incisors have serrated margins, prominent lingual and labial cuspules, and little or no diastema between the upper canine and the incisors (fig. 2.15B).

At full occlusion, the upper and lower incisors (if unworn) mesh so well that they may have been able to cut their way out of some prey ("cookie-cutter" ecomorph bite model; Naples and Martin, 2008), something our testing has shown the other incisor arcades cannot do. Martin and Naples' cookie-cutter bite would appear highly speculative, pending experimental verification, were it not for the existence of living proxies that successfully use similar tooth morphology in a similar fashion. Sharks such as *Carcharodon carcharias* have teeth with generally similar morphology to the more derived homothere incisors and produce devastating wounds on prey of varying size (Frazetta, 1988). Apart from size, an even closer proxy is *Pygocentrus nattereri* (piranha, fig. 2.16), which has bony jaws and is known to remove a bite-size bolus from prey much larger than itself (Fink, 1993; Pauly, 1994; Freeman et al., 2007).

At some level, we know that the cookie-cutter bite is feasible, but from previous bite model experiments, we also know that stringent morphological criteria apply. In both sharks and piranha, teeth are replaced, and bite capabilities presume unworn, undamaged teeth. The piranha is the better living proxy, having bony jaws that do not project during biting (Alfaro et al., 2001). In piranha, the teeth do not occlude but have an "underbite," with the jaw positioned to slide or shear the lower teeth past the uppers (Berkovitz, 2000), as do carnassials (scissor blades). The condition in cats with bladed incisors is for uppers and lowers to meet blade to blade, without overlap, and is more similar to wire or bolt cutters.

An unfortunate aspect of both experiments and direct observations of living proxies is the difficulty in scaling relative size up or down. Until full-scale bite tests are performed, with incisor batteries at various hide/tissue depths, with and without the upper canines included, we cannot determine the precise forces and capabilities of the bite. At the piranha scale, the cookie-cutter bite works, and the better question is how big a mouthful the cat will achieve.

A strong advantage of the scimitar-tooth bite is the capability of a default bite, using the incisor arcade alone. This would have afforded scimitar-tooth cats in general, and *Homotherium* and *Xenosmilus* in particular, the flexibility of taking smaller prey when necessary or taking large prey of a size that was beyond the reach of their upper canine killing bite by inflicting a "wounding bite." If there are enough hyenas, extant hyena packs, working in unison, can kill large prey in less time than it takes a lion to use a throat clamp (Kruuk, 1972).

Antón and Galobart (1999) have applied the dirk-tooth canine shear-bite model of Akersten (1985) to scimitar-tooth machairodonts such as *Homotherium*. Robocat experiments

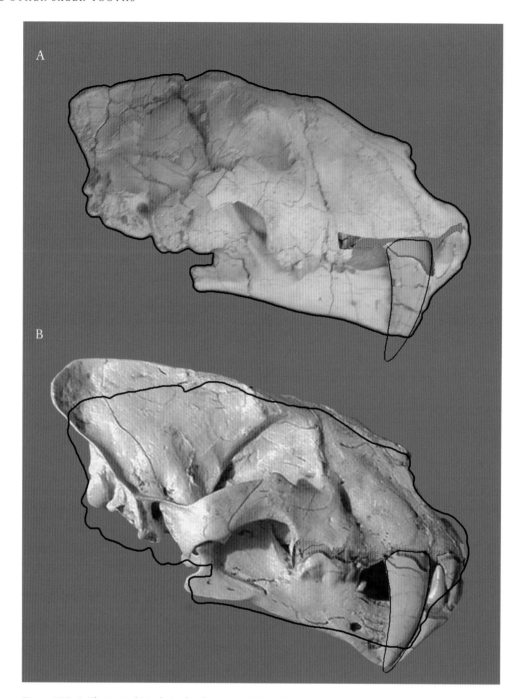

Figure 2.15. *A,* The typical *Machairodus* diastema, CEJ, and pre-sumed gingival relationship of the saber and incisor arcade (Wheeler, 2004). *B,* A more derived "cookie-cutter" diastema, CEJ, and presumed gingival relationship of saber and incisor arcade (Wheeler, 2004). Note difference in the step height region in the CEJ in otherwise similar saber profiles. (courtesy Vicki Wheeler)

have shown that, for any saber-tooth with a remotely con-ventional incisor row, once the incisors are engaged into the skin of the prey, the cat cannot free itself without reopening its mouth. By process of elimination, the bite model for scimitar-tooths starts with biting down with the upper ca-nines, which have coarse serrations that facilitate the draw-

cut effect on the downstroke. Once the cat has closed its mouth as far as possible, it has reached what I term a "point of inflection" and must move the upper canines in a different direction to withdraw. If the cat pulls back as it opens its mouth, the posterior margins of the serrated teeth would also draw-cut through the tissues of the prey on the upstroke,

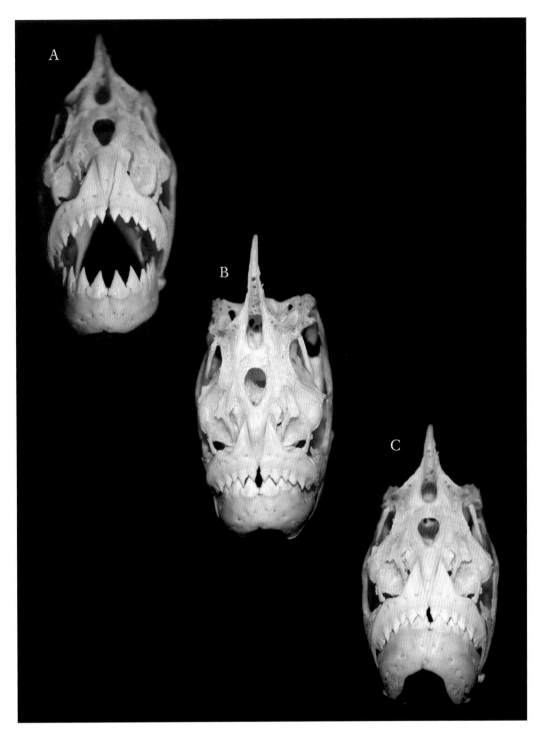

Figure 2.16. The dentition of *Pygocentrus nattereri* (piranha):
A, at full gape; B, the upper and lower dentition just meeting;
C, in the full closed position. Note that the lower teeth have slid
past the uppers, producing a scissor- or carnassial-like shearing
effect. (courtesy Vicki Wheeler)

potentially leaving a wound half again to double the antero-posterior width of the tooth. Properly sited on the neck of suitable prey, this would be almost as devastating as the dirk-tooth bite, but would have the advantage of not requiring the same high degree of tactile input and precision.

BITE MODELS AND FORCES

Substantial bite force is required to effect the scimitar-tooth bite. Unlike dirk-tooths, scimitar-tooth cats employed mandibular opposition throughout their killing bite in addition to using head-depressing force on the downstroke. This was followed by application of forces generated by the musculature of the forelimb and neck to widen the wound on the upstroke. Modern practical bite force studies began with basic "in vivo" studies on opossums, with low gape, and maximum force required at or close to occlusion (Thomason et al., 1990; Thomason, 1991). Thomason's opossums, and most extant carnivores, apply force with the upper canines against the prey that is equal to and opposite from the occluding force applied to the lower canines. This is not the case for saber-tooths in general and dirk-tooths in particular. The skull and mandible are not an isolated free body, and there is no need for the force on the maxillary dentition to be equal and opposite to the force on the mandibular dentition. The cat and prey are interconnected by the powerful forearms and body of the cat. Unlike conical-tooth cats, saber-tooths have an asymmetrical dentition, with bladed upper canines opposing a mandible with reduced incisor-like lower canines. The force that can be applied to the lower dentition is limited by the strength of the mandible and connective tissue, but the upper limit of force applied to the upper canines is only limited by the strength of the skull and body of the cat and the strength of its grasp on the prey.

A primary issue with "bite force club" thinking (Wroe et al., 2005) arises when you start to compare saber-tooth bites to conical-tooth carnivore bites, using maximum bite force values. The use of simplified lever arm models fits very well at or near full occlusion. However, bite force varies with gape, and it is the shape of the gape-versus-force curve that distinguishes conical-, scimitar-, and dirk-tooth bite models. Unlike conical-tooth cats, scimitar-tooths have a requirement for high bite force from full gape to occlusion. The problem with lever arms is that the mechanical advantage of the tension element (muscle) goes away as a function of the sine of the angle between the muscle and the lever arm. Ideally, this angle is 90° at the point where maximum force is required. For an herbivore, this is practical, in which case the cosine of the gape angle has the same value as the sine of the lever arm angle, which would be 1.0° at 0° or full occlusion.

Expanding this model to larger conical-tooth carnivores such as *Panthera leo* (African lion) raises concerns because of gape increase and uncertainty about the force-versus-gape requirements. In a carnivore such as *P. leo,* maximum force

for the carnassial would be required at a gape sufficient to clear the carnassials plus a fold of hide and any included flesh that the lion wants to swallow. Maximum force for the canines would be at a gape angle that clears the canines, plus a vertebra and associated musculature for a neck bite, but a secondary requirement is for sufficient force to pinch the wind pipe at close to maximum gape during a throat bite. Extant carnivores typically have at least 60° of gape (Radinsky, 1982). If the imaginary lion, when executing a bite, is at 30° of gape with its canines engaging the vertebra it wants to crush or separate, and the lever arm angle is set at 90°, the resultant trigonometric function (sin 90°) is 1.0. As the gape angle increases to 60°, the lion will have 0.866 (sin 60°) as much force available for the throat clamp. Therefore, the functional lion bite isn't that difficult to achieve.

If we had been drafting a blueprint for a bear, and started with a maximum force requirement at occlusion of essentially 0° gape (bears being omnivorous have bunodont crushing molars, not trenchant slicing carnassials), then at 60° of gape the available force would be 0.5 (sin 30°). You can see where this argument is leading, and it is the reason we don't have any saber-tooth bears. Because of the initial gape angle required just to clear the canine overlap, saber-tooths start with a gape requirement of 90°, which increases to 120° for some of the more derived nimravids. For our hypothetical saber-tooth bear, a gape of 90° would result in the lever arm angle being parallel to the tension element (muscle), and sin 0° is 0. This is also the reason dirk-tooths, with gapes exceeding 90°, need some different muscle attachment points just to close their mouths. Even for generalized carnivores with smaller gapes, the muscle fibers are oriented at multiple angles with multiple attachment points. The extant bears provide an example of some of the adaptations that match morphology with gape-versus-force curves.

If we look at the different skull shapes of an omnivorous brown bear *(Ursus arctos)* compared to a truly herbivorous panda *(Ailuropoda melanoleuca)* in dorsal view, the greater size of the temporalis musculature is apparent in the latter. However, there is a greater proportional width in the panda. Not only does the bamboo diet require a high bite force, the force requirement is at occlusion. For saber-tooths, a variety of additional compromises in the attachment points and angles are required. These variations are as different and complicated as seen in the multiple types of fossil cats and cat-like carnivores, but they are all different from extant cats. The cookie-cutter saber-tooth cats require high bite forces at occlusion, without loss of bite force at full gape, resulting in a unique morphology.

POTENTIAL PREY

The idea of a dirk-tooth such as *Smilodon* killing healthy full-grown *Mammuthus* (mammoth) has popular appeal, but little practicality. There is no fossil evidence for their having taken mature mammoths (Naples et al., 2000; Martin et al., 2001). As a weapon for harvesting bison, the dirk-tooth, as embodied by *Smilodon,* given sufficient cover, was unmatched until the development of the Sharps rifle. However, for any prey larger than a mammoth calf, dirk-tooth cats had no default means of inflicting bite wounds.

Extant lions, given sufficient time and enough lions, kill juvenile and adult elephants (Joubert and Joubert, 1994; 1997) that would be beyond the reach of *Smilodon* individually or in groups. What is notable about these accounts is that the targeted animal has to be debilitated (reduced to the brink of collapse) by drought or starvation. Because lions are not able to use a killing bite on juvenile or larger elephants, they essentially eat them to death, using a killing method more like that of hyenas and canids (Van Lawick and Goodall, 1971; Kruuk, 1972). This method is atypical for felids. It therefore seems reasonable that *Homotherium* could have used similar tactics and taken even healthy juvenile mammoths by attrition. There is direct evidence at Friesenhahn Cave, Texas (Graham, 1976), that *H. serum* took calves or even juvenile mammoths, and many other sites indicate (Martin et al., 2001) juvenile mammoth predation by *Homotherium.* However, mature mammoths would have been unsuitable as scimitar-tooth prey, absent some severely debilitating condition.

SUMMARY

Cats and cat-like carnivorans remain the ultimate hypercarnivores, and the shape of their upper canines defines their functional morphology. Each of the three basic types of canine morphologies—conical-tooth, scimitar-tooth, and dirk-tooth—is associated with a unique killing bite. Extant conical-tooth cats appear still to be using their traditional, and certainly proven, bite and crush technique. It has carried them through repeated extinction events and continues to support their current success. Bite model simulation experiments on full-size cadavers have confirmed the assertion that saber-tooths use a draw-cut to produce wounds with their upper canines. An essential feature of the saber-tooth killing bite, this draw-cut action is associated with the curved arc of the saber blade, in which the blade movement parallel to the axis facilitates a slicing movement perpendicular to the blade. Saber-tooth upper canines are not sharp enough to cut animal hide and tissue unless there is some associated slicing movement along the tooth axis. Each of the three upper canine morphologies achieves this differently. Conical-tooth big cats have a well-developed point and aspect ratio, permitting a spine-severing nape bite or an anchored suffocating throat clamp.

Scimitar-tooth cats bite into prey with an inflected arc trajectory and must open their mouths to withdraw or, in the special case of the cookie-cutter cat, fully occlude a bladed incisor arcade to cut out of prey. Scimitar-tooths use their short, broad, serrated upper canines to bite down in line with the tooth axis as they close their mouths. This action produces an entry wound the width of the tooth. Additional injury occurs as these sabers are withdrawn. When releasing the mandible and simultaneously pulling in the posterior direction, the serrated upper canines further enlarge the wound on the exit stroke. More derived scimitar-tooth cats could also inflict wounds with their incisor battery as a default bite when the prey was too large or otherwise unsuited to their specialized killing bite.

Dirk-tooth cats, with longer and relatively slender, less serrated upper canines, use another method to cut through hide and tissue. They carve through prey in a continuous arc, keeping the incisors free as they cut their way out. They apply force perpendicular to the tooth axis as the bite progresses, pulling the tooth into a slicing arc posterior to the tooth axis. This produces a wound several times the anteroposterior width of the tooth, and the trajectory results in the upper canines cutting their way out of the prey without any need to open the mouth. Unlike the scimitar-tooth, the dirk-tooth bite was so size- and prey-specific that this cat lacked an alternative or default means of seriously wounding prey unsuited to their primary killing bite.

The reason for Robocat experiments in a virtual FEA world is that computer analyses tend to reflect the initial assumptions, both explicit and implicit, without real world limitations.

Physical experiments not only provide answers to the questions posed, but frequently suggest better questions. They tell you when your assumptions are wrong or when essential issues have been overlooked in unambiguous ways, such as having the experimental bison specimen lift off the ground.

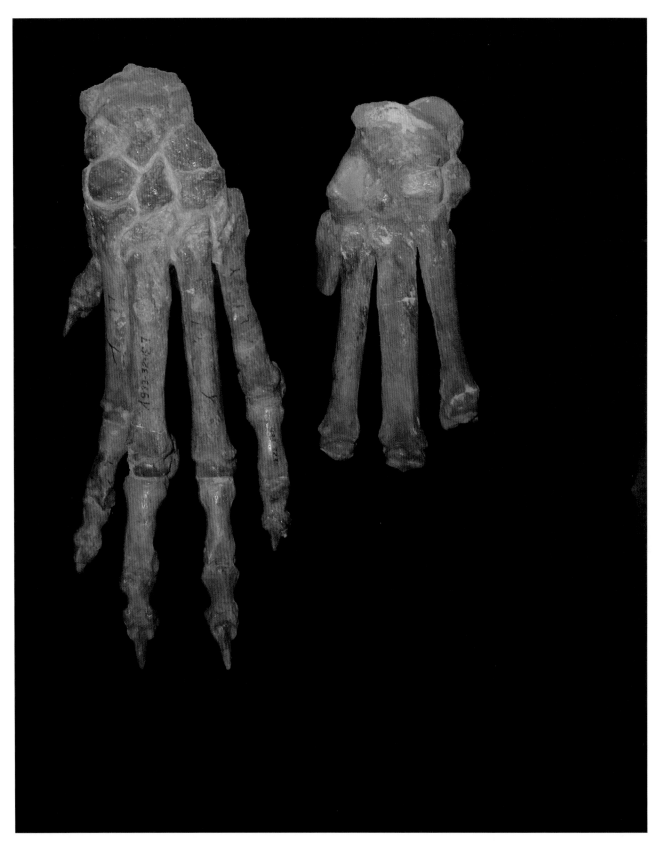

Dorsal views of the right and left manus of *Homotherium serum* (TMM 933-3231). Note the focal periosteal elevation identifying a stress fracture.

3

Pathology in Saber-tooth Cats

BRUCE M. ROTHSCHILD

LARRY D. MARTIN

PALEOPATHOLOGY, THE STUDY of ancient disease and trauma, provides insight into the behavior of extinct animals. In that sense, these studies overlap with ichnology, the study of trace fossils. It might even be thought of as a subset of that field, as the expression of disease in the fossil record results from the activity of organisms, as do other trace fossils. Behavior ceases at the time of death and is therefore hard to associate with any given individual, even if evidence of behavior in the form of trace fossils is abundant. In the end, we are usually forced to make arguments of analogy, using morphologically similar modern examples. Some animals, such as saber-tooth carnivores, have no similar modern analogues, and controversy surrounds interpretations of their behavior. Because pathology is directly associated with the skeletal remains of fossils organisms, it becomes an especially powerful way of filling in the gaps resulting from the lack of suitable modern examples. We propose to use paleopathologic analyses to gain insight into the behavior of saber-tooth cats.

PREDATORY BEHAVIOR

Two basic models of saber-tooth predatory behavior seem reasonable. In one, the prey is captured after a short pursuit, probably from the rear, with the predator grappling with the buttocks of prey and then pulling the animal down. From that position, it moves forward and delivers a killing bite to the abdomen (Akersten, 1985) or the throat (Antón and Galobart, 1999). In either case, the bite would be facilitated by rolling the prey onto its back. In the other model, the predator runs alongside the prey and grapples with the front end and shoulder, halting its forward progression. The predator stands firmly planted on its hind feet in this process, with one front paw over the shoulder and the other on the head of the prey, bending the head around so that the curvature of the neck can be bitten (Martin, 1980). Much of this behavioral scenario is analogous with a cowboy bulldogging a steer and might be referred to as the bulldogging model. A version of this model was proposed by Kurtén (1952), who considered bears to be examples of modern carnivore predators that use this kind of killing behavior. The two models are not mutually exclusive. Both might have been used at one time or another by the same individual, especially *Homotherium*. Martin et al. (2001) suggested that abdomen bites, if they were used at all, would have been more characteristic of the scimitar-tooth predators such as

Homotherium, while dirk-tooth predators like *Smilodon* were more likely to use the throat bite.

Anatomical studies reveal clues suggesting two behavior patterns. In contrast to pursuit predators, Andersson (2004) suggests that ambush predators would also grapple with their prey. Elbow joints of grapplers are wide and those of non-grapplers are narrow and box-like. Elbow morphology pits load transfer against stability (Jenkins, 1973). Elbow supination allows manipulation of prey but is reduced in cursorial animals, which require more stable and restricted joint movement (Andersson and Werdelin, 2003).

The two postulated predatory behavioral patterns result in different muscle insertion stresses, which at times overcame the elastic modulus of the insertion (muscle-bone interphase). The result is calcification/ossification of muscle insertions to form exostoses. The location of the exostosis identifies the affected muscle group and the direction and nature of the repetitively applied force.

Hoffman (1993, p. 52) described a *Smilodon* femur with a "nasty, knobby growth" from repeated adductor muscle attachment injury. This is exactly the anticipated stress response for an animal that repeatedly stands on its hind legs. Strong adduction would be required for firm leg "planting" in an ambush attack. Accentuated deltoid muscle attachment on the scapula in the form of an exostosis was also noted by Hoffman (1993) in *Smilodon* and is also the anticipated response to repetitive use as a forearm stabilizer for bending prey necks. Reported exostosis in *Smilodon,* affecting the flexor tendons of the upper limb and extensor tendons of the lower limb (Moodie, 1923; Heald, 1989; McCall et al., 2003), shows the anticipated distribution for grappling while standing on the hind feet. Adaptation in the feet for plantigrade posture also supports a standing posture during attack. Stress fracture in the radius of *Megantereon cultridens* suggests similar behavior patterns (fig. 3.1A). The presence of a healed greenstick fracture in that species suggests an injury early in life and supports a similar grappling behavior.

Additional paleopathologic evidence supports ambush behavior by *Smilodon.* Hoffman (1993) noted crushed limbs. Heald and Shaw (1991, p. 24) stated that "severe impact injuries and chronic re-injuries to anterior chest, transmitted through the ribs to the spinal column, attest to the frequent violent impact that *Smilodon* inflicted upon its prey." Chest contact is more likely to result from ambush behavior. Crushed, fused metatarsals (Hoffman, 1993) are parsimonious with the risk of standing on the hind legs while wrestling with the forequarters of a prey animal.

Homotherium crenatidens and *Homotherium latidens* also show prominent humeral shaft exostosis (Turner and Antón, 1997), with osteoarthritis of the elbow in *Homotherium serum.* This pattern also indicates repetitive grappling activity. However, there may be variation among the *Homotherium* species. *Homotherium serum* also had hind limb exostosis, but this is unreported in *Homotherium crenatidens. Homotherium*

serum may have been changing its attack strategy, a possibility further suggested by its slightly shortened tibia (Rawn-Schatzinger, 1992). Metatarsal stress fractures in *Homotherium serum* also support at least episodic ambush behavior.

CONSPECIFIC ANTISOCIAL INTERACTIONS

Some of the injuries noted in *Smilodon* have been interpreted as evidence of intraspecific fights. Shaw (1992) attributed shoulder and spine wounds and perforations to bites. Moodie (1930b) attributed a "necrosed," infected sacrum with draining sinuses caused by a stab wound. Hoffman (1993) documented a *Smilodon* skull with a "hole-accepting saber" wound.

Shermis (1983) reported osseous defects in flat bones from cat fangs as well as broken-off embedded fang tips. These were likely the result of territorial battles. Elongate anterior-posterior perforation of a left frontal (LACMHC 2001-24; Miller, 1983) appears also to represent conspecific interaction. A rib (LACMRLP R25876) with the tip of a *Smilodon* canine (LACMRLP R25877) embedded in the anterior surface further documents such behavior. Hooijer (1972) suggested that the bite occurred from a posterior approach.

There are a few cases of cross-specific combat between two different saber-tooths. The most famous is a *Nimravus bumpensis* skull with a partially healed saber wound, attributed to *Eusmilus* (Scott and Jepson, 1936; Macdonald and Sibley, 1969).

CONSPECIFIC SOCIAL INTERACTIONS

Was *Smilodon* a solitary hunter or a social animal that lived in packs? Heald (1989) speculated that the location and severity of some injuries (e.g., fractures) would have severely limited food acquisition and that some form of provisioning would have been required. This suggested a sort of social support system. However, consider modern cats. They can survive long enough (as long as five weeks) without food for fractures to heal (McCall, 1997). Most fractures in cats can heal on their own (Hill, 1977; Phillips, 1979). McCall et al. (2003) suggest that the injuries suffered by individual *Smilodon* at Rancho la Brea would not have prevented successful foraging (hunting/scavenging or water acquisition) by a solitary individual.

HANDEDNESS

The five instances of sacralization of *Smilodon* lumbar vertebrae at Rancho La Brea are all right-sided (Moodie, 1930a), and cervical and thoracic vertebral transverse process asymmetry is common. Thirty-five percent of late Pleistocene saber-tooth carnivores from Rancho La Brea have asymmetrical lateral depressions ranging from single nuchal crest shallows, to "coalescing apertures spread across the posterior

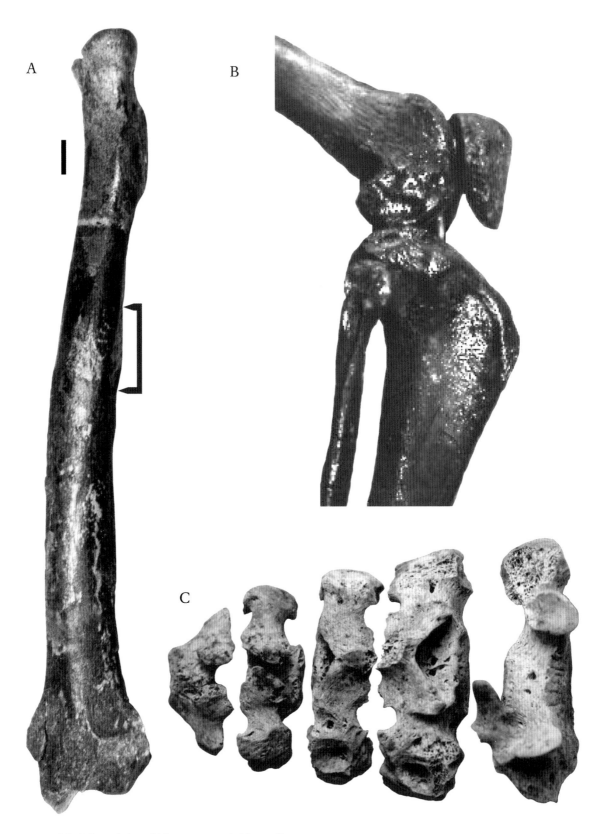

Figure 3.1. *A*, Dorsal view of *Megantereon cultridens* radius
(SE 311). A focal periosteal elevation (bump) identifies a stress
fracture. *B*, Lateral view of the knee of *Smilodon fatalis* from the
George C. Page Museum pathology collection. A ridge on the
proximal tibia identifies an osteoarthritic osteophyte. *C*, Dorsal
view of caudal, sacral, and lumbar vertebrae from a kitten, possibly belonging to *Megantereon cultridens*, from the Basel Museum collection. Disrupted posterior elements of sacral vertebrae
identify spina bifida.

brain case" (Moodie, 1930a). These appear to represent tendon avulsions. I have observed asymmetrical exostoses on
the posterolateral aspect of the occiput that suggest handedness or at least directionality in activities (Merriam and
Stock, 1932). Similar asymmetry also occurs in *Homotherium
serum* (TMM 3443).

DIRK-TOOTH CATS

Impaction with or without infection, tooth loss, and tooth
breakage with resorption of bone around the tooth are noted
in *Smilodon* (Shaw, 1992). One might assume that tooth damage would end predatory interactions (Shermis, 1983). However, Hoffman (1993) reported a *Smilodon* skull with a saber
fracture from rotational stress and a large saber fragment
thrust into the nasal cavity. This skull also has asymmetrical
deformation of facial bones accompanied by deformation of
front teeth positioning. It is also quite clear that the breakage
did not end the life of that *Smilodon* nor did it necessarily
affect tooth usage. McCall et al. (2003) reported seven *Smilodon* skulls with only one saber. They attributed this condition
to congenital agenesis and trauma. Hoffman (1993) reported
two cases at Rancho La Brea with canine teeth worn down
after breaks. Binder and Van Valkenburgh (2010) reported
variable frequency of tooth fracture among the various
Rancho La Brea pits, independent of animal age. This con-
trasted with Van Valkenburgh's (2001) notation of increased
tooth breakage of recent carnivores with age.

Van Valkenburgh and Hertel (1993) found the frequency
of tooth breakage in Rancho La Brea *Smilodon* was three
times that found in extant carnivores. They attributed this
to more rapid feeding, more aggressive guarding of kills, and
more complete utilization of carcasses because of limited
food availability. Specifically, Van Valkenburgh and Hertel
(1993) reported that 127 of 1,775 (7.15%) of *Smilodon* teeth
were broken. This included 52 of 544 (9.56%) of *Smilodon* incisors and 26 of 251 (10%) canines. Their frequency analysis
suggested a different killing technique than that used by
modern cats. Canine breakage in extant cats was related to
"large, unpredictable loads" incurred in the killing process
(Van Valkenburgh and Hertel, 1993). The elongated *Smilodon*
canines are more susceptible to bending moments. Akersten
(1985), Van Valkenburgh and Ruff (1987), and Emerson and
Radinsky (1980) suggested that *Smilodon* avoided bone in the
feeding process. Emerson and Radinsky (1980) noted that
"masseter fibers are more vertically oriented in saber-tooths,
as additional compensation for a weakened temporalis."

Predators can only assume slight risk in catching and
killing prey, given multiple prey encounters, as severe injury
usually precludes survival. Given the limited number of individuals with canine pathology, they must have devised an
effective strategy to protect their canines.

Epidemiologic analysis of the fossil saber-tooth record is,
with dramatic exception (e.g., Rancho La Brea tar pits), lim-

ited by small sample size. Although Hemmer (1978) thought
that there was a high frequency of skeletal element pathology, further analysis revealed only 1%. According to Moodie
(1930c), only 1% of *Smilodon* femurs had "osteopathology."

Shaw (1992) reported hundreds of traumatic injuries in
Smilodon. Among these, he included fractures, dislocations,
infections, and osteoarthritis (fig. 3.1B). He considered infections and osteoarthritis as post-traumatic events. In 1985, he
described "minor subluxations," a fracture, a septic hip, a
"pressure atrophy," three with "arthritis deformans," and ten
with "subperiosteal exostoses." Prevalence increased closer
to the extinction horizon (Duckler, 1997).

Hoffman (1993) reported a "mashed-up" pelvis, fractured
lower limbs, dislocated hips, and crushed chests. Reports of
vertebral compression, torsion and flexion fractures, and rib
fractures are prominent (Moodie, 1923; Heald, 1989; McCall
et al., 2003).

Traumatic damage, for example, fractures, was often
secondarily infected (Moodie, 1926b; Macdonald and Sibley,
1969). There was one infection in nine damaged *Smilodon*
sacra among the 1,086 examined (Moodie, 1930b). Fused,
infected vertebrae reported by Macdonald and Sibley (1969)
were attributed to post-traumatic complications. Seven percent of pathologic bones represented developmental anoma-
lies (Heald and Shaw, 1991). Shaw (1992) reported scoliosis
and mild spina bifida in *Smilodon*, as was also noted in vertebrae from the Basel Museum collection possibly belonging
to *Megantereon cultridens* (fig. 3.1C). Seven *Smilodon* specimens
had only one saber, which McCall et al. (2003) attributed in
some instances to agenesis.

Individual variation is another issue: when is it really
pathology? The *Smilodon* ectocuneiform is an intriguing
example. A hook-like tuberosity is broadly fused to the body
at the plantar (posterior) aspect in 70.8%–76.2% of Rancho
La Brea *Smilodon*, depending on author (Merriam and Stock,
1932; Shaw and Tejada-Flores, 1985). The latter authors found
that six of 11 specimens (54.5%) showed the same feature in
the Talara, Peru, asphalt deposits. A facet that articulated
with a small nodule of bone, the missing plantar process, was
present in 30.1% of 123 specimens from Rancho La Brea.
Shaw and Tejada-Flores (1985) attributed this variation to
muscular forces exerted on the bone. Nine of 123 specimens
(7.3%) "exhibited unexplained pathologic bone growth."

Vertebral pathology is addressed separately because of a
limited data base. With the exception of notation of spondylosis deformans, analysis to date is limited to a single aspect
of vertebral pathology: fusion or ankylosis (Moodie, 1926a;
b; c; Bjorkengren et al., 1987).

Merriam and Stock (1932) used the very appropriate terminology, spondylosis deformans, to define the vertebral
body osteophytes occasionally present in *Smilodon*. These
are marginal bony overgrowths extending perpendicular to
the vertebral centra. Based on its occurrence in humans,
these overgrowths appear to be an aspect of aging rather

than a pathological condition, and are distinguished from the disease of peripheral joints called osteoarthritis (Rothschild, 2008b).

Hoffman (1993) reported a fractured, fused first cervical vertebra in *Smilodon,* suggesting another injury due to violent impact. This would have been a severe injury (broken neck) and the fact that it was healed demonstrates a remarkable capacity to deal with trauma.

Fusion/ankylosis of *Smilodon* vertebrae has been attributed to osteomyelitis (Moodie, 1926b; c) or erroneously (Moodie, 1923; 1927) to myositis ossificans, a condition where damaged muscle tissue calcifies (Rothschild, 2008a). Examination of Moodie's (1926c) drawing reveals clear evidence of trauma in one, and probable infection (on the basis of holes) in the other.

Bjorkengren et al. (1987) found 48 specimens of *Smilodon* with fusion of two or more adjacent thoracic or lumbar vertebral segments among the one thousand vertebral specimens recovered by that time. They reported three distinct patterns: trauma with possible infection was present in 17; anterior longitudinal ligament ossification was present in nine; spondyloarthropathy was present in 11. Additionally, nine had a combination of either anterior longitudinal ligament ossification and trauma or spondyloarthropathy and trauma. The pathology in two individuals could not be classified.

Anterior longitudinal ligament ossification is a manifestation of a phenomenon called diffuse idiopathic skeletal hyperostosis, or DISH (Rothschild, 2008a). It is an asymptomatic phenomenon, not a pathology. It is unassociated with any identifiable clinical manifestations in humans. There was a peculiarity, however, in the observations of Bjorkengren et al. (1987) on *Smilodon.* DISH is a phenomenon of mature individuals. They observed that two of nine individuals with anterior longitudinal ligament ossification had incomplete vertebral ring apophysis ossification, indicative of immature skeletons. There is a condition found in some younger individuals that mimics DISH: hypervitaminosis A (excessive vitamin A ingestion; Seawright and English, 1967). This condition can occur from excessive ingestion of liver. Modern cats eat the viscera before proceeding to the musculature (Schaller, 1967). A diet of very large animals might provide an opportunity for young individuals to overindulge in liver. This would suggest that the young were brought to the carcass and allowed to eat their fill. Could this be a reflection of a specific dietary preference among *Smilodon*?

Spondyloarthropathy is a form of inflammatory arthritis (Resnick, 2002; Rothschild and Martin, 2006). It is a trans-mammalian disease that can affect almost any joint (Rothschild and Martin, 2006). It is common in recent carnivores, and afflicts 5% of contemporary large cats (Rothschild and Martin, 2006). As vertebral fusion is only one of its manifestations, full examination of all vertebrae, and indeed of all axial and peripheral bone articular surfaces, would be required to determine its population frequency. Such analysis would allow *Smilodon* to be compared with contemporary carnivores.

The occurrence of spondyloarthropathy appears to be related to infectious agent diarrhea-food poisoning and fecal-oral contamination. Could the frequency of spondyloarthropathy provide a window to the degree of "sanitation" in the Pleistocene? It would be intriguing to learn of variation in disease frequency through time and to learn if prevalence of traumatic injuries increased closer to the extinction horizon, as suggested by Duckler (1997). Lastly, it is impressive that of 48 fusion pathologies only two were not clearly identifiable as to etiology. This shows the power of scientific paleopathology to recognize pathology with confidence across phylogenetic lines.

SCIMITAR CAT BODY FORMS

Evidence for trauma in *Homotherium serum* from the Texas Memorial Museum Friesenhahn Cave collection was limited to a single healed fracture, four stress fractures (see figure on page 34), and one compression fracture (fig. 3.2A). Posttraumatic changes were also analogous to the T7 compression fracture found in *Homotherium serum*.

Other than *Homotherium latidens* humeral shaft exostosis (Shermis, 1983; Turner and Antón, 1997), comments on *Homotherium* are limited to asymmetrical longus capitis muscle scar insertions in Incarcal IN-V 211 (Incarcal, Spain; Shermis, 1983). Osteoarthritis with pseudoarthrosis (new pelvic articulation) and a complicated hip and acetabulum fracture was noted in LACMHC 9854 (Shermis, 1983) and also in the elbows of two *Homotherium serum* specimens from the Texas Memorial Museum collections. Ischial and pubic avulsions were associated with the dislocation of the joint and the failure of the fracture components to fuse with each other. While Shermis (1983) referred to the avulsions as myositis ossificans traumatica, current terminology labels these alterations as exostoses. More importantly, acetabular fractures have major postural implications. The mechanism of such injuries implies occurrence during bipedal posture.

Examination of *Homotherium serum* specimens from Friesenhahn Cave revealed minimal osteoarthritis (elbow osteophytes in two of 16 humeri). Stress fractures were found in three of 13 metatarsal IIIs. There is one healed fracture among nine metacarpal IIIs and a bent second metacarpal with periosteal reaction (Meade, 1961). There is a left metacarpal V that is thin and bent (Meade, 1961). There are distal tendon attachment stress changes in five of 17 femora. A compression fracture was present in one of three T3 vertebrae from among 26 thoracic vertebrae and 14 lower thoracic vertebrae. Infection affected two of 14 calcanei; the pathology in one was classic for tuberculosis (fig. 3.2B–C). Vertebral facet joint asymmetry was present in five vertebrae and basal skull asymmetry was found in one individual. In addition, a

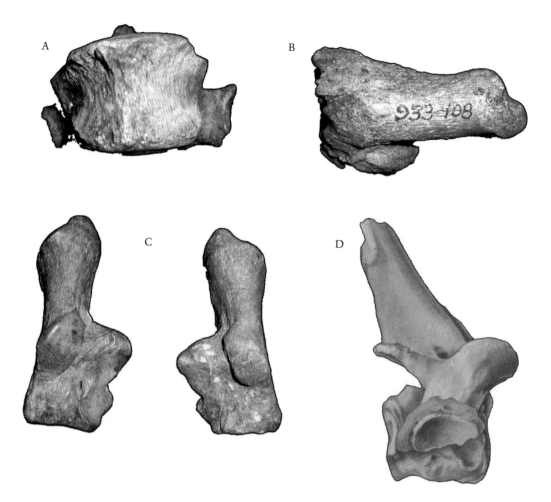

Figure 3.2. *A,* A ventral view of vertebra T10 of *Homotherium serum* (TMM 933-1382). Note the disparity of lateral vertebral height identifying a compression fracture. *B,* Lateral view of the calcaneus of *H. serum* (TMM 933-108). Disorganized and reactive bone identifies an infection. *C,* Dorsal view of the calcaneus of *H. serum* (TMM 933-1079). Undermining of the articular surface characteristic of tuberculosis (*left*). Normal conditions in a randomly selected specimen, also from the collections of the Texas Memorial Museum (*right*). *D,* An *en face* view of zygapophyseal facet joint of *Xenosmilus hodsonae* T1 (BIOPSI 101). Overgrowth of facet margins identifies osteoarthritis (green shading).

stress fracture was also found in metacarpal IV in the Idaho specimen (IMNH 900-11862), *Homotherium ischyrus.*

The pathologies noted above indicate a significant risk of injury to the manus in *Homotherium serum* and the use of the manus in some dangerous activity such as grappling with prey. Hind limb stresses were also more common in *Homotherium serum,* in contrast to the pathologies reported for *Homotherium crenatidens* (Shermis, 1983).

The evidence for tuberculosis in *Homotherium serum* is based on pathology noted in a calcaneus. Subchondral bone was undermined at the calcaneal tarsal joint, identical to undermining previously documented as of tubercular (*Mycobacterium tuberculosis*) origin in *Bison, Ovis, Symbos,* and mastodons. This suggests that *Homotherium serum* may have included as prey one or more of these taxa, from which it may have acquired the infection. The ecology and distribution of *Homotherium* specimens suggests that *Bison* was the most likely target.

Pathology analysis in *Xenosmilus hodsonae* was derived from examination of two specimens, BIOPSI 101 and UF

60,000. Examination revealed an assortment of trauma, congenital anomaly, and the only example to date in a saber-tooth carnivoran of a specific variety of arthritis—calcium pyrophosphate deposition disease. Osteoarthritis was noted in a cervical (C5) zygapophyseal facet and in a thoracic (T1) costovertebral joint (fig. 3.2D). Post-traumatic changes were noted in a thoracic vertebra (T6). Developmental anomaly (missing a capitular articulation facet) was present in a *Xenosmilus* thoracic (T11) costovertebral joint. There was slight thoracic costovertebral surface alteration at T8 characteristic of a disorder called calcium pyrophosphate deposition disease (CPPD). This is a form of crystalline arthritis (Rothschild and Bruno, 2008) that is occasionally observed in extant carnivores (Rothschild and Martin, 2006). Further examination revealed an exostosis of metacarpal V and erosions of costovertebral joints at T6, T7, and T10, and of the proximal fourth metacarpal. This documents the presence of spondyloarthropathy and suggests the likelihood of scavenging behavior, with its concomitant fecal-oral contamination.

The importance of making a full examination of collections is illustrated by the findings in *Xenosmilus hodsonae* and *Homotherium serum,* showing not only multiple forms of pathology, but also identification of a metabolic abnormality recognizable in one vertebra. As costovertebral involvement is an uncommon manifestation of CPPD, its recognition suggests that it may be usefully searched for in other saber-tooth cat collections, including the Page Museum Rancho La Brea collections.

SUMMARY

The pattern of pathology found in the dirk-tooth felid *Smilodon* is consistent with a model of predation in which the predator attacked from the side and grappled with its victim while standing. There is also evidence of injury during intraspecific combat, either in defending territories or prey. The scimitar-tooth *Homotherium serum* shows a similar pattern of stress-related bone modifications and injury, indicating that its mode of attack overlapped that of *Smilodon.* It may have been shifting from the pursuit mode used by the early homotheres to one more similar to that in the dirk-tooths, an evolutionary tract already tried by *Xenosmilus.* The discovery of tuberculosis in Pleistocene bison and the American mastodon (Rothschild and Laub, 2006) may explain its occurrence in one of the *Homotherium serum* specimens from Texas. Spondyloarthropathy is well represented in the large *Smilodon* sample, and we have now found it in *Xenosmilus.* We have not found it in the long-legged scimitar-tooth *Homotherium.* This may reflect the smaller sample size, or it might suggest that the short-legged *Smilodon* and *Xenosmilus* engaged in more scavenging.

Two individuals of *Xenosmilus hodsonae* drinking from a pool near the skeletal remains of a peccary they may have killed and eaten at an earlier time. Peccary remains were found associated with *Xenosmilus* and may have been their preferred prey. (drawing Mark Hallett)

4

The Osteology of a Cookie-cutter Cat, *Xenosmilus hodsonae*

LARRY D. MARTIN

JOHN P. BABIARZ

VIRGINIA L. NAPLES

THE LARGE CONICAL-TOOTH cats of the North American Ice Age included a relative of the lion *Panthera atrox*, a large northern subspecies of jaguar *P. onca augusta*, a cheetah-like cat *Miracinonyx trumani*, the mountain lion *Puma concolor*, and three saber-tooth cats. Of these, only the mountain lion presently resides north of the Mexico–United States border, although the jaguar may occasionally be found in southern Arizona (Lange, 1960). The remaining taxa are extinct. The mountain lion is a specialized predator on deer. Deer are medium-sized herbivores that flourished in wooded habitats at the end of the Pleistocene and still survive today. The large extinct cats were also specialized predators but preyed on larger animals such as horses, camels, bison, and possibly elephants. Of these larger prey species, only the bison avoided extirpation in North America; horses went extinct and were subsequently re-introduced by humans. The extant population of bison became smaller in body size and more adapted to open steppes than were their Pleistocene relatives. Treeless grasslands are difficult if not impossible habitats in which large cats can practice their traditional hunting strategy of a careful stalk followed by rapid closure with the prey. Bison may have made a suitable meal, but they would have been inaccessible to large cats because of the lack of ground cover. It is within this context that the extinction of the large cat population in North America at the end of the Pleistocene becomes not only explainable, but plausible.

Until recently, only two saber-tooth cats were recognized in the North American Pleistocene. One of these, *Smilodon fatalis*, is known from thousands of specimens recovered from the Rancho La Brea tar pits. For most people, *Smilodon* represents the archetype of a saber-tooth carnivore. Its long, slender upper canines and robust, short-limbed skeleton have intrigued scientists and fantasy authors alike. Contemporary with it, but rarely occurring in the same locality, is a more slender, longer-legged sabercat, with shorter, broader upper canines, identified as *Homotherium serum*. Before the skeletons of these two animals were described, there was considerable confusion between them. Initially, the North American *Homotherium* material had been described under the name *Dinobastis* by Cope (1893). This description of *Dinobastis* was largely based on an upper canine that was shorter and more coarsely serrated than that of *Smilodon*. The discovery of complete skeletons of *Dinobastis* in Friesenhahn Cave, Texas (Meade, 1961; Rawn-Schatzinger, 1992), demonstrated that there were considerable skeletal differences

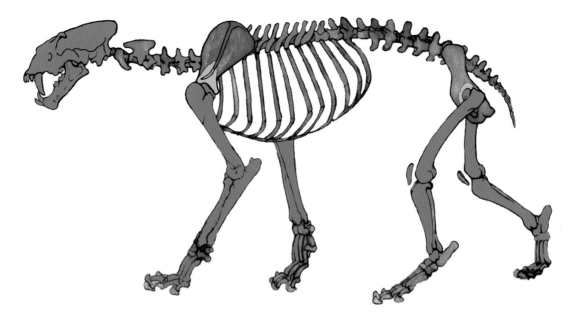

Figure 4.1. The combined skeleton of *Xenosmilus hodsonae*, BIOPSI 101 and UF 60,000, in left lateral view. The elements present are shown in deep blue. A fragment of the right scapula and right pelvis elements are indicated by a lighter blue. (drawing Mary Tanner)

between that form and *Smilodon*. Kurtén (1963) proposed that *Smilodon* and *Homotherium* represented different adaptive types of saber-tooth cats that could be recognized as tribes, the Smilodontini and the Homotheriini. Churcher (1966) suggested that *Dinobastis serus* from Friesenhahn Cave was, in reality, a *Homotherium* and that *Homotherium* could be placed in a separate subfamily, the Homotheriinae. *Smilodon*, along with *Megantereon*, would constitute the subfamily Smilodontinae. Kurtén (1963) had suggested the now popular term "scimitar-tooth" cat for the tribe Homotheriini and "dirk-tooth" cat for the tribe Smilodontini. He did not expect that these terms would be extended to other saber-tooth carnivores. However, Martin (1980) did exactly that. He suggested that dirk-tooth cats really represented an adaptive type, or ecomorph (White, 1984; White and Keller, 1984), which could subsequently be used to characterize a number of taxonomically unrelated saber-tooth carnivores. Also, "scimitar-tooth" cat could serve the same purpose. Martin also pointed out that these designations characterized animals that shared postcranial and cranial features as well as dental ones. This paradigm needed to be modified (Martin et al., 2000; Naples et al., 2002) because of the description of a third saber-tooth cat from the early Pleistocene of North America that belongs to the Homotheriini but differs in striking and important ways from typical scimitar-tooth cats.

This animal, *Xenosmilus hodsonae* or "cookie-cutter" cat (Martin et al., 2000; figs. 4.1, 4.2), was described from two

partial skeletons from an Early Pleistocene cave deposit in Florida (University of Florida collecting site Haile 21A). This site was discovered by quarry operators who were in the process of destroying it. Initially, these fossils were salvaged by a local commercial collector, Larry H. Martin, and later by the staff of the University of Florida at Gainesville. Partial skeletons of two individuals were recovered. One partial skeleton was donated to the University of Florida by the collector, Larry H. Martin. The other remained in a private collection (fig. 4.3). The abundance of a large peccary (*Platygonus vetus*, fig. 4.4A) encouraged commercial and amateur collectors to give the locality the colloquial name "hog heaven." The entire site has since been removed through mining operations. Along with the peccaries, there was an upper canine of a smaller saber-tooth cat, *Smilodon gracilis* (fig. 4.4B), and fragmentary remains of a gomphothere. The presence of a gomphothere with *S. gracilis* indicates an early Irvingtonian age for this locality, or about one M.Y.A. (fig. 4.4C; modified from Hulbert, 2001). It seems likely that the site was used as a den by the sabercats, and much of the other osteological material was introduced by them. Initially, the skeletons were thought to belong to *H. serum* and were largely ignored. The remaining partial skeleton was obtained by BIOPSI and later became the subject of a joint study by the authors (Martin et al., 2000), who showed that it represented a unique new form of saber-tooth cat.

Figure 4.2. A life reconstruction of *Xenosmilus hodsonae: A,* how the animal might be camouflaged in a tall grass savanna; *B,* reconstruction showing the striped-coat pattern typical of animals that live in grasslands, where the light is interrupted in vertical stripes by tall blades of grass. (drawing Mary Tanner)

METHODS AND MATERIALS

The two specimens of the cookie-cutter cat *X. hodsonae* (BIOPSI 101 and UF 60,000; fig. 4.1) were compared with homologous elements of the other large contemporary felids from North America, in addition to the brown bear *Ursus arctos* (KUVP 23034) and the giant panda *Ailuropoda melanoleuca* (Davis, 1964; CNHM 47432 and CNHM 74269). (The acronym CNHM used by Davis has been changed, and the present designation for these specimens is FMNH, Field Museum of Natural History.) The felid genera include: lion *Panthera leo* (KU 638), mountain lion *Puma concolor* (KU 96904), American lion *P. atrox* (KUVP 44409), dirk-tooth cat *Smilodon fatalis* (KUVP 98124), Pliocene scimitar-tooth cat *H. ischyrus* (IMNH 900-11862), and Pleistocene scimitar-tooth cat *H. serum* (TMM 933-3582; TMM 933-3231). The authors took all table measurements directly from original material when possible or from the published literature as noted. All measurements were obtained using digital or dial calipers. A list of institutional abbreviations can be found in appendix A. A measurement point location reference can be found in appendix B.

Figure 4.3. Photograph of the holotype, *Xenosmilus hodsonae*. (courtesy Jerry Stark)

DESCRIPTION AND COMPARISONS
THE SKULL

The skull (fig. 4.5; tables 4.A.1, 4.A.2; appendix B) is about the size of that of a small African lion *P. leo,* or that of a small example of the Pleistocene saber-tooth cat *Smilodon fatalis.* The skull is dolichocephalic, contrasting with the extremely brachiocephalic skull of *Homotherium.*

The nasals in *Xenosmilus* are rectangular, in contrast to wedge-shaped nasals, with a more pointed posterior (naso-frontal) suture as in *Panthera.* They are short, broad, and more arched than in *Panthera.* The nasal aperture in *Xeno-smilus* is narrower and more rectangular than in *Panthera* and *Smilodon.* In anterior view (fig. 4.6A), the frontal bones of *Xenosmilus* are depressed at their juncture with the nasals and are noticeably narrower than in *Panthera* and *Smilodon,* while the frontals of *Homotherium* (fig. 4.6B) are relatively wider than in all the other taxa discussed. In *Smilodon,* the frontal-maxillary suture is straight laterally and sinuously

curved. The posterior premaxillary process ends just above the front of the infraorbital foramen. The suture between the jugal and the zygomatic process of the squamosal is more vertical in *Xenosmilus* and *Smilodon* than in *Panthera,* in which it is more horizontal. In comparison to the other three taxa, and corresponding with the narrowness of the skull, the postorbital processes of *Xenosmilus* are extremely small, being reduced to thick, bony knobs with small, pointed tips that barely penetrate the space of the orbit. In *Panthera,* the postorbital processes are broad, somewhat sharp-edged, with ventrally curled, pointed tips, and extend almost half way across the orbit, where they approach the most anterior edge of the zygomatic arches. In *Homotherium,* the postorbital processes are also thick and blunt, but they extend over the orbit, aligning with the zygomatic arches in anterior view. *Smilodon* also has large, prominent postorbital processes that form a ridge on the posterior edges and are more robust and pointed than in *Homotherium, Panthera,* and *Xenosmilus.* In *Smilodon,* this feature extends out to about half the distance

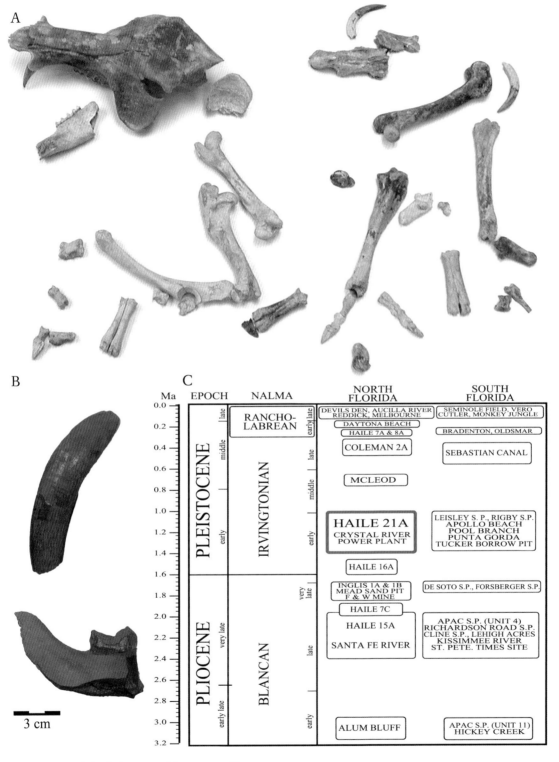

A

B

C

Ma	EPOCH		NALMA		NORTH FLORIDA	SOUTH FLORIDA
0.0	PLEISTOCENE	late	RANCHO-LABREAN	early late	DEVILS DEN, AUCILLA RIVER REDDICK, MELBOURNE	SEMINOLE FIELD, VERO CUTLER, MONKEY JUNGLE
0.2					DAYTONA BEACH	
					HAILE 7A & 8A	BRADENTON, OLDSMAR
0.4		middle	IRVINGTONIAN	late	COLEMAN 2A	SEBASTIAN CANAL
0.6				middle		
0.8					MCLEOD	
1.0		early		early	HAILE 21A CRYSTAL RIVER POWER PLANT	LEISLEY S. P., RIGBY S.P. APOLLO BEACH POOL BRANCH PUNTA GORDA TUCKER BORROW PIT
1.2						
1.4						
1.6					HAILE 16A	
1.8	PLIOCENE	very late	BLANCAN	very late	INGLIS 1A & 1B MEAD SAND PIT F & W MINE	DE SOTO S.P., FORSBERGER S.P.
2.0					HAILE 7C	
2.2				late	HAILE 15A	APAC S.P. (UNIT 4) RICHARDSON ROAD S.P. CLINE S.P., LEHIGH ACRES
2.4					SANTA FE RIVER	KISSIMMEE RIVER ST. PETE. TIMES SITE
2.6						
2.8		early late		early		
3.0					ALUM BLUFF	APAC S.P. (UNIT 11) HICKEY CREEK
3.2						

3 cm

Figure 4.4. *A*, Skeletal elements of the peccary, *Platygonus vetus*, BIOPSI 0175, a species that has been found in association with *Xenosmilus*. *B*, Elements of *Smilodon gracilis*, BIOPSI 0176, a smaller dirk-tooth cat that shared the environment with *Xenosmilus hodsonae*. *C*, Pliocene-Pleistocene timescale indicating the relationship of the Haile 21A locality (outlined in red), where these specimens were found, compared to other nearby Florida localities. The timescale is modified from Hulbert (2001).

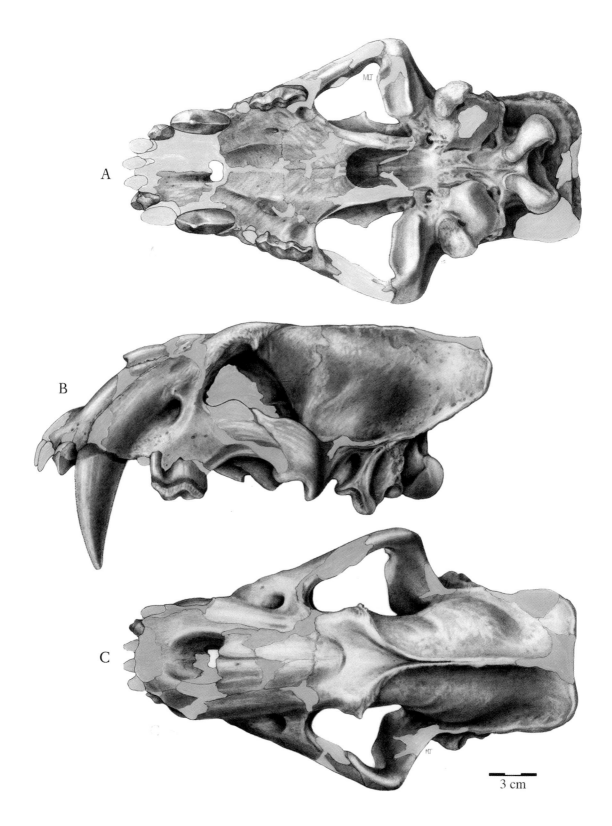

Figure 4.5. The skull of *Xenosmilus hodsonae*, BIOPSI 101:
A, ventral view; B, lateral view; C, dorsal view. Cranial mea-
surements are listed in tables 4.A.1 and 4.A.2. (drawing
Mary Tanner)

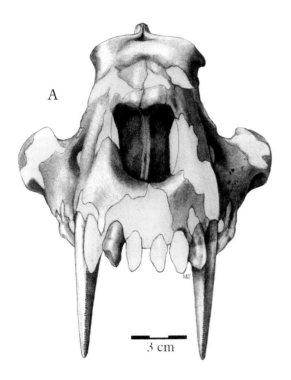
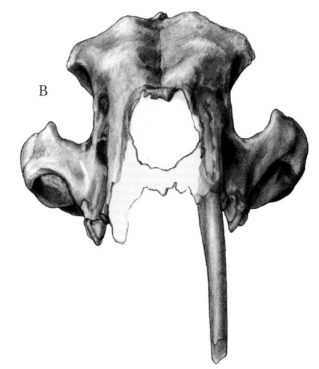

3 cm

Figure 4.6. *A*, The skull of *Xenosmilus hodsonae*, BIOPSI 101, anterior view. *B*, The skull of *Homotherium ischyrus*, IMNH 900-11862, anterior view. This comparison shows how much narrower the skull of *Xenosmilus* is than that of a more typical homothere. (drawing Mary Tanner)

toward the zygomatic arch in anterior view and helps to form the dorsal aspect of the orbit. The anterior half of the infratemporal fossa forms an infraorbital fossa, leading to the infraorbital foramen. It is smaller and rounder than in the other three felid taxa to which it was compared. The infraorbital foramen in *Xenosmilus* serves as the passage for both the sphenopalatine and pterygopalatine arteries and nerves and is both absolutely and relatively smaller than in either *Panthera* or *Smilodon*. In all three species it is oval, with the long axis oriented dorsoventrally and inclined slightly medially at the dorsal end. Although the infraorbital foramen in *Homotherium* is also oval and slightly larger than in *Xenosmilus*, its long axis is oriented dorsolaterally. There is a shallow sulcus above the infraorbital foramen for the lip retractor muscle, M. levator labii superioris. This feature is more pocketed in *Homotherium*, broader in *Panthera,* and more oval dorsoventrally in *Smilodon*. The incisor row in *Smilodon* is more procumbent than in *Panthera*, but in both of these genera the incisor arc is relatively flat. In contrast, the incisor arc of *Xenosmilus* is more procumbent and curved, as in *Homotherium*. The narrowness of the muzzle in both *Xenosmilus* and *Homotherium* is emphasized in anterior view (fig. 4.6A–B). While the width of the frontal bones, measured across the postorbital processes, is much wider in the Idaho homothere at 124 mm, compared to *Xenosmilus* at 75 mm, the muzzle of

the homothere is actually narrower, measuring 64 mm, compared to the muzzle of *Xenosmilus* at 72 mm. Both *Panthera* at 106 mm and *Smilodon* at 102 mm have overall larger, wider, and longer muzzles. These proportions may correlate with the acuity of the sense of smell, as in bears and compensate for the larger and wider roots of the canine teeth. The incisor row is medial to, and separated by a diastema from, the canine. The canine has a more lateral position in *Smilodon* and *Panthera*. In contrast, in *Xenosmilus* and *Homotherium*, the incisor row overlaps the canine teeth in anterior view. Furthermore, in *Xenosmilus*, the diastema between the upper canine and the upper I^3 is absent.

The zygomatic arch in *Xenosmilus* shows markedly less lateral flare than in either *Panthera* or *Smilodon* and slightly less than in *Homotherium,* again illustrating the narrowness of the skull. The zygomatic arch is short and arched upward, with a broad attachment on the lateral surface for M. masseter superficialis, which extends all the way posteriorly to the glenoid (fig. 4.5B). The dorsal surface of the squamosal portion of the zygomatic arch in *Xenosmilus* shows a slight concavity, as does that in *Panthera,* although the region is wider anteroposteriorly in *Xenosmilus*. It is narrower in both *Homotherium* and *Smilodon* and inclined ventrally toward the anterior edge, providing a more anterior support for the origin of M. temporalis. The zygomatic arches of both *Smilodon* and *Panthera* enclose a greater space for the temporal fossa than does that of *Homotherium* and *Xenosmilus*, although the latter is even more reduced.

The temporal region in *Xenosmilus* is stretched anteroposteriorly so that the nuchal crest greatly overhangs the occipital condyles. This contrasts with the normal condition found in saber-tooth cats, in which the occiput tends to be more vertical, thereby shortening the excursion for the temporalis muscle and giving it a more vertical line of action in closing the jaw. The sagittal crest extends anteriorly, with the ridges converging just behind the orbits, partially as a result of the frontals being extraordinarily narrowed. In fact, they are narrower than in any other felid that we have examined. The sagittal crest is high and continues into a large but not deeply recessed nuchal crest. Lack of a recessed nuchal crest is also characteristic of *Homotherium*, but not of *Smilodon*, in which the occipital crest region is deeply depressed just dorsal to the expansion for the braincase. This lack of recession in both *Xenosmilus* and *Homotherium* suggests that the action of the temporalis musculature differed from that in the smilodontins. The surface of the parietals in the area of the occipital crest is pockmarked by small foramina. These are nutrient foramina supplying the temporalis musculature.

The frontoparietal suture in *Xenosmilus* lies just posterior to the convergence of the sagittal crest, much as it does in *Panthera*. The cranium does not expand toward the orbital area as it does in *Panthera*, more closely resembling *Homotherium* in this respect. In *Homotherium*, the postorbital processes are broad, massive structures that extend ventrally and partially enclose the back of the orbit, protecting the eye from the movements of the temporalis muscle. This condition is absent in *Xenosmilus*. In *Xenosmilus*, the orbit is small and relatively vertical. The eye position is not rotated outward as far as in *Panthera*, and the position is more similar to that in *Smilodon*. There is a tiny ethmoid foramen. In *Xenosmilus*, a narrow groove measuring approximately 5 mm wide, leading to the orbital fissure, arises immediately ventral to the orbital process. This feature is narrower than in *Panthera*, *Smilodon*, and *Homotherium*. The ethmoid foramen is anterior to, and not encompassed by, this distinct groove. The optic foramen is broader and more lateral than in *Panthera* or *Smilodon*. The orbital fissure in *Xenosmilus* is rounder and larger than in *Panthera* and *Homotherium*. The foramen rotundum and foramen ovale in *Xenosmilus* are small compared to both *Panthera* and *Homotherium*.

All saber-tooth cats exhibit some cranial flexion. Unlike *Panthera*, but similar to *Homotherium* and *Smilodon*, the juncture between the glenoid fossa and the pterygoid wings is deeply incised. In *Xenosmilus*, the foramen ovale is separated from the auditory tube by a broad, bony bar, and is no longer on the ventral surface but faces anteriorly into the orbit. This probably results from cranial flexion, with the face tilted downward in *Xenosmilus*, as described by Emerson and Radinsky (1980) for other saber-tooth carnivores. The glenoid

fossa is not as broad as it is in *Panthera* and is shifted forward in relationship to the pterygoid wings, partially reflecting a shorter anteroposterior extension of the space enclosed by the zygomatic arches.

The posterior ventral processes of the glenoids in *Xenosmilus* are shorter and broader than in either *Homotherium* or *Panthera*, but the anterior processes are similar in all three genera. There is no postglenoid foramen. The glenoid surface is dorsal to the tooth row, as it is in *Panthera* and *Homotherium*.

In *Xenosmilus*, the mastoid process is larger than in *Homotherium* and is divided by a notch into an anterior flat surface and a larger posterior surface that is tilted forward. In *Homotherium*, this division is visible, but the posterior tilted portion is not as large or as strongly developed. The notch separating the two mastoid facets in *Xenosmilus* extends the entire mediolateral width of the process, ending just anterior to the stylomastoid foramen. The mastoid processes of *Xenosmilus* and *Homotherium* differ from those of *Smilodon*, being longer anteroposteriorly than wide mediolaterally. This is partially due to the mastoid process being entirely anterolateral to bullae that tilt toward the midline in *Homotherium*. The mastoid processes stretch across the anterior surface of the auditory bulla in *Smilodon*. The mastoid process in *Xenosmilus* is grooved posteriorly, and this groove is much more prominent than in *Homotherium*. In *Homotherium*, the mastoid process is separated posteriorly from the bulla by a broad, shallow groove that terminates posteriorly in a small paraoccipital process, widely separated from the mastoid process. The groove is missing from *Xenosmilus*, and although the region is damaged, a separate paraoccipital process, if present, was small. The paraoccipital process surface is the origin of the M. digastricus, and the lateral, tilted surface, along with the ventral borders of the mastoid, is the origin of the M. sternomastoideus.

In *Xenosmilus*, there is a single large stylomastoid foramen. The posterior lacerate and condylar foramina are enclosed in a single opening, as in advanced feline cats. The occipital condyles continue across the ventral surface of the basioccipitals. Homotheres have a keel extending anteriorly from the pharyngeal tubercle, which is absent in the pantherines. The pharyngeal tubercle in *Xenosmilus* is larger and more prominent than in the other cats examined. In *Xenosmilus*, there is a distinct depression posterior to the pharyngeal tubercle and between the anterior extensions of the condyles. This feature is absent in the three other genera examined. The external auditory opening is small and dorsal to the glenoid, in contrast to *Panthera*, reflecting the greater cranial flexion in the saber-tooths. The glenoid articulation in *Xenosmilus* is broad and rectangular with a long, low postglenoid process, while the articulation in *Homotherium* is narrower, with a larger and more triangular postglenoid process. In *Xenosmilus*, the anterior glenoid process is low, although probably better developed than in *Smilodon*. In general, the

A

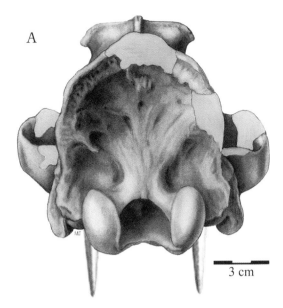

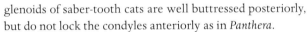

3 cm

B

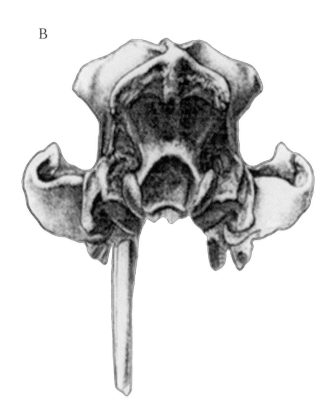

Figure 4.7. *A,* The skull of *Xenosmilus hodsonae,* BIOPSI 101, posterior view. *B,* The skull of *Homotherium ischyrus,* IMNH 900-11862, posterior view. Although the skull in *Xenosmilus* is narrower than that of a typical homothere, the posterior occipital region is wider as well as longer anteroposteriorly. (drawing Mary Tanner)

glenoids of saber-tooth cats are well buttressed posteriorly, but do not lock the condyles anteriorly as in *Panthera.*

In *Xenosmilus* (fig. 4.7A), the size of the occipital condyles correlates to the absolute size of the animal, yet the head is surprisingly small for the body size. The occipital condyles project farther posteriorly than do those in *Panthera* and are broader and more robust, with a large dorsal surface. The occiput is constricted above the foramen magnum, and the overhanging occipital surface is roughened, ridged, and has deep depressions, possibly from the presence of large nutrient foramina. Posteriorly, the rectus capitis muscles of *Xenosmilus* attach as a broad sheet, as in the pantherine cats, rather than into two distinct pockets, as in *Homotherium* (fig. 4.7B) and *Smilodon.* The elongated occipital crest extends forward almost horizontally to the zygomatic arch.

The Palate

The palatine ridges are accentuated, and the palate is depressed anteriorly in the direction of the incisive foramina (fig. 4.5A). The premaxillaries are rounded and very heavily built, with the incisive foramina more posterior than in *Panthera.* The incisive groove and foramina in *Xenosmilus* are almost confluent, whereas they are separate in *Smilodon* and *Panthera.*

In *Xenosmilus,* the pterygoid ridges converge posteriorly toward each other, and the internal nares are relatively small and taller than they are wide. There are embrasure pits, located just medial to the carnassials, and the M^1 is not developed. The posterior palatine foramina lie just anterior to a

maxillary-palatine suture that is more posteriorly situated than in *Panthera.* The external narial opening is taller than it is wide and is almost rectangular. The posterior border of the internal nares is considerably posterior to the tooth row. The anterior margin of the nasals lies at the back of the canines, and the premaxillaries project anteriorly.

In *Xenosmilus* (BIOPSI 101), the incisors have been worn. They are large and recurved posteriorly, but still show lateral serrated ridges and are arranged in a semicircle with no diastema separating them from the canines. In *Xenosmilus,* the alveoli show that the upper incisors were graded in size, increasing from front to back, with the third incisor about twice the size of the first. The canines are flattened mediolaterally, keeled, and serrated front and back as in other scimitar-tooth cats. The serrations are coarse. There is a short diastema between the posterior edge of the canine and the upper P^3. The P^3 is single-rooted, compared to the double-rooted P^3 of the Idaho homothere and *Smilodon.* The protocone of P^4 is absent, as in other homotheres, with a slight protoradix present. The carnassial is highly worn, and the wear pattern consists of vertical striations. The M^1 is absent and does not appear ever to have been present, whereas in *Homotherium, Smilodon,* and *Panthera* the M^1 is retained.

THE MANDIBLE

The lower jaw (fig. 4.8, BIOPSI 101; fig. 4.9, UF 60,000; tables 4.A.3, 4.A.4; appendix B) has a flattened anterior symphysis.

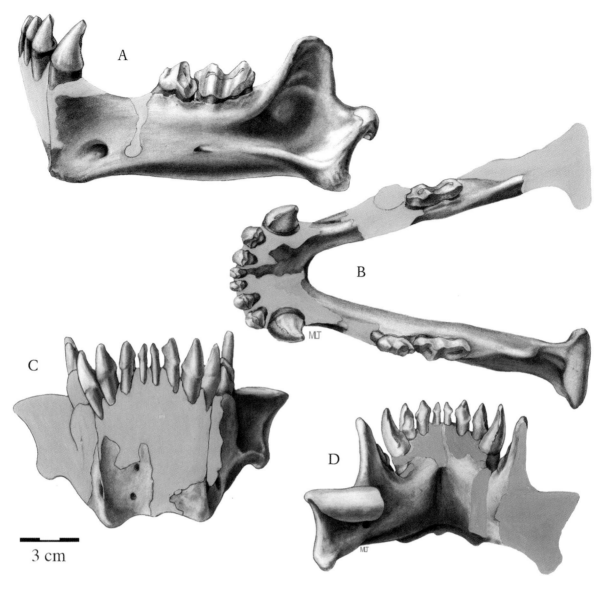

Figure 4.8. The mandible of *Xenosmilus hodsonae*, BIOPSI 101:
A, lateral view; *B*, dorsal view; *C*, anterior view; *D*, posterior view.
Mandibular measurements are listed in tables 4.A.3 and 4.A.4.
(drawing Mary Tanner)

Anteriorly, there are two or possibly three nutrient foramina at the bottom of the jaw surface, just lateral to the symphyseal connection. In *Xenosmilus,* there is a very slight dependent flange, whereas this flange is more pronounced and rounded in *H. serum* and much more flared anteriorly, forming a sharp ridge in *Smilodon*. There are two laterally facing mental foramina just below the diastema between C and P₄. *Homotherium* generally has a single large mental foramen, sometimes accompanied posteriorly by another small mental foramen in the body below P₃. *Smilodon* usually has a single lateral mental foramen below the diastema, but this feature, as in all other cats, tends to vary within individual specimens. *Panthera* lacks the anterior flange and usually has

three to four small lateral mental foramina. In *Xenosmilus* (BIOPSI 101), there is a third small mental foramen, located in the body of the mandible, just ventral to the junction of P₄ and M₁. Another similar small mental foramen occurs in UF 60,000, just anterior to the P₄. The mandibular fossa in *Xenosmilus* extends anteriorly to about the middle of the carnassial. This extension is about the same in *H. ischyrus* and *Smilodon*. In *Panthera* and *H. serum,* it ends abruptly, just below the posterior edge of the M₁. In *Xenosmilus, Smilodon,* and *H. ischyrus,* there is no anterior masseteric pocket, as is present in *H. serum* (Martin, Schultz, and Schultz, 1988). A low, blunt, and vertical coronoid process terminates at about the same level as the tip of the lower canine in both *Xeno-*

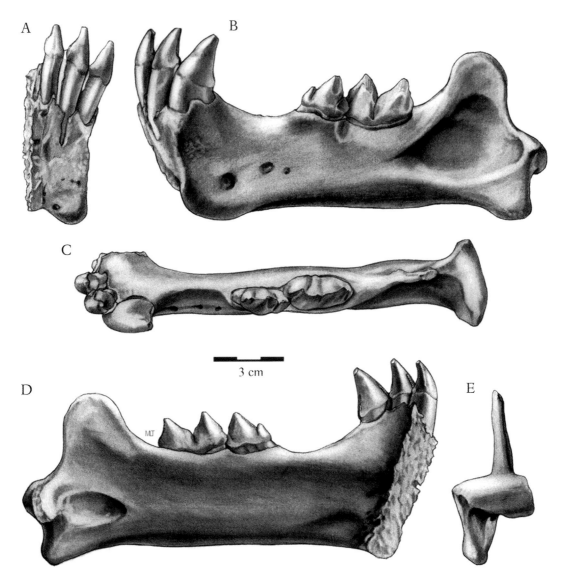

Figure 4.9. The left hemimandible of *Xenosmilus hodsonae*, UF 60,000: *A*, anterior view; *B*, lateral view; *C*, dorsal view; *D*, medial view; *E*, posterior view. (drawing Mary Tanner)

smilus and *H. ischyrus*. In *Smilodon*, the coronoid process angles posteriorly toward the condyle and is slightly lower than the tip of the lower canine. This reduction in height of the coronoid process helps avoid overstretching of the superficial masseter muscle (Emerson and Radinsky, 1980). In *Panthera*, the large, tall coronoid process angles posteriorly and ends about even with the anterior edge of the condyle, extending about 21 mm above the tip of the lower canine. In *Xenosmilus* and *H. ischyrus,* the masseteric fossa undercuts the lateral margin of the mandibular condyle. This feature is present, but not as obvious in *Panthera* and *Smilodon*. There is no undercutting by the masseteric fossa in *H. serum*. Ventral to the condyle, the angular process in *Xenosmilus* and *H. serum* is not separated by a notch, as in *Panthera, Smilodon,* and *H.*

ischyrus. The vertical blunt angular process of *Xenosmilus* does not inflect inward as it does in *Smilodon, Homotherium,* and *Panthera*. The dorsal surface of the mandibular condyle merges anteriorly and smoothly onto the mandibular body in *Xenosmilus*. It is not separated by the groove that divides the anterior condylar margin from the condylar notch in *Smilodon, Homotherium,* and *Panthera*. In *Xenosmilus,* the lateral mandibular surface below the condyle and ventral edge extends posteriorly to include the mandibular angle, which serves as the insertion of the M. masseter superficialis. It is similar to *H. serum* in this respect. The medial surface is the site of insertion of the M. pterygoideus medialis. Together, these muscles elevate the mandible, along with the M. temporalis, which inserts onto both the medial and lateral aspects of the coronoid process. The masseter and medial pterygoid muscles also act to maintain pressure while using the carnassials, keeping those teeth apposed during the cutting phase of the masticatory stroke. These muscles also control mediolateral deviation of the mandible, preventing misalignment of and damage to the canines during mandibular closure. In *Xenosmilus*, the masseteric fossa does not flare as far laterally as it does in both *Smilodon* and *Panthera*, being more similar to *Homotherium* in this feature. The degree of flare most likely reflects the relative thickness of the deep masseter muscle. In dorsal anterior view, the coronoid process of *Xenosmilus*, which serves as the insertion of the M. temporalis, is not as medial to the tooth row as it is in *Homotherium*, *Smilodon*, and *Panthera*. In *Xenosmilus*, the mandibular foramen is positioned directly below the coronoid process and opens posteriorly. This foramen is located slightly anterior to the vertical centerline of the coronoid process in *Homotherium* and *Smilodon*, and more so in *Panthera*, almost reaching the anterior boundary.

In *Xenosmilus*, the anterior symphysis of the lower jaw is heavily developed, suggesting that the mandibular rami could not shift while active and that the large forces generated during biting would be easily transmitted across the mandible. The lower incisors of *Xenosmilus* are larger than in *Smilodon*, *H. serum*, and *H. ischyrus*. They are recurved, cone shaped, with the anteroposterior length greater than the transverse. They are all graded in size, with I_3 about two-thirds the size of the canine. All the incisors show wear facets on both the lateral and medial sides from occlusion with the upper incisors. There are two small lateral cusps on each of the incisors, with the medial cusp being slightly elevated compared to the lateral. All the incisors have lateral and medial serrated ridges that terminate in serrated cuspules. In *Homotherium* and *Smilodon*, the lower canines are more laterally compressed, and they are almost conical in *Panthera*. In dorsal view, the large lower canine of *Xenosmilus* is triangular and closes the narrow gap between the upper canine and upper I^3 when the mouth is closed. This arrangement may prevent any tissue from bulging between the upper and lower canines and typifies the cookie-cutter bite.

The P_3 is absent in *Xenosmilus*, and, for the most part, in *Smilodon fatalis*, although some specimens retain this tooth. In *Xenosmilus*, P_4 is tilted posteriorly and overlaps M_1. The P_4 has a distinct parastyle and the paraconid is level with the paraconid of M_1. The P_4 is about two-thirds the size of M_1. On the posterior lateral surface of P_4 in BIOPSI 101, there is a large wear facet for occlusion with the upper P^4. The M_1 in *Xenosmilus* is slightly longer than in *H. ischyrus* and *H. serum*. In *Xenosmilus* and *Homotherium*, the carnassial notch of M_1 is broad, and the protoconid and paraconid are about equal in height in all three species. The paraconid of M_1 in *Xenosmilus* is about equal in length to the protoconid, and this is the same for *H. ischyrus*. The length of the protoconid

in *H. serum*, *Smilodon*, and *Panthera* is greater than the length of the paraconid. This is true for most of the saber-tooth cats (Merriam and Stock, 1932). The carnassial notch in both *Smilodon* and *Panthera* is more deeply incised than in *Xenosmilus* or *Homotherium*.

THE POSTCRANIAL SKELETON
The Axial Skeleton
The Cervical Vertebrae
THE ATLAS
The atlas in *Xenosmilus* was not recovered.

THE AXIS
The axis vertebra in *Xenosmilus* (BIOPSI 101, fig. 4.10A–C, table 4.1) is shorter and broader than in *Smilodon*. It is as short as in *Panthera* but broader. Most of the upper portion of the spinous process is missing. The articulation with the atlas is more forward facing and would allow less lateral movement than in any of the other cats including *Homotherium*. The odontoid process is about the same length as in *Panthera* but is much more robust and has a prominent dorsal knob. It is pointed, as in the Idaho homothere and *H. serum*. The centrum shape in *Xenosmilus* is triangular in anterior view, as opposed to rectangular in *Panthera*, and almost hourglass-shaped in *Smilodon* and *Homotherium*. *Homotherium* has an elongated centrum, as does *Smilodon*. The neural canal in *Xenosmilus* is much broader than in *Homotherium*, *Smilodon*, or *Panthera*. The floor of the neural canal is nearly flat in *Xenosmilus* and *Homotherium*, while a low median ridge is present in *Panthera* and a prominent one in *Smilodon*. The ventral surface of the centrum in *Xenosmilus* and *Homotherium* both have a high posteroventral medial keel, indicating unusual development of the M. longus coli for depression of the head. The transverse processes, which serve as attachment for muscles, are broadly spread in *Xenosmilus*. These processes are less spread in *Homotherium*, *Smilodon*, and *Panthera*, but in *Xenosmilus* they are more elongate than in the other saber-tooths. The vertebral foramina are large and oval in *Xenosmilus* and smaller and more slit-like in *Smilodon* and *Homotherium*. *Panthera* is intermediate in this respect. In *Xenosmilus* and *Panthera*, a groove continues from the posterior part of the vertebral foramen onto the posteromedial surface of the transverse process. This groove is absent in *Homotherium* and *Smilodon*. Anteriorly, the vertebral foramen opens into a deep pocket in *Xenosmilus*, *Smilodon*, and *Homotherium*. This pocket is absent in *Panthera*.

CERVICAL VERTEBRA THREE
The third cervical vertebra (BIOPSI 101, fig. 4.10D–F, table 4.2) is missing the dorsal arch but shows a large vertebral canal that inclines laterally, and the centrum is excavated ventrolaterally. This vertebra is shorter anteroposteriorly than

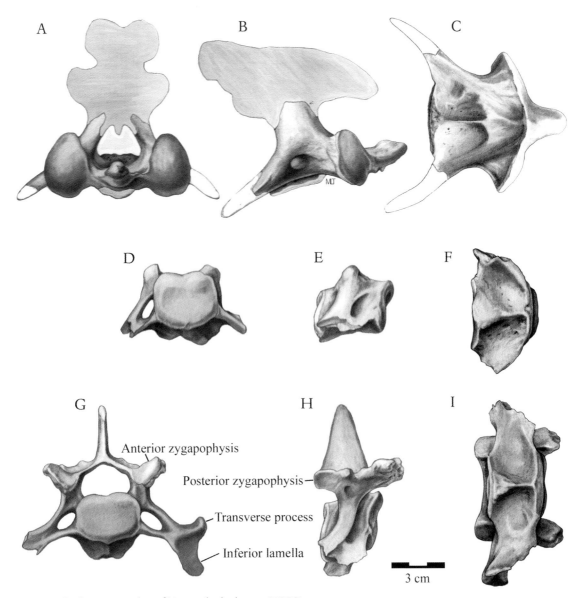

Figure 4.10. The axis vertebra of *Xenosmilus hodsonae*, BIOPSI 101: *A*, anterior view; *B*, lateral view; *C*, ventral view. Cervical vertebra three (C3) of *X. hodsonae*, BIOPSI 101: *D*, anterior view; *E*, lateral view; *F*, ventral view. Cervical vertebra five (C5) of *X. hodsonae*, UF 60,000: *G*, anterior view; *H*, lateral view; *I*, ventral view. (drawing Mary Tanner)

in *Smilodon* and slightly longer than in *Panthera*. The anterior surface of the centrum in *Xenosmilus* is divided into a dorsal half tilted ventrally and a ventral half facing straight anteriorly. This same flexure is present in the other two sabertooth cats, *Homotherium* and *Smilodon*, but is poorly developed in *Panthera*.

CERVICAL VERTEBRA FOUR

The fourth cervical vertebra was not recovered in *Xenosmilus*.

CERVICAL VERTEBRA FIVE

The fifth cervical vertebra (UF 60,000; fig. 4.10G–I; table 4.3) is reasonably complete. It has large vertebral arterial canals and a lower spinous process compared to *Smilodon*. The centrum is shorter than in *Smilodon* and lacks the two posterior ventral processes found in that genus. It is excavated ventrolaterally, with a prominent median keel. The centrum is slightly wider than in *Smilodon*, and the vertebral canal is larger and more inclined. In *Xenosmilus*, the dorsal surface of the neural arch lacks the lateral ridges found in *Smilodon*.

Table 4.1. Axis measurements (mm): *Xenosmilus hodsonae*. Compared specimens (observed range), *Smilodon fatalis*, *Panthera atrox* (Merriam and Stock, 1932). (*a* = approximate)

	X. hodsonae BIOPSI 101	Smilodon fatalis		Panthera atrox	
		2039-1	2039-10	2937-1	2937-8
Greatest length of neural spine	—	120.5	89.1	111.0	99.9
Greatest width of posterior end of neural spine	—	38.3	22.5	54.0	50.6
Length of centrum along median line measured parallel to lower surface from posterior	87.0	—	—	—	—
end to tip of odontoid process	31.5	102.9	86.9	99.8	103.7
Depth of centrum measured normal to floor of neural canal and across posterior epiphysis	46.0	26.0	21.8	29.0	30.7
Greatest transverse width across posterior epiphysis of centrum	86.8	41.8	36.9	47.0	46.0
Width across articulating surfaces for atlas	—	77.0	63.6	81.1	77.7
Width across outer ends of transverse processes	—	92.4	94.0	93.0*a*	94.0*a*
Width across posterior zygapophyses	—	65.0	45.1	68.8	65.4
Distance measured in median line from lower border of posterior epiphysis of centrum to top of posterior end of neural spine	25.5	90.0	81.4	97.3	97.8
Length of odontoid process	25.8	—	—	—	—
Width of odontoid process at base	15.8	—	—	—	—
Greatest depth of odontoid process	15.8	—	—	—	—

Table 4.2. Third cervical vertebra measurements (mm): *Xenosmilus hodsonae*. Compared specimens (observed range), *Smilodon fatalis* (Merriam and Stock, 1932). (*a* = approximate)

	X. hodsonae BIOPSI 101	Smilodon fatalis	
		2040-1	2040-10
Length from end of anterior zygapophysis to end of posterior zygapophysis	—	75.9	63.2
Greatest length from ends of anterior zygapophyses to ends of hyperapophyses	—	90.9	68.6
Length of centrum measured normal to posterior face and along median line	47.5	52.0	42.8
Width across anterior zygapophyses	—	69.6	53.8
Width across posterior zygapophyses	—	78.5	58.2
Greatest width of neural canal at anterior end	27.6	27.6	21.4
Greatest transverse width of posterior epiphysis of centrum	46.4	48.9	38.4
Greatest width across outer ends of transverse processes	—	148.6	108.3
Greatest length of transverse process from outer end to end of anterointeral projection of inferior lamella	—	77.7	66.3
Height from median posterior border of centrum to top of neural crest	—	76.0*a*	55.6
Depth of centrum measured normal to floor of neural canal and across posterior epiphysis	32.8	28.5	25.6

Table 4.3. Fifth cervical vertebra measurements (mm): *Xenosmilus hodsonae.* Compared specimens (observed range), *Smilodon fatalis* (Merriam and Stock, 1932).

	X. hodsonae BIOPSI 101	Smilodon fatalis 2042-1	Smilodon fatalis 2042-10
Length from end of anterior zygapophysis to end of posterior zygapophysis	57.8	64.8	54.9
Length of centrum measured normal to posterior face and along median line	35.6	45.7	38.3
Greatest width across anterior zygapophyses	76.5	86.4	71.5
Width across posterior zygapophyses	65.8	83.0	69.2
Greatest width of neural canal at anterior end	33.0	31.5	30.7
Greatest transverse width of posterior epiphysis of centrum	44.4	44.5	39.4
Depth of centrum measured normal to floor of neural canal and along median line of posterior epiphysis	35.1	30.2	25.0
Greatest width across outer ends of transverse processes	128.2	150.6	125.4
Greatest distance from end of transverse process to lower posterior end of lamella	36.0	51.4	41.0
Height from middle of ventral border to posterior epiphysis of centrum to top of neural spine	94.0	106.5	80.2

The neural spine is broader and inclined forward, while the neural spine of the corresponding *Smilodon* vertebra inclines posteriorly. The leading edge of the spinous process slopes anteriorly in *Xenosmilus* and posteriorly in *Smilodon.* The anterior zygapophyses of *Xenosmilus* are inclined upward at the lateral edges to a greater degree than in *Smilodon,* and the posterior zygapophyses are also more sharply inclined dorsally on the lateral edges. The transverse processes in *Xenosmilus* project directly laterally, while those of *Smilodon* are relatively longer and directed posteriorly, extending past the vertebral centrum. The ventral surface of the centrum of *Xenosmilus, Panthera,* and *Homotherium* lack the twin posteriorly projecting processes that are present in *Smilodon.* In *Xenosmilus,* the ventral surface of the centrum has a distinct keel, posteriorly widening into a triangular process. These features are also present in *Homotherium,* weakly developed in *Panthera,* and largely absent from *Smilodon.* The dorsal arch of the vertebral canal is anteroposteriorly narrower than in *Smilodon* or *Panthera* but resembles *Homotherium* in this respect. Furthermore, as in *Homotherium,* its anterior margin is posterior to the anterior edge of the centrum, and it is about even with the anterior edge in *Panthera* and *Smilodon.* In *Smilodon,* there is a large pocket ventral to the posterior zygapophysis that is absent in *Xenosmilus.* The transverse process in *Smilodon* sweeps posteriorly behind the centrum, but unlike *Smilodon,* this process in *Xenosmilus* is almost straight, ending at the posterior ventral border. The ventral wing of the transverse process in *Xenosmilus* extends ventrally so that it is taller than its anteroventral extent. In *Homotherium, Smilodon,* and *Panthera,* the converse is true. In *Xenosmilus,* the dorsal edges of the prezygapophyses and postzygapophyses lie on a line much as they do in *Panthera,* but in *Smilodon* and *Homotherium,* this margin is stepped,

with the prezygapophyseal edge distinctly higher than the postzygapophyses. The anterior edge of the neural arch in *Xenosmilus* is straight and deeply recessed in comparison to the edge of the centrum. In *Panthera,* the arch bulges anteriorly and is about even with the anterior edge of the centrum. *Smilodon* is intermediate between *Panthera* and *Xenosmilus* in this respect. The dorsal prominence on the neural spine is elongated anteroposteriorly in *Xenosmilus* and transversely in *Smilodon.* The posterior border of the neural arch in *Xenosmilus* is broadly triangular and about even with the posterior border of the centrum. In *Smilodon,* it is notched behind the spine, but otherwise straight, and overhangs the centrum. Viewed posteriorly, the dorsal surface of the neural arch is rounded in *Smilodon,* much as it is in *Panthera,* but is pointed in *Xenosmilus.* The posterior ventral surface of the centrum in *Smilodon* is deflected upward and is more nearly straight in *Xenosmilus, Homotherium,* and *Panthera.*

CERVICAL VERTEBRA SIX

For cervical vertebra six (BIOPSI 101, fig. 4.11A–C, table 4.4), the posterior margin of the neural arch is "V" shaped in *Xenosmilus* and *Homotherium* and straight in *Smilodon* and *Panthera.* The transverse process is shorter and tilted upward in *Xenosmilus* and *Panthera* and not so grooved posteriorly as in *Smilodon.* Unlike C5, the dorsolateral edges of the zygapophyses are stepped in *Xenosmilus* as in *Smilodon* and *Homotherium.* In *Xenosmilus,* there is also a thin anterior ridge on the transverse process of C6 that extends laterally to the vertebral foramen. The dorsal arch of the vertebral canal is narrow in both *Homotherium* and *Xenosmilus* and broad in *Smilodon* and *Panthera.* The vertebral foramen in *Xenosmilus* is relatively and absolutely larger than in *Smilodon,* and although both are ovate, that in *Xenosmilus* is more

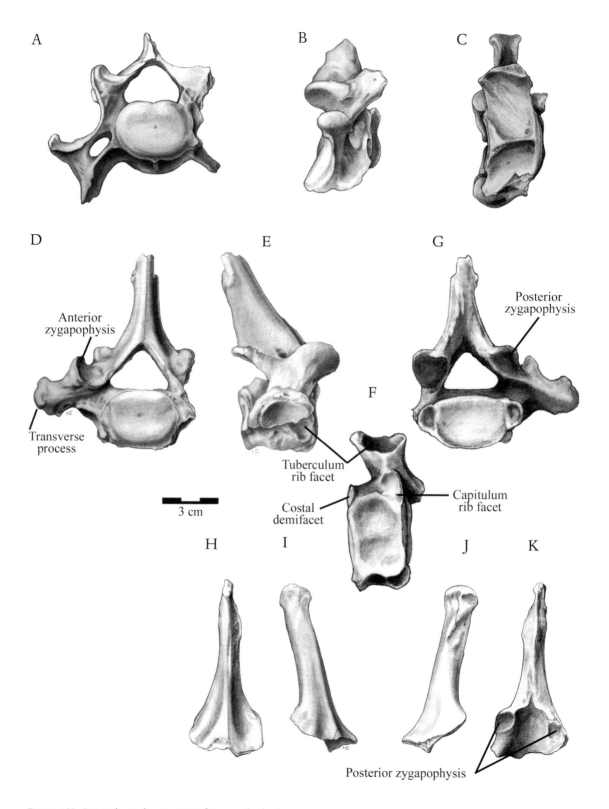

Figure 4.11. Cervical vertebra six (C6) of *Xenosmilus hodsonae*, BIOPSI 101: *A*, anterior view; *B*, lateral view right side; *C*, ventral view. Thoracic vertebra one (T1) of *X. hodsonae*, BIOPSI 101: *D*, anterior view; *E*, lateral view right side; *F*, ventral view; *G*, posterior view. Thoracic vertebra five (T5) of *X. hodsonae*, spinous process, UF 60,000: *H*, anterior view; *I*, lateral view right side; *J*, lateral view left side; *K*, posterior view. (drawing Mary Tanner)

Table 4.4. Sixth cervical vertebra measurements (mm): *Xenosmilus hodsonae*. Compared specimens (observed range), *Smilodon fatalis* (Merriam and Stock, 1932). (*a* = approximate)

	X. hodsonae BIOPSI 101	Smilodon fatalis 2043-1	2043-10
Length from end of anterior zygapophysis to end of posterior zygapophysis	52.1	62.5	49.1
Length of centrum measured normal to posterior face and along median line	34.4	39.9	34.0
Greatest width across anterior zygapophyses	77.2	89.5	73.8a
Width across posterior zygapophyses	70.8	86.0	68.9
Greatest width of neural canal at anterior end	32.7	33.9	33.0
Greatest transverse width of posterior epiphysis of centrum	45.2	40.0	35.2
Depth of centrum measured normal to floor of neural canal and along median line of posterior epiphysis	33.7	27.5	23.5
Greatest width across outer ends of transverse processes	125.5	141.7	111.3
Greatest width across inferior lamella	89.2a	98.0a	75.0
Greatest length of inferior lamella measured along lower border	55.5a	77.7	50.6
Height from middle of ventral border of posterior epiphysis of centrum to top of neural spine	—	116.0	89.4

Table 4.5. First thoracic vertebra measurements (mm): *Xenosmilus hodsonae*. Compared specimens (observed range), *Smilodon fatalis* (Merriam and Stock, 1932). (*a* = approximate)

	X. hodsonae BIOPSI 101	Smilodon fatalis 2045-1	2045-10
Greatest length from end of anterior zygapophyses to end of posterior zygapophyses	55.7	67.8	57.4
Length of centrum measured normal to posterior face along median line	33.6	40.4	34.0
Greatest width across anterior zygapophyses	77.2	81.4	71.7
Greatest width across posterior zygapophyses	61.1	63.1	51.3
Height of neural canal at anterior end	23.0	23.7	19.5
Greatest transverse width of posterior face of centrum across capitular facets	63.6	57.4	48.6
Depth of centrum measured normal to floor of neural canal and along median line of posterior epiphysis	33.3	30.2	27.6
Greatest width across outer ends of transverse processes	122.0	134.5	110.7
Greatest anterioposterior diameter of outer end of transverse process	33.7	32.0	27.0
Height from middle of ventral border of posterior epiphysis of centrum to top of neural spine	140.0a	152.5	115.6

horizontally oriented and located farther laterally within the transverse process. The centrum in *Xenosmilus* is larger and more rounded than in *Smilodon*, heart-shaped anteriorly, and ovoid posteriorly. The ventral surface of the centrum of C6 is keeled in *Xenosmilus* and *Homotherium* and unkeeled in *Smilodon* and *Panthera*. The centrum in *Xenosmilus* is shorter than in *Smilodon* and *Panthera* but is approximately equal to that in *Homotherium*. The posterior face of the centrum of *Xenosmilus* exceeds that of *Smilodon*, *Homotherium,* and *Panthera* in all dimensions. The ventral border of the anterior face of the centrum in *Xenosmilus* bends posteriorly to a greater degree than in *Homotherium*, whereas *Smilodon* and *Panthera* lack such a deflection.

CERVICAL VERTEBRA SEVEN

The seventh cervical vertebra was not recovered in *Xenosmilus*.

The Thoracic Vertebrae

THORACIC VERTEBRA 1–13

In anterior view the dorsal spine of *Xenosmilus* is thicker and more robust than in either *Smilodon* or *Panthera* but resembles *Homotherium*. There is a posterior ridge on the dorsal spine in *Smilodon* and *Panthera*, but this surface is flattened posteriorly in *Xenosmilus* and *Homotherium*. On the T1 of *Xenosmilus* (BIOPSI 101, fig. 4.11D–F, table 4.5), there is a bony exos-

Table 4.6. Fifth thoracic vertebra measurements (mm): *Xenosmilus hodsonae* (spinous process). Compared specimen (observed range), *Smilodon fatalis* (Merriam and Stock, 1932). (*a* = approximate)

	X. hodsonae UF 60,000	Smilodon fatalis table 21-5
Greatest length from end of anterior zygapophyses to end of posterior zygapophyses	—	60.7
Length of centrum measured normal to posterior face along median line	—	35.1
Greatest width across anterior zygapophyses	—	38.8
Width across posterior zygapophyses	41.6	39.9
Greatest height of neural canal at anterior end	—	15.9
Greatest width of neural canal at anterior end	—	—
Greatest transverse width of posterior face of centrum across capitular facets	—	50.3
Depth of centrum measured normal to floor of neural canal and along median line of posterior epiphysis	—	28.2
Greatest width across outer ends of transverse processes	—	85.2
Greatest anteroposterior diameter of outer end of transverse process	—	20.2
Height from middle of ventral border of posterior epiphysis of centrum to top of neural spine	—	119.8
Length of spine from middle of notch between anterior zygapophyses to top measured parallel to anterior end	96.0*a*	95.7

tosis on the right lateral side of the spinous process, midway along the length of the spine, which may reflect injury to the region. The prezygapophyses are tilted upward as in *Homotherium* and *Smilodon* and less so *Panthera*. The postzygapophyses expand laterally in all four genera. In *Xenosmilus* and *Homotherium*, the neural canal is triangular. It is oval in *Smilodon* and nearly circular in *Panthera*. The anterior centrum is wider than high in *Xenosmilus* and *Homotherium*, higher than wide in *Smilodon*, and nearly circular in *Panthera*. In *Xenosmilus,* the ventral capitular rib facet faces laterally and down as it does in *Homotherium*. In *Panthera*, it is tilted more anteriorly and in *Smilodon* more anteriorly and ventrally. In *Xenosmilus,* the posterior margin of the neural arch extends far behind the posterior edge of the centrum, while in *Smilodon* these margins are about even. *Homotherium* and *Panthera* resemble *Xenosmilus* in this respect. In *Xenosmilus,* the posterior margin of the ventral rib facet lies across from the middle of the dorsal spinous process. In *Smilodon*, the posterior margins are about even. *Panthera* is more like *Smilodon* in this respect. In *Xenosmilus,* the tubercular rib articulation on the transverse process is more lateral and in *Smilodon* more ventral. This articulation is nearly twice as large in *Xenosmilus* as in *Smilodon* and *Panthera* and close to the size of the articulation in *Panthera* and *Homotherium*. The ventral surface of the centrum is broad, flattened, and margined by a curved ridge, contrasting with all the other compared vertebrae.

THORACIC VERTEBRAE TWO THROUGH FOUR
Thoracic vertebrae two through four were not recovered.

THORACIC VERTEBRA FIVE
The centrum was not recovered in *Xenosmilus*, but the neural arch with the spinous process was recovered (UF 60,000; fig. 4.11H–K; table 4.6).

The spinous process is only slightly shorter than in *Smilodon* and *Panthera* but differs greatly in shape. In *Smilodon* and *Panthera*, the process is roughly triangular in lateral view, while it is rectangular in *Xenosmilus*. This rectangular feature resembles the spinous process of the giant panda. Interestingly, the shape of the ventral centrum in *Xenosmilus* is also rather bear-like as compared to the other cats. Dorsally, the tip of this process in *Smilodon* and *Panthera* is a rounded tubercle expanded transversely, while the tip in *Xenosmilus* terminates in a thickened anteroposterior ridge. The posterior edge has a distinct keel in *Xenosmilus* and is flattened and roughened in *Smilodon* and *Panthera*. In *Xenosmilus*, the dorsal spine is asymmetrical, with a distinct ridge on the left side, suggesting some preferential movement in the direction of bending of the spine and the possibility of handedness. The posterior zygapophyses are restricted to a rounded arch in *Xenosmilus*. They are inflected to form flat, outward-facing surfaces in *Smilodon* and *Panthera*.

Table 4.7. Sixth thoracic vertebra measurements (mm): *Xenosmilus hodsonae.*
Compared specimen (observed range), *Smilodon fatalis* (Merriam and Stock, 1932).

	X. hodsonae UF 60,000	Smilodon fatalis table 21-6
Greatest length from end of anterior zygapophyses to end of posterior zygapophyses	—	58.7
Length of centrum measured normal to posterior face and along median line	35.4	35.0
Greatest width across anterior zygapophyses	—	38.4
Greatest width across posterior zygapophyses	—	40.6
Height of neural canal at anterior end	—	15.0
Greatest transverse width of posterior face of centrum across capitular facets	61.2	50.5
Depth of centrum measured normal to floor of neural canal and along median line of posterior epiphysis	31.5	28.5
Greatest width across outer ends of transverse processes	—	85.5
Greatest anteroposterior diameter of outer end of transverse process	—	21.4
Height from middle of ventral border of posterior epiphysis of centrum to top of neural spine	—	117.6
Length of spine from middle of notch between anterior zygapophyses to top measured parallel to anterior end	—	95.2

Table 4.8. Seventh thoracic vertebra measurements (mm): *Xenosmilus hodsonae.*
Compared specimen (observed range), *Smilodon fatalis* (Merriam and Stock, 1932).
(*e* = estimate, + = damaged element)

	X. hodsonae UF 60,000	Smilodon fatalis table 21
Greatest length from end of anterior zygapophyses to end of posterior zygapophyses	—	58.2
Length of centrum measured normal to posterior face and along median line	35.5	34.5
Greatest width across anterior zygapophyses	—	37.5
Greatest width across posterior zygapophyses	—	38.2
Height of neural canal at anterior end	—	17.2
Greatest width of neural canal at anterior end	25.0e +	—
Greatest transverse width of posterior face of centrum across capitular facets	56.0	53.3
Depth of centrum measured normal to floor of neural canal and along median lien of posterior epiphysis	31.5	27.8
Greatest width across outer ends of transverse process	—	84.5
Greatest anteroposterior diameter of outer end of transverse process	—	21.3
Height from middle of ventral border of posterior epiphysis of centrum to top of neural spine	—	115.8
Length of spine from middle of notch between anterior zygapophyses to measured parallel to anterior end	—	96.5

THORACIC VERTEBRA SIX

The sixth thoracic vertebra in *Xenosmilus* (UF 60,000; fig. 4.12A–C; table 4.7) is missing the entire neural arch. The centrum in *Xenosmilus* is larger than in *Smilodon* and somewhat broader, but is otherwise similar.

THORACIC VERTEBRA SEVEN

The centrum of the seventh thoracic vertebra in *Xenosmilus* (UF 60,000; fig. 4.12D–F; table 4.8) is broader than long with respect to those of the other cats. The neural arch was not recovered. The anterior rib articulation is nearly circular in *Smilodon*, more oval in *Xenosmilus*, and more laterally situated

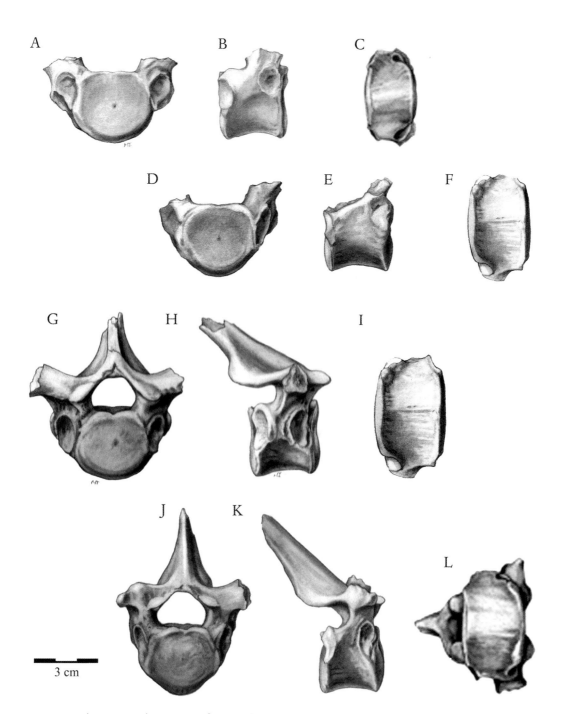

Figure 4.12. Thoracic vertebra six (T6) of *Xenosmilus hodsonae*, UF 60,000: *A*, anterior view; *B*, lateral view right side; *C*, ventral view. Thoracic vertebra seven (T7) of *X. hodsonae*, UF 60,000: *D*, anterior view; *E*, lateral view right side; *F*, ventral view. Thoracic vertebra eight (T8) of *X. hodsonae*, UF 60,000: *G*, anterior view; *H*, lateral view right side; *I*, ventral view. Thoracic vertebra nine (T9) of *X. hodsonae*, BIOPSI 101: *J*, anterior view; *K*, lateral view right side; *L*, ventral view. (drawing Mary Tanner)

Table 4.9. Eighth thoracic vertebra measurements (mm): *Xenosmilus hodsonae.*
Compared specimen (observed range), *Smilodon fatalis* (Merriam and Stock, 1932).

	X. hodsonae UF 60,000	Smilodon fatalis table 21-8
Greatest length from end of anterior zygapophyses to end of posterior zygapophyses	49.4	57.4
Length of centrum measured normal to posterior face along median line	36.1	34.9
Greatest width across anterior zygapophyses	41.8	35.9
Width across posterior zygapophyses	41.1	34.8
Greatest height of neural canal at anterior end	19.6	17.6
Greatest width of neural canal at anterior end	25.6	—
Greatest transverse width of posterior face of centrum across capitular facets	61.0	51.3
Depth of centrum measured normal to floor to neural canal and along median line of posterior epiphysis	33.8	27.7
Greatest width across outer ends of transverse processes	—	80.6
Greatest anteroposterior diameter of outer end of transverse process	—	23.0
Height from middle of ventral border of posterior epiphysis of centrum to top of neural spine	—	110.4
Length of spine from middle of notch between anterior zygapophyses to top measured parallel to anterior end	—	99.8

than in *Smilodon. Homotherium* is similar to *Xenosmilus* in this respect. *Panthera* has oval rib attachments as in *Xenosmilus,* but those of *Panthera* are more laterally placed. The ventral surface of the centrum is more keeled in *Smilodon,* and the anterior and posterior surfaces are more rounded.

THORACIC VERTEBRA EIGHT

The dorsal spine of the eighth thoracic vertebra in *Xenosmilus* (BIOPSI 101, fig. 4.12G–I, table 4.9) is pocketed on the lateral side of its anterior face so as to delineate a broad anterior ridge. This feature is not found in *Smilodon, Panthera,* or *Homotherium.* In *Xenosmilus,* there are distinct ridges going to the transverse processes and setting off the prezygapophyses. The prezygapophyses in *Xenosmilus* are broader and spread more than in the other cats. The neural canal in *Xenosmilus* is more oval than in the other felids. The centrum is broader for its length in *Xenosmilus* than in the other felids. The rib facets on the centrum in *Xenosmilus* are kidney bean shaped and closer together than in the other compared cats. The pedestal for the neural arch in *Xenosmilus* is also narrower than in the other cats. The postzygapophyses are expanded laterally more than in *Smilodon* or *Panthera* but resemble those in *Homotherium.*

THORACIC VERTEBRA NINE

The ninth thoracic vertebra in *Xenosmilus* (BIOPSI 101, fig. 4.12J–L, table 4.10) has a more robust neural spine than is found in *Smilodon* or *Panthera,* with shorter but broader prezygapophyses. The neural canal is large and not as rounded as it is in *Smilodon.* The anterior rib attachment on the centrum is kidney bean shaped as opposed to being nearly circular in

Smilodon. As in the preceding vertebra, the pedestal for the neural arch is narrower, and the posterior zygapophyses are more expanded laterally than in either *Smilodon* or *Panthera.*

THORACIC VERTEBRA TEN

The tenth thoracic vertebra in *Xenosmilus* (BIOPSI 101, fig. 4.13A–C, table 4.11) is missing the neural spine. The transverse process swings upward as in *Panthera,* rather than being straight as in *Smilodon.* The tubercular rib facet on the transverse process is nearly circular and faces almost laterally, while in *Smilodon* it is more oval and faces more ventrally. The pedestal for the neural arch is narrower than in *Smilodon* and *Panthera,* and the centrum is shorter and broader than in either *Smilodon* or *Panthera.* The neural arch is not as rounded as in either *Panthera* or *Smilodon. Smilodon* has a distinct keel on the ventral surface of the centrum that is missing in *Xenosmilus* and *Panthera* and weakly developed in *Homotherium.*

THORACIC VERTEBRA ELEVEN

Thoracic vertebra eleven (BIOPSI 101, fig. 4.13D–G, table 4.12) is the transition vertebra between the anterior thoracic and posterior thoracic series. The neural spine of the eleventh thoracic vertebra in *Xenosmilus* is broken but could not have been very tall. The prezygapophyses are smaller and broader than in *Smilodon* and *Panthera.* The ridge coming from the transverse process posterior to the zygapophyses in *Smilodon* is missing in *Xenosmilus* and *Panthera. Smilodon* and *Xenosmilus* differ from *Panthera* in having a high projection on the transverse process extending above the rib articulation and continuing into a posterior process that bifurcates at its ter-

Table 4.10. Ninth thoracic vertebra measurements (mm): *Xenosmilus hodsonae.*
Compared specimen (observed range), *Smilodon fatalis* (Merriam and Stock, 1932).
(*a* = approximate)

	X. hodsonae BIOPSI 101	Smilodon fatalis table 21-9
Greatest length from end of anterior zygapophyses to end of posterior zygapophyses	49.3	60.8
Length of centrum measured normal to posterior face along median line	36.5	34.8
Greatest width across anterior zygapophyses	40.8	33.0
Width across posterior zygapophyses	35.4	34.6
Greatest height of neural canal at anterior end	21.1	18.2
Greatest width of neural canal at anterior end	27.3	—
Greatest transverse width of posterior face of centrum across capitular facets	58.4	48.6
Depth of centrum measured normal to floor of neural canal and along median line of posterior epiphysis	33.7	28.7
Greatest width across outer ends of transverse processes	—	77a
Greatest anteroposterior diameter of outer end of transverse process	—	24.1
Height from middle of ventral border of posterior epiphysis of centrum to top of neural spine	—	101.6
Length of spine from middle of notch between anterior zygapophyses to top measured parallel to anterior end	—	92.5

Table 4.11. Tenth thoracic vertebra measurements (mm): *Xenosmilus hodsonae.*
Compared specimen (observed range), *Smilodon fatalis* (Merriam and Stock, 1932).
(*a* = approximate, *e* = estimate, + = damaged element)

	X. hodsonae BIOPSI 101	Smilodon fatalis table 21-10
Greatest length from end of anterior zygapophyses to end of posterior zygapophyses	—	57.6
Length of centrum measured normal to posterior face along median line	37.4	34.5
Greatest width across anterior zygapophyses	40.0a	32.7
Width across posterior zygapophyses	—	36.0
Greatest height of neural canal at anterior end	19.9	18.1
Greatest width of neural canal at anterior end	25.1	—
Greatest transverse width of posterior face of centrum across capitular facets	63.0	50.5
Depth of centrum measured normal to floor of neural canal and along median line of posterior epiphysis	35.6	28.9
Greatest width across outer ends of transverse processes	106e +	78.7
Greatest anteroposterior diameter of outer end of transverse process	20.2	23.1
Height from middle of ventral border of posterior epiphysis of centrum to top of neural spine	—	95.8
Length of spine from middle of notch between anterior zygapophyses to top measured parallel to anterior end	—	85.0

mination in *Xenosmilus* but not in *Smilodon*. *Homotherium* is more like *Smilodon* in those respects, but it lacks the posterior dorsal pocketing found on the transverse process in *Smilodon* but not found in *Xenosmilus* or *Panthera*. The rib articulation on the transverse process in *Homotherium* is large and rounded. In the BIOPSI 101 specimen of *Xenosmilus*, the T11 rib facet is rounded on the left side, as in *Homotherium*, and is ovate on the right side (fig. 4.13B), suggesting once again an asymmetry in the back of this individual. The anterior articulation on the centrum is small and elongated. The converse is true in *Smilodon*. The pedestal for the neural arch is narrower in *Smilodon* and *Panthera*. On the posterior

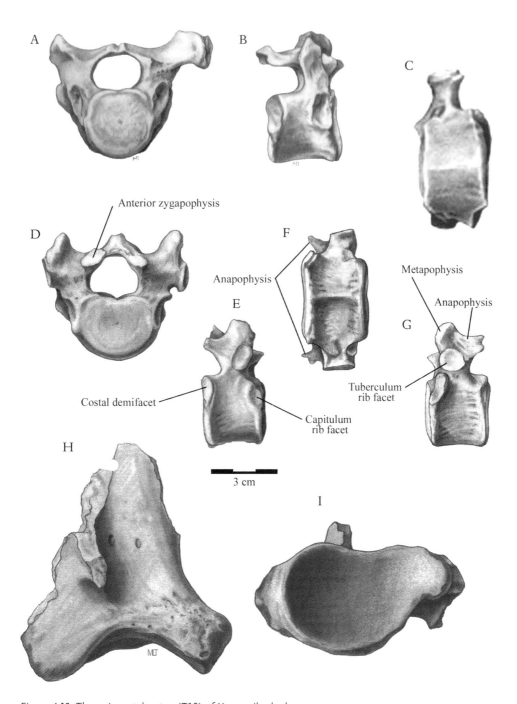

Figure 4.13. Thoracic vertebra ten (T10) of *Xenosmilus hodsonae*,
BIOPSI 101: *A*, anterior view; *B*, lateral view right side; *C*, ventral
view. Thoracic vertebra eleven (T11) of *X. hodsonae*, BIOPSI 101:
D, anterior view; *E*, lateral view right side; *F*, ventral view; *G*,
lateral view left side. This vertebra begins the transition to the
lumbar region. The left side shows a developmental anomaly
in having no posterior costal demifacet, although one of normal
appearance is present on the right side. The tuberculum facet on
the transverse process of the right side is oblong, while that on
the left side is circular. The partial right scapula of *X. hodsonae*,
UF 60,000: *H*, lateral view; *I*, posterior view of the distal end,
showing the glenoid cavity. (drawing Mary Tanner)

Table 4.12. Eleventh thoracic vertebra measurements (mm): *Xenosmilus hodsonae.*
Compared specimens (observed range), *Smilodon fatalis* (Merriam and Stock, 1932).

	X. hodsonae BIOPSI 101	*Smilodon fatalis* 2055-1	2055-10
Greatest length from end of anterior zygapophysis to end of posterior zygapophysis	—	67.4	49.0
Length of centrum measured normal to posterior face along median line	37.6	41.3	31.8
Greatest width across anterior zygapophyses	40.6	40.2	32.9
Width across posterior zygapophyses	—	36.4	20.8
Greatest height of neural canal at anterior end	23.1	—	—
Greatest width of neural canal at anterior end	26.5	—	—
Greatest transverse width of posterior face of centrum	57.5	45.4	34.5
Depth of centrum measured normal to floor of neural canal and along median line of posterior epiphysis	31.2	28.8	25.6
Greatest width measured from outer ends of facets for tubercle of rib	76.1	77.0	59.4
Height from middle of ventral border of posterior epiphysis of centrum to top of neural spine	—	111.4	64.2
Length of spine from middle of notch between anterior zygapophyses to top measured parallel to anterior end	—	87.7	41.0
Greatest anteroposterior diameter of plate above facet for tubercle of rib	30.6	42.2	27.1
Greatest height of neural canal at posterior end	22.7	—	—
Greatest width of neural canal at posterior end	28.1	—	—
Greatest width across metapophyses	67.1	—	—
Greatest width across anapophyses	73.7	—	—
Greatest length from end of metapophyses to end of anapophyses (right)	30.4	—	—
Greatest length from end of metapophyses to end of anapophyses (left)	32.5	—	—

surface of the centrum of the BIOPSI 101 *Xenosmilus* specimen, there is a rib facet on the right-hand side but not the left. In the other cats, no posterior rib demi-facets are present on the eleventh vertebrae. *Xenosmilus, Homotherium,* and *Smilodon* all have a low keel on the ventral surface of the centrum, but none is present in *Panthera.*

THORACIC VERTEBRAE TWELVE AND THIRTEEN
The twelfth and thirteenth thoracic vertebrae in *Xenosmilus* were not recovered.

The Ribs, Sternum, Lumbar Vertebrae, Sacrum, and Caudal Vertebrae
These elements were not recovered for *Xenosmilus.*

The Appendicular Skeleton
The Forelimb and Pectoral Girdle
THE SCAPULA
There is a fragment of the distal end of the right scapula (UF 60,000; fig. 4.13H–I; table 4.13). It is remarkable for its large size and massiveness, exceeding even *Smilodon* in these respects. The glenoid in *Xenosmilus* is very large and oval, decreasing more in size anteriorly than in *Homotherium.* This aspect suggests greater lateral movement of the forelimb for

grappling with prey. There is a prominent facet for the origin of the long head of the M. biceps brachii. *H. serum* (Rawn-Schatzinger, 1992, fig. 15B) shows the glenoid fossa to be essentially rectangular, with a distinct notch on the anteromedial margin, while in *Xenosmilus* this margin slants smoothly.

THE HUMERUS
The humerus in *Xenosmilus* (BIOPSI 101, fig. 4.14, table 4.14) is slightly more slender than in *Smilodon,* but is more massive than in either *Panthera* or *Homotherium.* The humeral head in *Xenosmilus* is larger and more rounded than in *Smilodon,* but it does not extend as far distally. The humeral head in *Panthera* scribes a greater number of degrees of arc toward the anterior surface of the bone shaft than in *Xenosmilus.* The humeral head *Homotherium* is significantly more compressed than in *Xenosmilus.* In lateral view in *Xenosmilus,* the greater tuberosity is both taller and broader than in *Panthera* or *Smilodon.* The corresponding tuberosity in *Homotherium* is also tall, but the dorsal aspect that serves as the origin of the M. supraspinatus is narrower than in *Xenosmilus.* In *Xenosmilus, Homotherium,* and *Smilodon,* there is a large, round depression encompassing almost the entire lateral face of the greater tuberosity for the insertion of the

Table 4.13. Right scapula measurements (mm): *Xenosmilus hodsonae.* Compared specimens (observed range), *Smilodon fatalis, Panthera atrox* (Merriam and Stock, 1932). (*a* = approximate)

	X. hodsonae UF 60,000	Smilodon fatalis 2004			Panthera atrox 2902	
		R-6	R-7	R-8	R-1	R-3
Greatest length from coracoid process to suprascapular border measured along axis of spine	—	343.0	266.0	358.0	354.0	290.0
Width of scapular blade measured obliquely across spine	—	186.0	148.0	191.0	215*a*	169.0
Greatest anteroposterior diameter of articulating end measured across glenoid cavity	68.4	79.8	67.0	87.1	85.0	65.1
Greatest transverse diameter across glenoid cavity (fossa)	42.8	49.5	41.5	57.9	56.8	44.0
Distance from inner border of glenoid cavity to top of spine	—	89.0	70.1	—	98.0	81.5
Least width of neck across articulating end	74.0*a*	76.4	55.1	73.2	77.8	58.7
Width of supraspinous fossa	—	—	—	—	—	—
Width of infraspinous fossa	—	—	—	—	—	—

M. infraspinatus. Even though the insertion is similarly shaped and positioned, that of *Smilodon* is less prominent than in either of the scimitar-tooth cats. In *Panthera*, the corresponding depression is smaller and restricted to the anterior one-half of the lateral aspect of the greater tuberosity. In *Xenosmilus*, the initiation of the prominent tricipital ridge is distal to the posterolateral "corner" of the scar for insertion of the M. infraspinatus. This ridge terminates in the deltoid tuberosity. The length of the ridge is shortest in *Xenosmilus*, slightly longer in *Homotherium* and *Smilodon*, and significantly more distally positioned in *Panthera*. The ridge is similarly located and as prominent in *Xenosmilus* and *Homotherium*, although less so in *Smilodon*, and in *Panthera*, it is more prominent than in the saber-tooths. The distal deltoid expansion of the ridge is sharp and prominent in *Xenosmilus*, slightly less so in *Homotherium*, and significantly broader but less sharply defined in *Smilodon*. The deltoid expansion is best defined in *Panthera*. In all species this ridge marks the attachments of several humeral muscles. In *Xenosmilus*, the insertion of the M. teres minor is a lens-shaped rugosity anterior to the tricipital ridge. In *Smilodon*, the scar occurs in a location similar to that in *Xenosmilus*, although it is less clearly demarcated. In *Panthera*, the scar for this muscle insertion is similar in location to that of the other cats, although it is less well marked than in *Xenosmilus*. The tricipital ridge serves as the origin for the M. triceps lateralis and is most clearly marked in *Xenosmilus*, slightly less so in *Homotherium*, rather vaguely defined in *Smilodon*, and strikingly sharp and prominent in *Panthera*. The insertion of the M. deltoideus in *Xenosmilus* is prominent and rugose, as is also the case in *Homotherium*, less so in *Smilodon*, but most prominent in *Panthera*.

In anterior view, the humerus shows a broad groove between the greater tuberosity and the humeral head. When viewed from the medial side, this groove is not well developed in *Xenosmilus* or the other felids. The medial edge of this bicipital groove is pocketed with nutrient foramina.

These foramina are more laterally positioned in *Xenosmilus*, *Smilodon*, and *Homotherium* than in *Panthera*. The shaft of the humerus is straighter in the homotheres and more arched in *Smilodon* and *Panthera*. The anterior face of the humeral shaft in *Xenosmilus* is narrower than in *Smilodon*.

On the medial aspect of the humerus, the lesser tuberosity that serves as the insertion of the M. subscapularis in *Smilodon* is larger and longer dorsoventrally than in *Xenosmilus*, *Homotherium*, and *Panthera*. This feature is not only shorter, but narrower in *Homotherium*, making it relatively smaller in this cat.

The anterior humeral surface proximal to the trochlea and capitulum is more broadly depressed in *Xenosmilus* and *Homotherium* than it is in *Panthera* and *Smilodon*. When viewed distally, the trochlea projects farther posteriorly in *Smilodon* than it does in *Xenosmilus*, which is similar to *Homotherium* and *Panthera*. At the distal end, the external condylar surface is wider and less bulbous in *Smilodon* than it is in *Xenosmilus* and *Homotherium*. In *Xenosmilus*, the medial trochlear surface is distinctly more distal than the lateral, differing from the other cats in the degree of the development of this feature. At the distal edge of the lesser tuberosity, there is a noticeable pocket for the origin of the M. triceps medialis. This feature is almost flat in *Homotherium*, and also present but less significant in *Panthera*. *Smilodon* is similar to *Homotherium* in this respect. The lateral epicondylar crest is more sharply developed in *Xenosmilus* than in *Smilodon*, resembling *Panthera* and *Homotherium*.

The entepicondylar bridge is inclined in *Smilodon* and oriented more vertically in *Xenosmilus*, the latter feature being more like the other cats. The bridge is also narrower in *Xenosmilus* and *Homotherium* but is broad in *Smilodon* and *Panthera*. The posterior opening of the entepicondylar foramen that allows passage of the median nerve and artery is pocketed in all the other felids, but the surface is nearly flush in *Smilodon*. The entepicondylar foramen is larger in *Smilodon*, *Panthera*, and *Xenosmilus* than it is in the Idaho

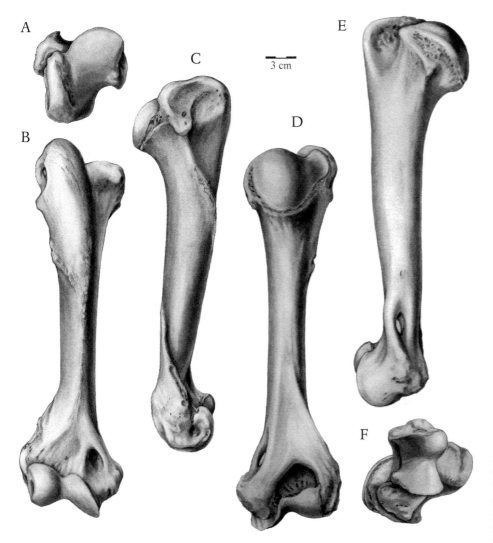

Figure 4.14. The right humerus of *Xenosmilus hodsonae*, BIOPSI 101: *A*, proximal end view; *B*, anterior view; *C*, lateral view; *D*, posterior view; *E*, medial view; *F*, distal end view. (drawing Mary Tanner)

Table 4.14. Right and left humerus measurements (mm): *Xenosmilus hodsonae*. Compared specimens (observed range), *Smilodon fatalis*, *Panthera atrox* (Merriam and Stock, 1932); Los Angeles County Museum, Hancock Collection (HC), *LACMHC-X-6910.

	X. hodsonae BIOPSI 101		Smilodon fatalis 2005		Panthera atrox 2903	
	R	L	R-1	R-10	R-1	R-9
Greatest length measured parallel to longitudinal axis	360.0	361.0	385.0	309.0	409.0	328.0–*324.8
Greatest transverse diameter (mediolateral) of proximal extremity	108.0	102.0	92.4	75.4	100.2	78.7
Greatest anteroposterior diameter of proximal extremity (head)	82.0	76.0	117.6	92.0	123.4	99.1
Transverse diameter at mid shaft	35.0	36.0	41.7	32.2	38.2	33.0–*29.6
Anteroposterior diameter at mid shaft	44.0	44.0	57.7	47.5	59.4	53.8–*48.9
Greatest width of distal extremity	106.0	107.0	125.0	98.8	107.6	85.7
Least anteroposterior diameter of articulating surface for ulna	30.0	29.0	35.4	27.5	35.0	30.0
Maximum width of entepicondylar foramen	9.0	8.8	—	—	—	—
Maximum length of entepicondylar foramen	17.5	17.6	—	—	—	—
Greatest breadth across distal articular surface	69.9	67.5	—	—	—	—

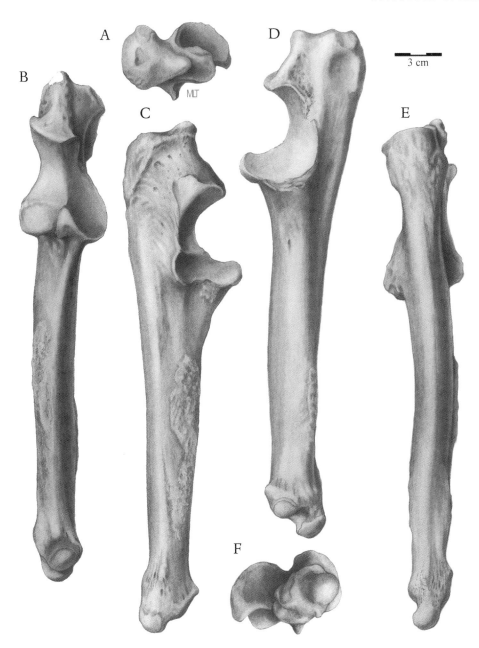

3 cm

Figure 4.15. The right ulna of *Xenosmilus hodsonae*, BIOPSI 101: A, proximal end view; B, anterior view; C, lateral view; D, medial view; E, posterior view; F, distal end view. (drawing Mary Tanner)

Homotherium. In *H. serum* the entepicondylar foramen is relatively longer and more inclined than it is in any of the other cats discussed here.

When viewed distally, the internal distal condyle projects farther posteriorly in *Smilodon* than it does in *Xenosmilus*, which is similar to *Homotherium* and *Panthera*. In *Panthera*, the olecranon fossa is deeply recessed, as it is in *Homotherium*. It is broader in *Smilodon* and *Xenosmilus*, and is pocketed dorsally in *Xenosmilus*. The lateral condylar crest is more sharply developed in *Xenosmilus* than in *Smilodon*, resembling *Panthera* and *Homotherium*. In *Xenosmilus*, the posterior aspect of the articular surface of the trochlea extends farther proximally than in the other cats in this study, creating an asymmetrical articulation that would allow more lateral move-

ment of the forearm in conjunction with flexion. This adaptation would permit a greater degree of pronation of the forearm, a movement necessary for grappling with large or strong prey or for climbing. This feature of *Xenosmilus* is reminiscent of bears in general and *Ailuropoda* more specifically. In the panda, however, this anatomical arrangement is probably important in climbing rather than in acquisition of prey.

THE ULNA

The ulna in *Xenosmilus* (BIOPSI 101, fig. 4.15, table 4.15) is even shorter than in *Smilodon*, and is shorter than the humerus. As in *Smilodon*, it is straight and massive. The olecranon processes in *Xenosmilus* and *Smilodon* are roughly vertical to

Table 4.15. Right and left ulna measurements (mm): *Xenosmilus hodsonae*. Compared specimens (observed range), *Smilodon fatalis*, *Panthera atrox* (Merriam and Stock, 1932); Los Angeles County Museum, Hancock Collection (HC), LACM-X-7403, *HC-1561.

	X. hodsonae BIOPSI 101		*X. hodsonae* UF 60000		*Smilodon fatalis* 2010		*Panthera atrox* 2908	
	R	L	R	L	R-1	R-10	R-1	R-7
Greatest length measured parallel to longitudinal axis of ulna	320.0	318.0	—	330.0	372.0	387.0	438.0	384.0–*375.8L
Greatest width of posterior surface of olecranon process	41.0	41.0	—	43.0	48.8	33.6	47.7	35.9
Greatest transverse width of greater sigmoid cavity	53.0	51.0	—	57.0	60.2	41.5	65.0	48.0
Length of radial notch	42.5	42.5	—	40.0	—	—	—	—
Height of trochlear notch (inside)	40.3	40.5	—	—	—	—	—	—
Width of trochlear notch	50.4	—	—	—	—	—	—	—
Anteroposterior diameter of shaft at proximal end of tendon scar	35.0	36.0	—	40.0	47.6	34.8	56.7	42.5
Transverse diameter (mediolateral) of shaft at proximal end of tendon scar	25.0	25.0	—	24.0	29.9	20.9	24.0	20.7
Greatest anteroposterior diameter of distal extremity	37.0	37.0	—	32.0	46.6	33.3	44.6	36.2
Greatest width of distal extremity	27.0	26.0	—	38.0	33.8	20.4	27.5	23.0–*18.5L
Anteroposterior diameter of shaft at mid shaft	34.8	35.0	—	—	—	—	—	—
Transverse diameter at mid shaft	25.9	26.2	—	—	—	—	—	—

the shaft, slightly inclined posteriorly in *Panthera*, and much more inclined in *Homotherium*. The olecranon process in the Idaho homothere and *Panthera* is roughly equal in height but shorter in *Xenosmilus* and *H. serum*. Rawn-Schatzinger (1992) suggests that the low, rounded olecranon of *H. serum* might be indicative of a plantigrade stance. However, long olecranon processes are commonplace in plantigrade animals that have powerful forelimbs, for example digging mammals. The olecranon process, which serves as the insertion of all of the parts of the M. triceps and the origin of the M. anconeus and M. flexor carpi ulnaris, is broader mediolaterally in distal view and more robust than in *Panthera* and *Homotherium*, but less so than in *Smilodon*. In *Xenosmilus* and *Homotherium*, the posterior edge of the ulnar shaft is rounded, and in *Panthera* and *Smilodon*, it is flattened. In *Smilodon*, the anterolateral tubercle is hypertrophic, so that it extends higher and more lateral to the dorsal surface of the olecranon. In *Xenosmilus*, it is also massive, but is low and rounded, and is at about the same level as the dorsal surface and points medially. This same process is prominent but smaller in *Homotherium* and *Panthera*. In *Panthera*, but not in the saber-tooths, the tubercle is separated from the medial ridge by a distinct groove. In both *Xenosmilus* and *Smilodon*, the posterior surface of the olecranon is greatly broadened and flattened as compared to the more rounded and smaller surfaces in *Homotherium* and *Panthera*.

In lateral view, just anterior to the sigmoid notch, M. anconeus attaches as a broad horizontal band in *Xenosmilus*. In *Smilodon*, *Homotherium*, and *Panthera*, the attachment is more vertical. The sigmoid notch is wider in *Smilodon* than it is in *Panthera* and *Homotherium*, but about equal in width to that of *Xenosmilus*. The bottom of the sigmoid notch is extended farther laterally in *Xenosmilus* than in *Smilodon*.

The lateral expansion in *Panthera* is less than in *Smilodon*, and in *Homotherium* it is even more reduced. The space behind the sigmoid notch is roughened in *Smilodon* and *Xenosmilus* on the lateral side, but not in *Homotherium* and *Panthera*. In anterior view, *Xenosmilus* has a median guiding ridge between the medial and lateral expansions of the notch. This is similar to *Panthera*, *Homotherium*, and *Smilodon* but is more exaggerated in *Xenosmilus*. The radial notch in *Xenosmilus* is very long, as in *Panthera*, but it is shorter in *Smilodon* and *Homotherium*. The ulnar shaft in *Xenosmilus* is extremely robust and marked by many rugose scars for muscle and ligamentous attachments. Distal to the radial notch in *Xenosmilus* is a vertically oriented, lens-shaped rugosity that served as the ulnar insertion of the M. biceps brachii. This feature is identifiable in all cats compared in this study, although it is more prominent and relatively larger in *Xenosmilus*. As in *Smilodon*, the origin for the M. flexor digitorum profundus in *Xenosmilus* encompasses a large area. This suggests a greater capacity for thumb and digit two opposition. In *Xenosmilus*, there is an enormous rugosity, almost one-third the length of the ulnar shaft, reaching distally almost to the radioulnar articulation, for origin of the M. pronator quadratus. This attachment surface is larger as well as more directly lateral in *Xenosmilus* than in *Smilodon*, allowing *Xenosmilus* a greater ability to pronate the forearm. In *Panthera*, this region is a sharp but narrow ridge. In *Xenosmilus*, the ulnar shaft is markedly shorter than that of *Homotherium* and slightly shorter than that of *Smilodon*, although of approximately the same degree of robustness. The shaft of the bone in *Homotherium* is also only about two-thirds the diameter of that of the other saber-tooths. The robustness of the ulnar shaft in *Panthera* is intermediate as compared to the saber-tooth cats. The medial face of the bone in *Xenosmilus* is occupied by a large broad roughened area that served as the attachment of the interosseus membrane and the muscles that arise partially from this ligament as well as from the ulnar shaft. This rugosity is nearly as prominent in *Smilodon*, but it is shorter and more proximal in *Panthera*. "The interosseus crest in *Homotherium serum* is low and rounded, giving the distal shaft an ovoid cross section" (Rawn-Schatzinger, 1992, p. 29). In *Xenosmilus,* the origin of the M. abductor pollicis longus arises from the most proximal portion of the posteromedial (extensor) surface of the ulnar shaft. Distal to this origin is that of the M. extensor pollicis longus, and farther down the origin is that of the M. extensor indicus. The large oval articulation between the distal radius and ulna in *Xenosmilus* emphasizes the capability of this animal to pronate the forearm, as well as to rotate the radius against the ulna to a greater degree than was possible for the other large cats. Enlargement of the pronator quadratus allows more forceful pronation and movement of the elbow. This is further facilitated by the shape of the articular surfaces at the elbow. This combination of features would have enhanced the ability of *Xenosmilus* to grapple with prey. Like *Xenosmilus*, although with a lesser differential, the ulna of *Smilodon* is shorter than the humerus. The ulna is longer than the humerus in the Idaho homothere, *H. serum,* and *Panthera.*

The ulna styloid process is short and rounded in *Xenosmilus* and longer and more pointed in *Smilodon*. It is rounded in *Homotherium* and *Panthera* but separated by a more distinct notch in those genera than it is in *Xenosmilus*. The articulation for the M. extensor carpi ulnaris is broader than in *Panthera* or *Smilodon* and resembles *Homotherium*, although in *Homotherium* it is less pronounced than in any of the other cats to which we have compared it. In *Xenosmilus*, the distal end of the ulna bears a large oval facet containing a smaller rounded depression that articulates with a corresponding strongly marked oval facet of the distal radius. These features are also distinct in *Smilodon* and *Panthera* but less distinct and relatively much smaller in *Homotherium*.

THE RADIUS

The radius (BIOPSI 101, fig. 4.16, table 4.16) is shorter and thicker in *Xenosmilus* than it is in *Smilodon*. In distal anterior view, the radius of *Xenosmilus* and *Smilodon* bow laterally, similar to the panda *Ailuropoda* (Davis, 1964). The radial shaft (diaphysis) is straighter in *Panthera* and *Homotherium*. The radial head in *Xenosmilus* is rounder and not notched as it is in *Smilodon*, being similar to *Homotherium*. The radial head in *Panthera* is oval and notched, and the bicipital tuberosity is an oval flat projection. It is similar but more elongate in *Homotherium* and *Smilodon*. In *Xenosmilus,* the bicipital tuberosity is more complex, forming a roughened ridge that continues proximally toward the radial head. This is the site of insertion of the M. biceps brachii. The proximal radial

shaft is not as flattened as in *Panthera* or *Homotherium*. The interosseus crest in *Panthera* and *Homotherium* is a sharp ridge but is broad and roughened in *Xenosmilus* and *Smilodon*. As in *Smilodon*, the nutrient foramen in *Xenosmilus* is more proximal than in *Panthera*. The posterior radial surface in *Xenosmilus* shows a broad but shallow groove for the origin of the M. extensor pollicis brevis. Proximally, a roughened ridge serves as the origin of the M. abductor pollicis longus. The prominence of these two muscles suggests that *Xenosmilus* had enhanced ability to extend and abduct the thumb. These features are more prominent on the right side of *Xenosmilus*, again suggesting handedness. These features were not as distinctly developed in the other cats examined. This suggests that *Xenosmilus* could generate more force for extension and abduction of the thumb.

In *Xenosmilus*, the facet for the distal articulation with the ulna is an oval depression with a deeper oval cup in the center. This feature is wider anteroposteriorly and resembles the facet of *Smilodon* and *Panthera*. The corresponding feature in *Homotherium* is approximately half the size, less distinct, and lacks the central recessed cup. The articulation for the scapho-lunar is broader in *Xenosmilus* and *Smilodon* than it is in *Panthera* or *Homotherium*. The styloid process in *Xenosmilus* is more pointed than in *Smilodon*, although that in the Idaho *Homotherium* is even more pointed. In *Panthera,* the styloid process is long but rounded. The carpal articular surface in *Xenosmilus* is deeply concave and dumbbell-shaped, with the medial side of the dumbbell larger than the lateral. This is also the case in *Smilodon*. Rawn-Schatzinger (1992) describes this surface as narrow, elongate, and key-shaped in *H. serum*. The carpal articular surface in the Idaho homothere is also dumbbell-shaped, but the lateral side is significantly smaller than the medial. Overall, this surface is smaller dorsoventrally than that of *Xenosmilus*, reflecting the more gracile radius. The articular surface in the Idaho homothere is also more compressed mediolaterally, allowing a smaller overall articulation for carpal bones and a more compact proximal manus. In contrast, the carpal articular surface of *Panthera* is wider mediolaterally than in the comparison species although it is also less deeply concave, less deep dorsoventrally, and the two ends of the dumbbell are more nearly equal in size than in the comparison species.

THE WRIST (CARPAL BONES)

None of the carpal bones of *Xenosmilus* were recovered.

THE MANUS

The metacarpals of *Xenosmilus*, like those of *Smilodon*, are short and broad, contrasting with the elongated metacarpals of *Homotherium* and modern cats (Rawn-Schatzinger, 1992). The metacarpals of *Smilodon* are more bowed and not as tightly apposed as in *Xenosmilus*. The anterior faces of the metacarpals are flattened in *Xenosmilus* and more rounded in

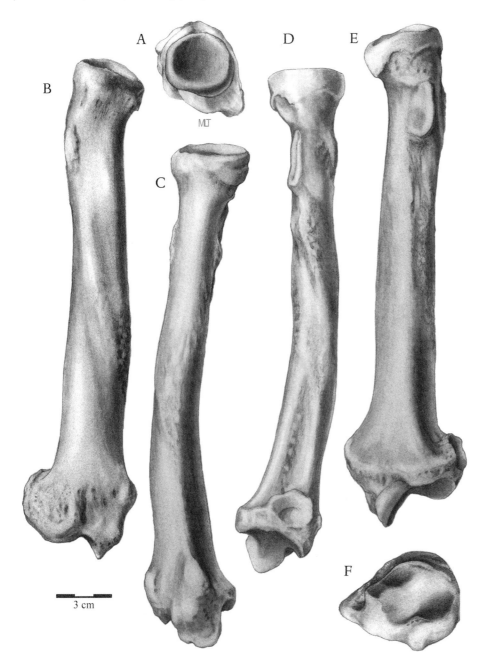

Figure 4.16. The right radius of *Xenosmilus hodsonae*, BIOPSI 101: *A*, proximal end view; *B*, lateral view; *C*, anterior view; *D*, posterior view; *E*, medial view; *F*, distal end view. (drawing Mary Tanner)

Panthera and *Smilodon*. This suggests that foot structure in *Xenosmilus* was derived from that of a more cursorial common ancestor. In contrast, in *Homotherium*, the foot shows greater adaptation toward cursoriality than was true of the common ancestor. Therefore, foot evolution in the common ancestor diverged toward a more elongated digitigrade foot in *Homotherium* and a shortened, plantigrade foot in *Xenosmilus*. It is possible that *Smilodon* descended from an ancestor that was never significantly more digitigrade.

METACARPAL I

Metacarpal I was not recovered for *Xenosmilus*, but metacarpal II has two distinct surfaces for its articulation.

METACARPAL II

In *Xenosmilus*, the second metacarpal (BIOPSI 101, fig. 4.17A–F, table 4.17) is straight and robust, with an oval shaft, contrasting with the elongate metacarpal II in *H. serum* and the Idaho homothere. This element is similar in degree of robustness and shape in *Smilodon* and *Panthera* to that of *Xenosmilus*. In *Smilodon*, metacarpal II diverges distally from metacarpal III. This makes the second finger semi-opposable. In *Xenosmilus*, *Homotherium*, and *Panthera*, metacarpals II and III are tightly pressed together. This would suggest that *Smilodon* had better grasping hands than did the other cats. The proximal articular surface of metacarpal II in

Table 4.16. Right and left radius measurements (mm): *Xenosmilus hodsonae.* Compared specimens (observed range), *Smilodon fatalis, Panthera atrox* (Merriam and Stock, 1932); Los Angeles County Museum, Hancock Collection (HC), *LACMHC-H-1560. (*a* = approximate)

	X. hodsonae BIOPSI 101		X. hodsonae UF 60,000		Smilodon fatalis 2006		Panthera atrox 2904	
	R	L	R	L	R-1	R-10	R-3	R-10
Length measured along internal border	259.0	259.0	265.0	260.0	289.0	235.0	366.0	*303.8–317.0
Long diameter of proximal end (mediolateral diameter)	40.0	41.0	41.1	41.0	49.8	41.8	55.7	44.0a
Greatest diameter taken at right angles to long diameter of proximal end	33.0	34.0	35.0	34.0	42.0	32.2	38.8	33.0a
Width of shaft at middle (mediolateral diameter)	30.0	30.0	30.0	30.0	38.8	26.0	39.6	*34.0–38.1
Thickness of shaft at middle (anteroposterior diameter)	22.0	23.0	25.0	24.0	22.0	16.5	27.5	*18.7–20.8
Greatest width at distal end taken normal to internal face	64.0	65.0	66.0	66.0	62.8	50.8	77.6	61.5
Greatest thickness of distal end	41.0	42.0	—	—	46.7	35.6	48.7	38.3
Greatest breadth of distal articular surface (facet)	46.0	45.7	—	—	—	—	—	—

Xenosmilus has two triangular facets for the trapezoid that are broader anteriorly than in *Panthera* and *Smilodon,* and in this respect, resembles *Homotherium. Smilodon* has an hour-glass-shaped (upper portion larger) anteromedial facet for the trapezium that is raised above the rest of the articular surface (Merriam and Stock, 1932). In *Xenosmilus,* this feature is also raised, oval, and has a wider horizontal axis. In the Idaho *Homotherium,* the facet is similarly shaped, although relatively smaller. In *Panthera,* the facet is also raised, but is in the shape of a rounded triangle. In *Xenosmilus,* the proximal end of metacarpal II has a pointed posterior process that is more flattened in *Smilodon.* The medial facet for the trapezium is broad and rounded in *Xenosmilus.* In *Panthera,* it is similar in shape but is impressed rather than raised, as in *Xenosmilus.* In *Smilodon,* it is narrow and depressed. The groove for the radial artery is prominent in *Smilodon* and *Panthera,* but not in *Xenosmilus* and *Homotherium.* In *Panthera, Smilodon,* and *Homotherium,* the proximal end is distinctly indented into the shaft, while in *Xenosmilus,* it extends medially past the edge of the shaft. The articular surface for the dorsal interosseus muscle in *Xenosmilus* extends across the face of the shaft and is much more strongly developed than in the other cats. The articular facets for metacarpal III are continuous for *Xenosmilus* and are separated by a groove in *Smilodon* and *Panthera.* The anterior facet extends to the articular surface in *Smilodon* so that the lateral facet for the magnum found in *Panthera* (Merriam and Stock, 1932) and *Xenosmilus* is lacking in *Smilodon.* The facets for metatarsal III show less overhang in *Smilodon* than they do in *Panthera* and *Xenosmilus.* In *Xenosmilus,* the proximal ligamental attachment surface with metacarpal III extends farther distally than in *Smilodon* and *Panthera,* and the anterolateral corner of the articular surface also extends farther in *Xenosmilus.*

The shaft of metacarpal II in *Xenosmilus* expands distally so that there is a large distolateral process abutting against metacarpal III and separating their shafts. This process is not as well developed in the other cats. The proximal edge of the anterodistal articular surface in *Xenosmilus* is not as depressed as in the other cats, being flatter and less round. On the palmar side, the medial ridge extends farther proximally in *Smilodon* and *Panthera* than in *Xenosmilus.*

METACARPAL III

In *Xenosmilus,* the metacarpals are unusual in that II, III, and IV are of similar length. In all of the cats under comparison (*Smilodon, Panthera,* and the Idaho *Homotherium*), metacarpal III is the longest (BIOPSI 101, fig. 4.17G–L, table 4.18). In *H. serum* (Meade, 1961), metacarpal III is also the largest, and in *Xenosmilus* it is also the most robust, although metacarpals II and IV are nearly as robust. The magnum facet is concave in *Xenosmilus* and *Smilodon* but convex in *Homotherium* (Rawn-Schatzinger, 1992) and *Panthera.* It is broader in *Xenosmilus* than in the other cats. The surface for the magnum in *Xenosmilus* is separated from the lateral facet for the unciform as in *Smilodon* (Merriam and Stock, 1932), and the facet for metacarpal IV overlaps, contrasting with the flat and more lateral facet in *Panthera.* The dorsal articular facet for metacarpal II is large and raised in *Smilodon,* yielding an unusually diverged metacarpal II. The attachment for the M. extensor carpi radialis is strongly developed in *Smilodon* and *Xenosmilus.* On the distal end, the articular surface is more strongly keeled in *Smilodon* than in *Xenosmilus.*

METACARPAL IV

Metacarpal IV in *Xenosmilus* (BIOPSI 101, fig. 4.17M–R, table 4.19) is larger and more robust than in *Smilodon.* The proximal end is broader and less triangular than in *Smilodon, Homotherium,* or *Panthera.* The articular facet for metacarpal III is elongate, continuous, and semilunate. In *Smilodon* and *Panthera,* it is shorter and distinctly notched. The facet extends anteriorly beyond the facet for the unciform in *Xenosmilus,* while ending at about the same level in *Smilodon, Panthera,*

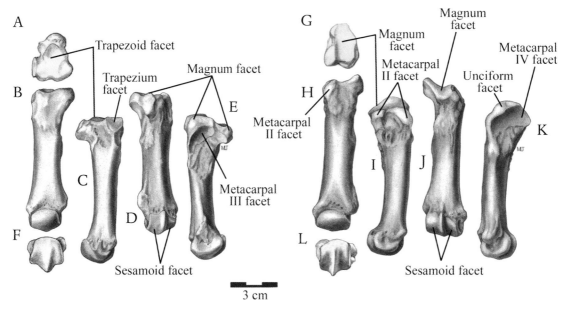

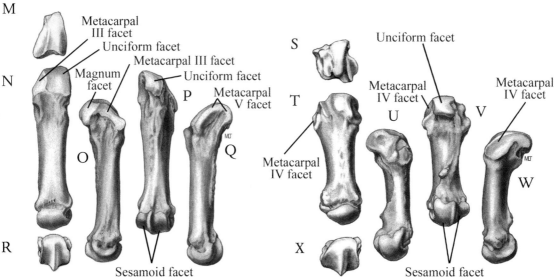

Figure 4.17. The left metacarpal II (Mc II) of *Xenosmilus hodsonae*, BIOPSI 101: *A*, proximal end view; *B*, dorsal view; *C*, medial view; *D*, ventral view; *E*, lateral view; *F*, distal end view. The left metacarpal III (Mc III) of *X. hodsonae*, BIOPSI 101: *G*, proximal end view; *H*, dorsal view; *I*, medial view; *J*, ventral view; *K*, lateral view; *L*, distal end view. The left metacarpal IV (Mc IV) of *X. hodsonae*, BIOPSI 101: *M*, proximal end view; *N*, dorsal view; *O*, medial view; *P*, ventral view; *Q*, lateral view; *R*, distal end view. The right metacarpal V (Mc V) of *X. hodsonae*, BIOPSI 101: *S*, proximal end view; *T*, dorsal view; *U*, lateral view; *V*, ventral view; *W*, medial view; *X*, distal end view. (drawing Mary Tanner)

and *Homotherium*. The anteroproximal surface is deeply notched in *Smilodon*, and more flattened in *Xenosmilus*. The scars for the palmar interosseus muscles are large and paired in *Smilodon*. They are joined and central in *Xenosmilus*. The keel on the distal end in *Xenosmilus* is not as well developed as in *Smilodon*.

METACARPAL V

Metacarpal V (fig. 4.17S–X, table 4.20) in *Xenosmilus* is a little larger and more robust than the same bone in *Smilodon*, and the facet for metacarpal IV makes a small anterolateral shelf that is narrower and extends farther distally. In *Smilodon*, the

Table 4.17. Left metacarpal II measurements (mm): *Xenosmilus hodsonae.* Compared specimens (observed range), *Smilodon fatalis, Panthera atrox* (Merriam and Stock, 1932).

	X. hodsonae BIOPSI 101		Smilodon fatalis 2014		Smilodon fatalis 2014		Panthera atrox 2912		Panthera atrox 2912	
	L-A	L-B	R-1	R-3	L-1	L-3	R-1	R-3	L-1	L-3
Greatest length	89.7	94.8	103.8	76.4	105.5	76.4	124.4	104.6	122.8	100.4
Greatest transverse diameter (mediolateral) of proximal end	26.1	27.7	22.1	20.6	22.1	21.6	29.0	23.8	29.0	22.9
Greatest dorsoventral (dorsopalmar) diameter of proximal end	31.1	31.7	34.0	28.0	34.0	27.2	28.5	35.8	36.4	32.0
Transverse diameter at mid shaft	17.2	17.5	18.7	15.0	17.9	15.7	19.3	16.4	19.8	16.7
Dorsoventral diameter at mid shaft	14.9	15.3	16.5	13.9	17.8	14.8	20.8	17.1	20.8	17.4
Greatest transverse diameter at distal end of shaft	26.4	25.4	27.6	22.0	26.4	22.6	29.0	26.6	27.7	23.0
Greatest dorsoventral diameter at distal end	21.7	20.5	—	—	—	—	—	—	—	—

Table 4.18. Left metacarpal III measurements (mm): *Xenosmilus hodsonae.* Compared specimens (observed range), *Smilodon fatalis, Panthera atrox* (Merriam and Stock, 1932).

	X. hodsonae BIOPSI 101	Smilodon fatalis 2015		Smilodon fatalis 2015		Panthera atrox 2913		Panthera atrox 2913	
		R-1	R-3	L-1	L-3	R-1	R-3	L-1	L-3
	L	Max.	Min.	Max.	Min.	Max.	Min.	Max.	Min.
Greatest length	97.8	107.8	83.0	109.6	83.2	139.4	113.9	138.8	116.0
Greatest transverse diameter (mediolateral) of proximal end	25.6	30.7	24.6	29.6	24.2	34.2	30.6	34.0	27.8
Greatest dorsoventral (dorsopalmar) diameter of proximal end	27.2	28.8	23.0	29.2	22.8	35.2	28.3	33.2	25.8
Transverse diameter at mid shaft	18.8	19.5	14.6	20.0	14.8	19.4	16.0	21.3	14.4
Dorsoventral diameter at mid shaft	14.1	15.7	13.2	15.1	12.3	17.7	14.8	18.0	13.0
Greatest transverse diameter at distal end of shaft	28.3	29.4	23.7	29.3	22.4	29.0	24.0	29.2	23.0
Greatest dorsoventral diameter at distal end	22.5	—	—	—	—	—	—	—	—

Table 4.19. Left and right metacarpal IV measurements (mm): *Xenosmilus hodsonae.* Compared specimens (observed range), *Smilodon fatalis, Panthera atrox* (Merriam and Stock, 1932).

	X. hodsonae BIOPSI 101		X. hodsonae UF 60,000	Smilodon fatalis 2016		Smilodon fatalis 2016		Panthera atrox 2914		Panthera atrox 2914	
				R-1	R-3	L-1	L-3	R-1	R-3	L-1	L-3
	R	L	R	Max.	Min.	Max.	Min.	Max.	Min.	Max.	Min.
Greatest length	96.1	100.8	—	105.6	79.4	107.4	79.4	136.4	112.0	142.7	112.1
Greatest transverse diameter (mediolateral) of proximal end	22.3	24.4	24.5	24.2	20.4	26.6	18.9	27.6	22.2	25.7	22.9
Greatest dorsoventral (dorsopalmar) diameter of proximal end	31.8	28.9	35.0	27.4	22.8	25.8	20.6	31.7	26.9	29.3	27.0
Transverse diameter at mid shaft	17.5	17.7	16.9	16.0	15.3	15.9	12.3	18.9	15.1	17.4	15.0
Dorsoventral diameter at mid shaft	15.9	16.0	14.8	14.9	14.4	14.4	12.2	17.2	14.8	16.9	14.6
Greatest transverse diameter at distal end of shaft	25.7	24.9	—	24.6	19.6	22.7	20.2	27.4	21.6	26.3	21.7
Greatest dorsoventral diameter at distal end	22.9	—	—	—	—	—	—	—	—	—	—

lateral component of the facet continues as two broad strips posteriorly, but not in *Xenosmilus.* The unciform facet is composed of two broad ridges in *Xenosmilus* but widens and flattens posteriorly in *Smilodon.* The entire proximal end is broader and more symmetrical in *Xenosmilus.* The palmar surface of the shaft is triangular in *Smilodon* and flattened in *Xenosmilus.* The keel on the distal end is more proximal and strongly developed in *Smilodon,* although in *Xenosmilus* there is a large mediodistal prominence not seen in *Smilodon.*

THE PHALANGES

THE PHALANGES INDIVIDUAL DESCRIPTIONS None of the phalanges for *Xenosmilus* were found articulated, and of those associated, sequence identification is difficult.

The left proximal phalanx for digit V (fig. 4.18H–L, table 4.21) is preserved. It is more slender and longer than the corresponding metatarsal phalanx, and the distal end is more constricted. The shaft is broader than in *Panthera,* but the

Table 4.20. Left and right metacarpal V measurements (mm): *Xenosmilus hodsonae.* Compared specimens (observed range), *Smilodon fatalis, Panthera atrox* (Merriam and Stock, 1932).

	X. hodsonae BIOPSI 101	*X. hodsonae* UF 60,000	*Smilodon fatalis* 2017		*Smilodon fatalis* 2017		*Panthera atrox* 2915		*Panthera atrox* 2915	
	R	L	R-1 Max.	R-3 Min.	L-1 Max.	L-3 Min.	R-1 Max.	R-3 Min.	L-1 Max.	L-3 Min.
Greatest length	82.2	82.6	85.0	62.8	86.6	61.6	115.3	90.6	113.0	91.7
Greatest transverse diameter (mediolateral) of proximal end	30.5	30.5	28.8	20.3	27.9	20.1	27.7	25.1	30.0	24.9
Greatest dorsoventral (dorsopalmar) diameter of proximal end	27.6	28.3	31.0	25.2	32.2	23.3	30.8	27.0	33.2	27.3
Transverse diameter at mid shaft	17.4	15.8	17.2	13.3	17.9	12.9	15.8	14.2	17.6	14.2
Dorsoventral diameter at mid shaft	14.5	12.6	14.2	11.7	16.1	10.8	15.2	13.2	15.8	12.2
Greatest transverse diameter at distal end of shaft	24.5	24.6	25.7	20.4	26.6	19.6	25.2	21.2	26.7	22.2
Greatest dorsoventral diameter at distal end	21.3	21.4	—	—	—	—	—	—	—	—

Table 4.21. Manus first phalanx digits II, IV, V measurements (mm): *Xenosmilus hodsonae.* (suggested positions)

	X. hodsonae BIOPSI 101		
	II	IV	V
Greatest length	40.9	45.4	45.1
Greatest dorsoventral (dorsopalmar) diameter of proximal end	17.6	19.2	16.6
Greatest transverse diameter (mediolateral) of proximal end	22.8	22.9	20.0
Dorsoventral diameter at mid shaft	10.0	11.5	11.5
Transverse diameter at mid shaft	16.1	16.4	14.5
Greatest transverse diameter at distal end of shaft	16.6	17.4	14.3
Greatest dorsoventral diameter of distal end	10.4	11.5	11.0

distal end is more constricted, with the internal ridge of the distal articulation bent inward. The internal corner of the proximal articulation is thrown medially, suggesting that the digit had little tendency toward lateral rotation. This would suggest the hand was more capable of gripping than the hand in *Panthera.* Two distal phalanges that bore the keratinous claws are preserved (fig. 4.18A–G). These were presumably for the right and left digit III. The claws are hooded and similar to those of *Panthera,* but the articulation is smaller and less depressed. The ventral tendinous articulation is pronounced but smaller.

The Hindlimb Skeleton and Pelvic Girdle
THE INOMINATE
In *Xenosmilus,* the inominate (UF 60,000; fig. 4.19A; table 4.22) is represented by a fragment of the right ilium showing a large acetabulum. The ilium is not twisted medially as it is in *Ursus.* There was a large area for the origin of the M. rectus femoris. The inominate was very thick and robust.

THE FEMUR
In addition to the comparisons with other felids, some features of the femur (BIOPSI 101, fig. 4.20, table 4.23) were compared with the femur of the bear *Ursus arctos,* and the giant panda *Ailuropoda melanoleuca.* All of the bone markings, both proximal and distal, are more robust and prominent in *Xenosmilus* than in our comparison genera, reflecting a greater muscle mass in this predator. The femur of *Smilodon* is the next most robust and is bowed anteroposteriorly in comparison to that of *Panthera. Homotherium* has the least robust femur of the group. The femur of *Xenosmilus* is the shortest among those of the four felid genera. The mediolateral diameter at the midpoint of the femoral shaft is relatively greater in *Xenosmilus,* slightly less so in *Smilodon,* even more compressed in *Panthera,* and most compressed in *Homotherium.* This compression seems to reflect progressively increasing cursoriality. The femoral head in all four genera is approximately round, but that in *Xenosmilus* is of greater diameter. The femoral head diameter of both *Smilodon* and *Panthera* are similar in size, but in *Homotherium,* it is significantly smaller. In addition to being of greater diameter, the femoral head in *Xenosmilus* shows a deeper concavity where the femoral neck merges with the anterior face of the femoral shaft. This margin in *Panthera* is less concave, that of *Smilodon* farther reduced, and in *Homotherium* is the least concave among the felids. When viewing the anterior faces of the femora, the head in *Homotherium* is angled anteriorly to the least extent; *Panthera* and *Smilodon* share similar slightly increased angulation, with *Xenosmilus* showing the greatest angulation among the felids. *Ursus arctos* shows a concavity only slightly less deep than that in *Xenosmilus,* as does *Ailuropoda,* although the femoral neck in the latter is of greater robustness than in all genera studied here other than *Xenosmilus.* The orientation of the femoral head with respect to the shaft in *Xenosmilus* most closely resembles that of *Ailuropoda,* although the femoral neck in *Xenosmilus* and the other felids is much shorter than that of *Ailuropoda*

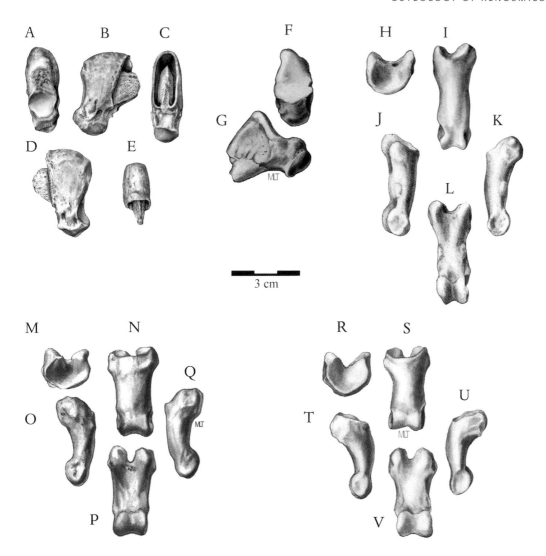

3 cm

Figure 4.18. The left terminal phalanx (claw) of digit III of *Xenosmilus hodsonae*, UF 60,000: *A*, proximal end view; *B*, medial view; *C*, ventral view; *D*, lateral view; *E*, distal end view. The right terminal phalanx (claw) of digit III of *X. hodsonae*, UF 60,000: *F*, anterior view of proximal end; *G*, medial view. The first phalanx of left digit V (manus) of *X. hodsonae*, UF 60,000: *H*, proximal end view; *I*, dorsal view; *J*, medial view; *K*, lateral view; *L*, ventral view. The first phalanx of left digit IV (manus) of *X. hodsonae*, UF 60,000: *M*, proximal end view; *N*, dorsal view; *O*, medial view; *P*, ventral view; *Q*, lateral view. The first phalanx of left digit II (manus) of *X. hodsonae*, BIOPSI 101: *R*, proximal end view; *S*, dorsal view; *T*, medial view; *U*, lateral view; *V*, ventral view. This specimen is housed in a private collection. (drawing Mary Tanner)

and *Ursus*. The fovea capitis in the femoral head of *Xenosmilus* is relatively larger than in the other three felid genera and strikingly larger than in *Ailuropoda* and *Ursus*. It is positioned similarly in *Ailuropoda*, *Ursus*, and the other felids. The femoral neck in *Xenosmilus* is thicker than in the other felids and *Ursus*, although similar to the diameter in *Ailuropoda*. This is primarily because the neck between the head and the greater

trochanter in *Xenosmilus* has a shallower notch, and the proximal tip of the greater trochanter is wider mediolaterally than in *Panthera*, the felid with the next most robust femoral neck. The femoral neck is of slightly decreased diameter in *Smilodon* and greatly decreased in *Homotherium* compared to *Panthera*, because in these two cats the greater trochanteric notch is increasingly larger. This region in *Ailuropoda* and *Ursus* cannot

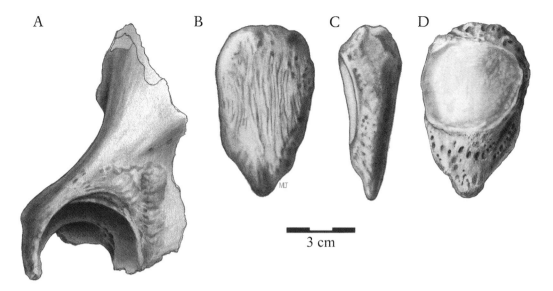

Figure 4.19. *A*, The inominate (pelvis) of *Xenosmilus hodsonae*, UF 60,000, right ilium fragment, lateral view. The left patella of *X. hodsonae*, UF 60,000: *B*, anterior view; *C*, medial view; *D*, posterior view. (drawing Mary Tanner)

Table 4.22. Right inominate measurements (mm): *Xenosmilus hodsonae.* Compared specimens (observed range), *Smilodon fatalis*, *Panthera atrox* (Merriam and Stock, 1932). (*a* = approximate)

	X. hodsonae UF 60,000	Smilodon fatalis 2008		Panthera atrox 2006	
		R-1	R-10	R-1	R-5
Length from anterior end of ilium to posterior border of ischium	—	368.0	283.0a	370.0	347.0
Greatest depth of ilium	—	94.0	73.8	90.6	83.0
Length of pubic symphysis	—	125.0	—	146.2	128.0
Width of ischium measured from ischial tuberosity to posterior end of ischial symphysis	—	117.7	88.6	111.1	101.3
Diameter of acetabulum fossa measured at right angles to long axis of notch	54.3a	54.7	46.4	54.8	47.3
Long diameter of obturator foramen	—	74.2	69.9	85.0	79.4
Greatest diameter of obturator foramen taken normal to long diameter	—	52.8	45.0	46.9	46.9
Greatest width at acetabulum measured between anteroposterior margins of inominate	97.3	—	—	—	—

be compared with the felids because it is significantly different in shape, with a greater trochanter that does not extend as far proximally.

The greater trochanter of *Xenosmilus* shows the greatest convexity and has the relatively largest mediolateral breadth among the felids. The robustness of the greater trochanter in *Panthera* and *Smilodon* is approximately equal to that in *Xenosmilus,* although it is more gracile in *Homotherium.* The greater tuberosity continues as a crest on the lateral femoral shaft more distally in *Xenosmilus* and *Smilodon* than in *Panthera* and *Homotherium.* In all genera this crest continues as

a raised ridge merging into the lateral epicondylar crest. Proximally, the ridge is more posterior in *Panthera* and more anterior in *Homotherium* than in either *Smilodon* or *Xenosmilus.* The insertion of the M. gluteus medius is similar in shape and orientation among the felids, although more robust in *Xenosmilus.* The distal tip of the femoral crest serves as the site of the insertion of the M. piriformis. It is more expanded and pointed in *Xenosmilus* than in the other felids. In *Homotherium,* this process is more convex and projects less far proximally than in the other species and has a more convex anterolateral surface.

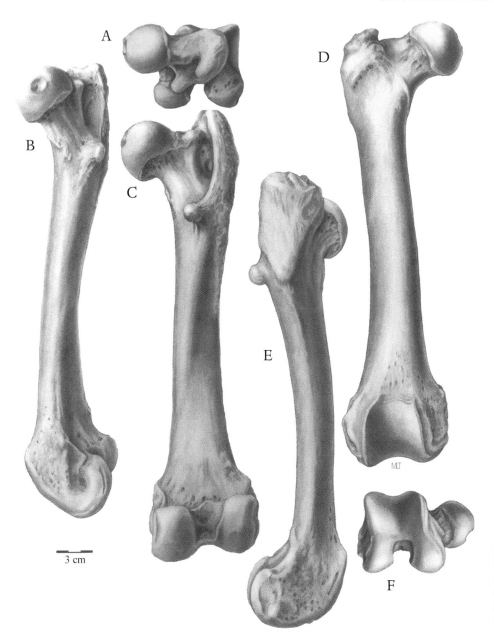

Figure 4.20. The right femur of *Xenosmilus hodsonae*, BIOPSI 101: *A*, proximal end view; *B*, medial view; *C*, posterior view; *D*, anterior view; *E*, lateral view; *F*, distal end view. (drawing Mary Tanner)

In *Xenosmilus,* the anterior femoral face is flattened and shows the same roughening for attachment of the muscles of the rectus femoris group as in all of the animals with which it has been compared. The trochanteric fossa is a deep elongated ovate depression in all species examined. The oval shape of the trochanteric fossa is more distinct in *Xenosmilus, Ailuropoda,* and *Ursus* and is bordered proximally by a bony strut crossing toward the greater trochanter. The fossa in *Panthera* is also oval but lacks the proximal bridge. In *Smilodon,* the oval depression widens proximally, and in *Homotherium* this trend is carried even farther.

Medial to the trochanteric fossa, the femoral neck is slightly concave and smooth in *Ursus, Ailuropoda,* and *Xenosmilus,* while there is a slight process in *Panthera* and *Homotherium* and a large process in *Smilodon.*

In *Xenosmilus,* the lesser trochanter is greatly enlarged as a prominent rounded bony knob for attachment of the M. psoas major. The unusually large size of this muscle reflects a correspondingly greater capacity for stabilizing the inominate when the animal was standing bipedally and pulling with the forelimbs. This feature is the next largest in *Panthera* but is more proximal and slightly more medial. The process is also large in *Smilodon,* but comes to a point rather than being rounded, and is more distal and medial than in *Xenosmilus.* In *Homotherium,* the process is both smaller and more proximal and lateral on the posterior face of the femoral shaft than in the other felid genera. In *Ursus,* the lesser trochanter is about equal in size and pointed in shape as in *Smilodon* and more mediodistally positioned. The raised area encompassed by the linea aspera in *Xenosmilus* is wider

Table 4.23. Right femur measurements (mm): *Xenosmilus hodsonae.* Compared specimens (observed range), *Smilodon fatalis, Panthera atrox* (Merriam and Stock, 1932); Los Angeles County Museum, Hancock Collection (HC), *LACMHC-15292, **LACM-PMS23 75-19L; Page Museum Salvage (PMS), PMS-2375-19L.

	X. hodsonae BIOPSI 101		X.hodsonae UF 60,000		Smilodon fatalis 2009		Panthera atrox 2907	
	R	L	R	L	R-1	R-10	R-1	R-10
Greatest length from top of greater trochanter to distal condyles measured parallel to long axis of femur	364.0	360.0	375.0	—	408.0	317.0	*462.0	391.0–**383.3
Transverse diameter of proximal end outer face of greater trochanter to inner side of head, taken normal to median longitudinal plane	103.0	102.0	100.0	—	108.8	83.0	*113.0	97.9
Greatest anteroposterior diameter of head	50.0	50.0	50.0	—	50.7	39.4	47.0	44.2
Transverse diameter (mediolateral) at mid shaft	33.0	35.0	—	—	40.4	30.1	37.0	**29.6–36.2
Anteroposterior diameter at mid shaft	38.0	38.0	32.0	—	35.4	26.8	32.5	**34.0–33.1
Greatest width of distal extremity	84.0	87.0	81.0	—	90.2	65.2	87.1	79.9
Greatest anteroposterior diameter of the distal extremity at right angles to longitudinal axis of femur	83.0	83.0	84.0	—	80.3	63.9	85.4	82.6
Greatest width of patellar groove (trochlea)	43.0	42.3	—	—	60.0	43.4	53.0	40.7
Greatest width of intercondylar notch	23.0	24.0	24.0	—	21.7	14.5	24.7	18.6
Greatest width of articular surface of medial (inner) condyle	33.0	33.0	—	—	35.7	26.5	32.1	31.3

than in the other genera, with that in *Smilodon* almost as broad, and in *Panthera* it is less broad but raised more prominently. It is the narrowest in *Homotherium.* In *Ursus,* the area is not raised but is flat and marked by prominent medial and lateral ridges and encompasses an area almost as broad as that in *Xenosmilus.* The nutrient foramen in *Xenosmilus* is located on the posterior face of the shaft. On the right femur of *Xenosmilus,* the nutrient foramen is centrally located, while on the left it is immediately medial to the linea aspera and approximately 24 mm distal from mid-shaft. The other species either show centrally located nutrient foramina (*Homotherium, Smilodon*) or laterally positioned foramina (*Panthera, Ursus*).

The patellar groove in *Xenosmilus* is wider than in the other five genera. This allows space for a relatively larger patella, reflecting an increased area of insertion for the muscles of the quadriceps group. Larger quadriceps muscles permit more forceful extension and greater stability of the knee joint. The patellar groove is rectangular as is that in *Panthera, Ailuropoda,* and *Ursus.* It narrows anterodorsally in *Smilodon* and *Homotherium.*

In *Xenosmilus,* the femoral condyles are block-like in form, shorter dorsoventrally, and wider than those of *Smilodon.* In *Panthera,* the femoral condyles have a greater dorsoventral extent than in either of the previously described saber-tooths. In *Homotherium,* the femoral condyles are shaped similarly to those of *Xenosmilus* but extend a shorter dorsoventral distance along the posterior face of the bone, indicating less flexion of the knee.

In lateral view, *Xenosmilus* and *Panthera* have femoral condyles that project posteriorly to a greater extent than in the compared genera. This anatomy allows *Xenosmilus* and *Panthera* a greater ability to flex the knee, a movement pattern that is essential for crouching and that would facilitate

stalking prey. The posterior edges of the distal femoral condyles are slightly less elongated in *Smilodon.* In *Homotherium* and *Ailuropoda* they are even shorter.

The posterior intercondylar groove in *Xenosmilus* is relatively narrower than in *Panthera* and closest in size to that of *Smilodon.* Overall, it is similar in proportion to that of *Ailuropoda* and *Ursus,* which have relatively shorter and broader posterior condylar margins. In *Xenosmilus, Ailuropoda,* and *Ursus,* the distal end of the femoral shaft widens farther laterally than medially, while the converse is true in *Panthera* and *Homotherium.* In *Smilodon,* the distal expansions are about equal. The lateral epicondyle of *Xenosmilus* differs from all other compared felid taxa, exhibiting a distinct shoulder proximal to the distal epicondyle. The scar for the lateral fabella, a sesamoid bone, is dorsomedial to it but not as strongly developed in *Xenosmilus* and *Smilodon* as in *Panthera* and *Homotherium.* In *Xenosmilus,* the pitted areas that serve as the origins for the anterior and posterior cruciate ligaments of the knee are relatively larger and more distinct than in the other genera. In *Xenosmilus,* the scar for the anterior cruciate ligament is only slightly lateral to the more anteriorly located scar for the posterior cruciate ligament, an arrangement that would minimize the difference in orientation of the two ligaments.

THE PATELLA

The patella in *Xenosmilus* (UF 60,000; fig. 4.19B–D; table 4.24) is large and relatively flat, with a long triangular distal portion. The femoral attachment is rounded but not convex as in *Homotherium.* The proximal portion shows roughening for attachment of the quadriceps tendon. The overall shape of the patella is more similar to that of *Panthera* than of *Smilodon.* In contrast, the patellae in *Ailuropoda* and *Ursus* are rounder and rugose.

Table 4.24. Right and left patella measurements (mm): *Xenosmilus hodsonae.* Compared specimens (observed range), *Smilodon fatalis, Panthera atrox* (Merriam and Stock, 1932).

	X. hodsonae UF 60,000		Smilodon fatalis 2012		Panthera atrox 2910	
	R	L	R-1	R-4	R-1	L-3
Greatest proximodistal diameter (length)	78.5	—	64.2	44.7	75.7	63.8
Greatest transverse (mediolateral) width	47.6	48.5	47.0	33.6	46.6	42.7
Anteroposterior diameter through middle of articulating surface	22.9	22.5	29.6	25.0	27.6	26.4

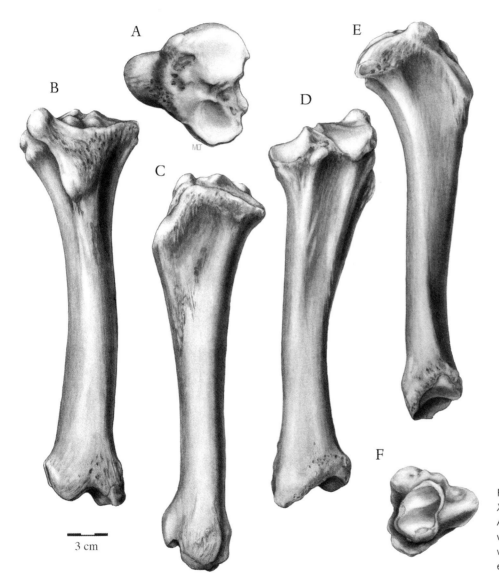

Figure 4.21. The right tibia of *Xenosmilus hodsonae,* BIOPSI 101: *A,* proximal end view; *B,* anterior view; *C,* medial view; *D,* lateral view; *E,* posterior view; *F,* distal end view. (drawing Mary Tanner)

3 cm

THE TIBIA

The tibia (BIOPSI 101, fig. 4.21, table 4.25) is about three-quarters the length of the femur in *Xenosmilus,* a ratio it shares with *Smilodon.* In *Homotherium,* the ratio is slightly larger, about 84%, but both the tibia and the femur are absolutely longer and more slender. The smallest anteroposterior shaft diameter on the homothere was 27 mm, while the smallest transverse is 30 mm. In *Xenosmilus,* it is roughly round, with both diameters at about 32 mm. Merriam and Stock (1932) report a range of 25–32 mm for the smallest transverse shaft diameter in *Smilodon* and 32–43.4 mm in *Panthera.* The tibial lengths for *Panthera* range from 338–400 mm, but the length in

Table 4.25. Right and left tibia measurements (mm): *Xenosmilus hodsonae.* Compared specimens (observed range), *Smilodon fatalis, Panthera atrox* (Merriam and Stock, 1932); Page Museum Salvage (PMS), *PMS-2338-19R.

	X. hodsonae BIOPSI 101		X. hodsonae UF 60,000		Smilodon fatalis 2010		Panthera atrox 2908	
	R	L	R	L	R-1	R-10	R-1	R-10
Greatest length measured parallel to long axis	286.0	283.0	290.0	—	305.0	241.0	*400.8	*322.9
Greatest transverse diameter (mediolateral) of proximal end	84.0	85.0	80.0	77.0	84.4	74.8	107.0	86.7
Transverse diameter at mid shaft	30.0	30.0	30.0	31.0	26.8	25.7	40.0	*28.5–33.0
Greatest transverse diameter of distal end	66.0	67.0	63.0	—	63.6	50.2	74.8	59.1
Greatest anteroposterior diameter at distal end	44.0	43.0	47.0	—	40.8	33.3	52.3	42.4
Greatest anteroposterior diameter at proximal end	83.0	83.1	—	—	—	—	—	—
Greatest anteroposterior diameter at mid shaft	38.7	35.6	—	—	—	—	—	*32.0

Smilodon is 241–305 mm. The length for *Homotherium* is 323 mm. The length of the tibia in *Xenosmilus* is 286 mm.

In *Xenosmilus,* the proximal lateral articular surface of the tibia is slightly more elongated posteriorly than in *Smilodon,* and it curves down distally, giving a rounded surface and the potential for a greater range of motion in retraction of the upper leg. In *Panthera,* this feature is slightly less exaggerated, and it is the least developed in *Homotherium.* At the middle of the posterior face of the tibia there is a sharp raised ridge, the soleal line, which runs distally and trends medially along the medial edge of the bone. This ridge is less elevated in *Smilodon* and even less so in *Panthera.*

In *Xenosmilus,* the articular surface of the medial condyle is relatively broader than in *Smilodon* and *Homotherium* but equal to that in *Panthera.* In *Xenosmilus,* anterior to the articular surface of the tibial condyle is a large tubercle just anterior to the depression that serves as the distal attachment for the lateral collateral ligament of the knee. This tubercle is also present in *Homotherium,* but in *Smilodon* the process is significantly smaller and more distal. The tibial tuberosity for the attachment to the patellar ligament is much broader in *Xenosmilus* than it is in *Homotherium* and even broader than in *Smilodon.* The anterior face of the tibia in *Xenosmilus* bears a relatively broader surface for attachment of the patellar tendons than do the tibiae of the other felids and *Ursus,* although it is of similar relative proportion in *Ailuropoda.*

The tibial shaft in *Xenosmilus* is twisted so that the anterior tibial tubercle is oriented more laterally than in *Homotherium,* but this is compensated by a greater medial bowing of the midshaft, resulting in the distal tibial articular surfaces showing a similar orientation in both cats. *Xenosmilus* differs from the other cats in having the medial proximal surface concave and roughened for the attachment of relatively larger muscles and ligaments. The articular surface of the medial malleolus is extended distally beyond the other articular surfaces in *Smilodon* but lie at about the same level with the other articular surfaces in *Homotherium. Xenosmilus* resembles *Smilodon* in this respect, but the notch between it

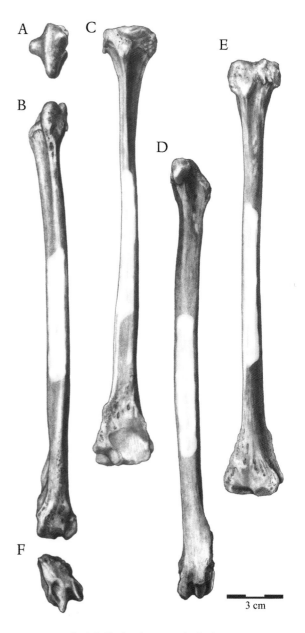

Figure 4.22. The left fibula of *Xenosmilus hodsonae,* UF 60,000: *A,* proximal end view; *B,* anterior view; *C,* medial view; *D,* posterior view; *E,* lateral view; *F,* distal end view. (drawing Mary Tanner)

Table 4.26. Fibula measurements (mm): *Xenosmilus hodsonae*. Compared specimens (observed range), *Smilodon fatalis*, *Panthera atrox* (Merriam and Stock, 1932); Page Museum Salvage (PMS), *PMS 2374-19R. (*a* = approximate)

	X. hodsonae UF 60,000		*Smilodon fatalis* 2011		*Panthera atrox* 2909
	R	L	R-2	R-10	R-1
Greatest length	243.0*a*	242.5*a*	284.7	212.7	369.0–*298.8
Greatest anteroposterior diameter of proximal end	—	29.7	45.9	32.5	35.4
Greatest mediolateral diameter of proximal end	—	20.9	—	—	—
Greatest anteroposterior diameter of distal end	34.4	33.2	26.4*a*	17.2	39.4
Greatest transverse diameter (mediolateral) of distal end	18.4	19.4	—	24.4	21.0
Anteroposterior diameter at mid shaft	10.7	10.9	11.9	9.2	16.1
Transverse diameter at mid shaft	13.2*a*	12.3*a*	13.5	9.4	10.9

and the lateral articular surface is shallower, and the process for the medial malleolar surface is broader than in *Smilodon*, resembling *Homotherium*. The external articular surface is larger than in *Smilodon*, once again resembling *Homotherium*, but in *Homotherium* the articular surfaces are nearly level, while in *Xenosmilus* they are highly inclined proximally. When the astragalus and calcaneum are fitted into this articular surface, due to the reduced angle of the calcaneum, it becomes obvious that *Xenosmilus* was plantigrade as were *Ailuropoda* and *Ursus*. This contrasts to the digitigrade foot in *Homotherium* and *Panthera* and the semidigitigrade foot in *Smilodon*. There is a deep groove, the sulcus maleolaris, for the tendon of M. tibialis posterior in *Xenosmilus*. This groove is broader than in *Homotherium* but resembles *Smilodon*.

THE FIBULA

In *Xenosmilus*, the fibula (UF 60,000; fig. 4.22; table 4.26) is even more robust than in *Smilodon*. The proximal end of the fibula has a large expansion in *Smilodon* that is not present in *Xenosmilus*, although the proximal end in *Xenosmilus* is absolutely as wide as it is in *Smilodon*. Proximally, the shaft in *Xenosmilus* has a large sharp ridge that is absent from *Smilodon*, while the *Smilodon* fibula has a spoon-shaped depression on the anterolateral surface. On the distal end, the digital notch in *Xenosmilus* is interrupted by an expansion of the medial articular surface. The distal articular surface is almost 30% larger than it is in *Smilodon*.

ANKLE (TARSAL BONES)

CALCANEUM

Only the right calcaneum (BIOPSI 101, fig. 4.23A–E, table 4.27) was recovered. The two astragular facets of *Xenosmilus* are more broadly connected than in both *Panthera* and *Smilodon*. The inner process, or sustentacular facet, of *Xenosmilus* is not turned medially as much as in either *Panthera* or *Smilodon*. The facet is also larger in *Xenosmilus*. The groove for the interosseus ligament is long in *Homotherium, Smilodon,* and

Xenosmilus, strengthening the joint in the saber-tooth cats, but it is short in *Panthera*. *Xenosmilus* and *Homotherium* have a distinct facet for the navicular that is even more prominent than that in *Smilodon*. This facet is absent from *Panthera, Xenosmilus,* and *Homotherium*. It is more oval in *Smilodon* than in *Xenosmilus*. On the lateral side of the calcaneum of *Panthera*, there is a distinct process separating two grooves. The dorsal groove is for the tendon of the fibularis brevis, and the ventral is for the tendon of the fibularis longus muscle. In *Xenosmilus*, the process and grooves are indistinct. The plantar view of the calcaneum shows it to be broader and more massive than in *Panthera*, with a more extensive articular surface for the calcaneal tendon. The neck leading to the articular facet for the cuboid in *Xenosmilus* and *Homotherium* resembles that of *Smilodon* and is much shorter than it is in *Panthera*. Although *Smilodon* and *Xenosmilus* are among the least cursorial of cats, they do not show a well-defined area of origin for the accessory flexor muscle, which is associated with lack of cursoriality by Turner and Antón (1997). Contrary to Turner's observation, a well-defined area of origin can be seen in a number of modern cats including *Panthera* and *Puma*. It is noted here that the term cursorial strictly applies to those animals that are capable of running fast and for long distances (Christiansen and Adolfssen, 2007). Certainly *Xenosmilus* and *Smilodon* were capable of a short burst of speed before pouncing on their prey, and *Homotherium* and *Panthera* could exceed this short burst. We suggest using the term semi-cursorial for *Panthera* and *Puma* and non-cursorial for *Xenosmilus* and *Smilodon*.

ASTRAGALUS

Both the left and right astragalus (BIOPSI 101, fig. 4.23F–J, table 4.28) were recovered. In *Xenosmilus*, the trochlear surface is broad and shallow as it is in *Smilodon*, while in *Panthera*, it is narrower and more deeply grooved. The neck is short and ursid-like in *Xenosmilus*. It is longer in *Smilodon* and *Homotherium* and longest in *Panthera*. The juncture between the neck and the trochlea is distinctly notched in *Homotherium*,

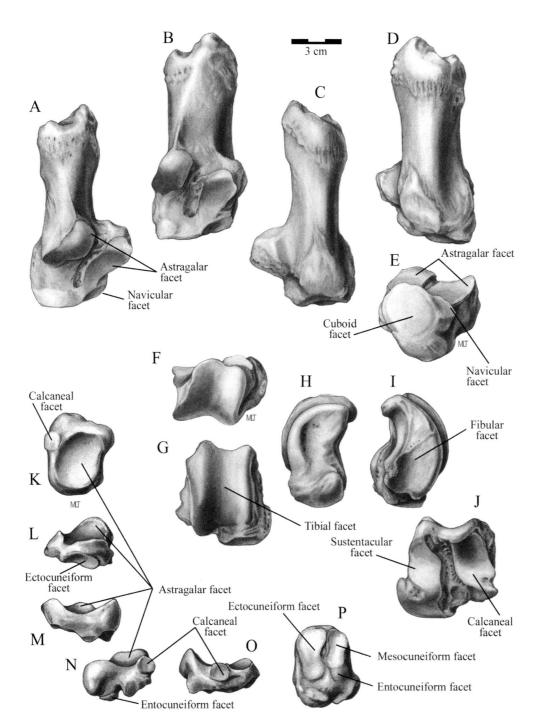

Figure 4.23. The right calcaneum of *Xenosmilus hodsonae*,
BIOPSI 101: *A*, anterolateral view; *B*, dorsal view; *C*, medial view;
D, posterolateral view; *E*, distal end view. The right astragalus
of *X. hodsonae*, BIOPSI 101: *F*, proximal end view; *G*, dorsal view;
H, medial view; *I*, lateral view; *J*, distal end view. The right navicular of *X. hodsonae*, BIOPSI 101: *K*, proximal end view; *L*, dorsal
view; *M*, medial view; *N*, ventral view; *O*, lateral view; *P*, distal
end view. (drawing Mary Tanner)

Table 4.27. Right calcaneum measurements (mm): *Xenosmilus hodsonae*. Compared specimens (observed range), *Smilodon fatalis, Panthera atrox* (Merriam and Stock, 1932).

	X. hodsonae BIOPSI 101	X. hodsonae UF 60,000	Smilodon fatalis 2031 R-1	R-3	Panthera atrox 2929 R-1-*2	R-10
Greatest length	102.5	104.1	106.8	79.4	142.0	109.0
Greatest width measured across astragalar facets As^1 and As^2	41.5	42.0	50.8	40.4	56.0	41.5
Greatest width across cuboid surface measured from astragalar facet As^3 to lateral (outer) side	37.7	36.7	44.2	32.7	37.8	26.0
Greatest anteroposterior diameter of lateral face measured normal (perpendicular) to plantar border and astragalar facet As^1 (neck)	46.6	44.3	48.9	36.2	*60.3	46.5
Greatest width across cuboid facet measured from medial (inner) border of facet (As^3) to lateral side	26.8	26.0	—	—	—	—
Greatest mediolateral (width) of calcaneal tuberosity measured normal to plantar surface	31.4	30.1	—	—	—	—

Table 4.28. Right astragalus measurements (mm): *Xenosmilus hodsonae*. Compared specimens (observed range), *Smilodon fatalis, Panthera atrox* (Merriam and Stock, 1932).

	X. hodsonae BIOPSI 101 R	L	Smilodon fatalis 2030-R1	2030-R4	Panthera atrox 2928-R1	2928-R10
Greatest length	55.1	51.1	61.2	44.6	74.1	61.8
Greatest width	51.6	52.0	61.2	46.0	66.0	51.8
Least width of neck	27.2	23.9	29.5	27.7	35.4	27.8
Greatest diameter of head	34.9	34.8	35.1	29.0	43.0	36.4
Length of neck	4.6	4.5	—	—	—	—
Width of groove of interosseous ligament	4.3	5.6	—	—	—	—

Panthera, and *Smilodon* but less so in *Xenosmilus*. *Panthera, Homotherium,* and *Xenosmilus* have a distinct lateral process for the articulation for the fibula that is not developed in *Smilodon*. Although the articulation with the navicular is rounded, it is not rounded as much as in *Smilodon*. It is also broader, more closely resembling *Panthera* in that respect. The inner (sustentacular) calcaneal facet is elongated in *Xenosmilus* and is nearly circular in *Panthera*. It is also elongated in *Homotherium* and *Smilodon,* although that in *Smilodon* more clearly resembles that in *Panthera*. The groove marking the attachment of the interosseus ligament between the calcaneal facets is longer in *Xenosmilus, Smilodon,* and *Homotherium* than it is in *Panthera,* suggesting that the anklebones in the saber-tooth cats needed to resist higher torsional stresses than in animals for which motion was more restricted to an anterior-posterior plane. The lateral calcaneal facet is narrower in *Xenosmilus* and *Homotherium* than it is in *Panthera* or *Smilodon*. In *Xenosmilus,* the trochlear ridges terminate in a distinct notch at the end of the groove for the interosseus ligament. In *Smilodon,* they terminate in a rounded process that joins the groove for the interosseus ligament. In *Panthera,* the notch is present, but the groove for the interosseus ligament does not approach it. In both *Smilodon* and *Xenosmi-*

lus, the neck is more turned medially than it is in *Panthera*. The medial side of the navicular facet has an overhanging lip in *Xenosmilus* and *Homotherium* that is not present in *Panthera*. The navicular facet is not as angled in *Smilodon* and *Xenosmilus* as it is in *Panthera*.

NAVICULAR

The tubercle of the navicular in *Xenosmilus* (BIOPSI 101, fig. 4.23K–P, table 4.29) is large and round, as it is in *Panthera* and *Homotherium*. It is reduced in *Smilodon*. In *Smilodon,* the articulation for the astragalus is oval, while in *Xenosmilus* and *Panthera* it is more pear-shaped. In *Smilodon* and *Panthera,* the facet for the ectocuneiform is broad and short, while in *Xenosmilus* and *Homotherium* it is elongated. It extends nearly as far as the facet for the mesocuneiform. In *Smilodon* and *Panthera,* the facet for the ectocuneiform is smaller and more sloping than in *Xenosmilus*. In *Xenosmilus,* the navicular is deeply notched at the edge of the facet for the ectocuneiform, while this region terminates in a ridge in *Smilodon* and *Panthera*. Just lateral to the astragalar facet is a small round calcaneal facet, approximately 12.5 mm in diameter. This facet in the Idaho *Homotherium* is oblong and reduced in width, measuring 12 mm long and 10 mm wide.

Table 4.29. Navicular measurements (mm): *Xenosmilus hodsonae.* Compared specimens (observed range), *Smilodon fatalis, Panthera atrox* (Merriam and Stock, 1932).

| | *X. hodsonae* BIOPSI 101 | | *Smilodon fatalis* 2036 | | *Panthera atrox* 2934 | |
	R	L	R-1	R-3	R-1	R-10
Greatest dorsoplantar length	45.0	45.6	48.0	37.0	52.5	43.0
Greatest transverse diameter (mediolateral)	38.0	37.4	37.4	28.5	38.2	33.7
Greatest proximodistal diameter	16.7	16.8	—	—	—	—

THE PES (HIND FOOT)

METATARSAL I

The first metatarsal was not recovered, but the proximolateral surface of metatarsal II is roughened in such a way to imply a vestigial metatarsal I in *Smilodon.* The proximal facet for the mesocuneiform extends at least as far back as it does in *Panthera* and *Puma,* terminating in a long, curving process. In *Xenosmilus* and *Homotherium,* this articulation and process is even more elongated.

METATARSAL II

Metatarsal II in *Xenosmilus* (BIOPSI 101, fig. 4.24A–F, table 4.30) is as short as in *Smilodon,* and, if anything, slightly more massive, contrasting with the elongate metatarsal II in *Homotherium* and *Panthera.* In *Xenosmilus,* metatarsal II broadens more distally than in *Smilodon* and has a larger roughened area for the ligaments binding it to metatarsal III. In *Xenosmilus,* the posterior trochlear surface has a nearly straight margin as it does in *Homotherium.* In *Smilodon* and *Panthera,* the medial trochlear ridge extends proximally beyond this margin. In *Xenosmilus,* metatarsal II is slightly shorter and less robust than metacarpal II. In *Smilodon,* metacarpal II and metatarsal II are about the same length, although metacarpal II is slightly more robust or massive. In *Xenosmilus,* the muscle scar on the plantar side of the shaft is located on the midline as in *Homotherium* (Rawn-Schatzinger, 1992).

METATARSAL III

The proximal end of the third metatarsal in *Xenosmilus* (UF 60,000; fig. 4.24G–L; table 4.31) is broader and expands more posteriorly than in either *Panthera* or *Smilodon.* The lateral proximal articular surfaces are separated by a groove in *Xenosmilus* but not in *Smilodon* or *Panthera. Xenosmilus* resembles *Homotherium* in this respect. The muscle attachment on the plantar side of the shaft extends farther distally and is much larger than in either *Panthera* or *Smilodon,* but resembles *Smilodon* more closely. The shaft is more bowed in *Smilodon* than in *Xenosmilus,* and it is even less bowed in *Panthera.* The middle trochlear ridge on the distal end does not extend as far proximally as in *Panthera* or *Smilodon.* The

anterolateral facet for the fourth metatarsal is more concave in *Panthera* and *Smilodon* than it is in *Xenosmilus.* The trochlear keel in both *Smilodon* and *Panthera* is broader than in *Xenosmilus.* The proximal articular surface in *Xenosmilus* extends down slightly onto the anterior face, in contrast to the sharp straight edge in *Smilodon* and *Panthera.*

METATARSAL IV

Metatarsal IV of *Xenosmilus* (BIOPSI 101, fig. 4.24M–R, table 4.32) resembles that of *Smilodon* in overall size and proportions compared to those of *Panthera* and *Homotherium.* As with the other metatarsals, metatarsal IV of *Smilodon* is more bowed than that of *Xenosmilus.* The muscle insertion on the plantar side of the shaft is about the same length in *Smilodon* and *Xenosmilus* but more centrally located in *Xenosmilus* as it also is in *Homotherium.* The muscle attachment in *Panthera* is laterally situated as it also is in *Smilodon.* In *Xenosmilus,* the proximal articular surface is more rounded than in either *Smilodon* or *Panthera,* and it extends farther onto the anterior face of the shaft. In *Smilodon,* the posterior one-third is stepped downward from the rest of the surface. In *Xenosmilus,* the proximal facets that articulate with metatarsal III are elongated, while in *Panthera* they are more rounded. The posterior surface of these facets is more laterally situated in *Xenosmilus.* These two medial facets are separated from each other by a groove that is more narrow and deep in *Panthera.* The proximolateral articular surface is not as pocketed as in *Panthera* and *Smilodon.* In *Smilodon,* there is a distinct lateral facet that is less pronounced in *Panthera* and not developed in *Xenosmilus.* The distal articular surface is narrower and more rounded in *Panthera* than in *Smilodon* and *Xenosmilus,* with a deeper and broader sulcus at the proximal margin. In *Panthera,* the sulcus is bordered by large ligamental prominences; these are not so prominent in *Smilodon* or *Xenosmilus.*

METATARSAL V

The overall size of metatarsal V in *Xenosmilus* (UF 60,000; fig. 4.24S–X; table 4.33) is about the same length as in *Smilodon,* but broader, and in particular, the distal end is much wider than in *Smilodon.* In *Xenosmilus,* there is a distinct

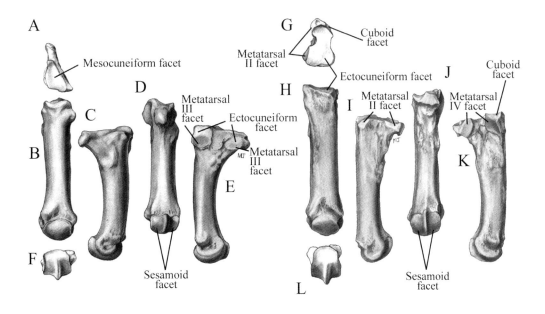

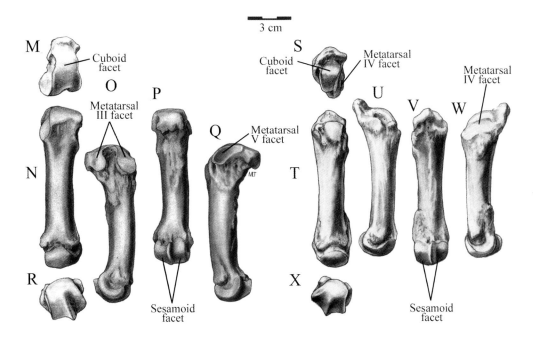

Figure 4.24. The left metatarsal II (Mt II) of *Xenosmilus hodsonae*, BIOPSI 101: *A*, proximal end view; *B*, dorsal view; *C*, medial view; *D*, ventral view; *E*, lateral view; *F*, distal end view. The right metatarsal III (Mt III) of *X. hodsonae*, BIOPSI 101: *G*, proximal end view; *H*, dorsal view; *I*, medial view; *J*, ventral view; *K*, lateral view; *L*, posterior distal view. The left metatarsal IV (Mt IV) of *X. hodsonae*, BIOPSI 101: *M*, proximal end view; *N*, dorsal view; *O*, medial view; *P*, ventral view; *Q*, lateral view; *R*, distal end view. The right metatarsal V (Mt V) of *X. hodsonae*, UF 60,000: *S*, proximal end view; *T*, dorsal view; *U*, lateral view; *V*, ventral view; *W*, medial view; *X*, distal end view. (drawing Mary Tanner)

Table 4.30. Left and right metatarsal II measurements (mm): *Xenosmilus hodsonae.* Compared specimens (observed range), *Smilodon fatalis, Panthera atrox* (Merriam and Stock, 1932).

	X. hodsonae BIOPSI 101	X. hodsonae UF 60,000	Smilodon fatalis 2019		Smilodon fatalis 2019		Panthera atrox 2917		Panthera atrox 2917	
			R-1 Max.	R-3 Min.	L-1 Max.	L-3 Min.	R-1 Max.	R-3 Min.	L-1 Max.	L-3 Min.
	R	L								
Greatest length	86.9	86.6	95.4	73.7	96.5	74.1	148.5	119.7	147.6	120.3
Greatest transverse diameter (mediolateral) of proximal end	21.3	22.3	16.7	14.6	17.0	14.8	22.0	18.4	21.8	18.8
Greatest dorsoventral diameter (dorsoplantar) of proximal end	35.0	36.2	29.4	26.2	31.7	26.6	38.0	30.8	37.7	32.5
Transverse diameter at mid shaft	16.1	16.9	13.8	13.2	16.0	13.4	19.4	16.2	18.8	15.7
Dorsoventral diameter at mid shaft	13.3	12.9	15.8	15.0	15.0	15.0	21.0	15.9	19.3	17.4
Greatest transverse diameter at distal end of shaft	23.9	22.8	22.0	20.3	23.7	20.5	27.9	23.7	27.8	23.6
Greatest dorsoventral dimension of distal end	19.4	19.3	—	—	—	—	—	—	—	—

Table 4.31. Right metatarsal III measurements (mm): *Xenosmilus hodsonae.* Compared specimens (observed range), *Smilodon fatalis, Panthera atrox* (Merriam and Stock, 1932). (P.C. = private collection)

	X. hodsonae UF 60,000	X. hodsonae P.C.	Smilodon fatalis 2020		Smilodon fatalis 2020		Panthera atrox 2918		Panthera atrox 2918	
			R-1 Max.	R-3 Min.	L-1 Max.	L-3 Min.	R-1 Max.	R-3 Min.	L-1 Max.	L-3 Min.
	R	R								
Greatest length	100.9	95.1	111.5	85.5	112.6	85.6	161.0	131.2	163.4	132.3
Greatest transverse diameter (mediolateral) of proximal end	23.8	24.6	36.2	28.8	36.4	29.7	43.0	35.1	42.7	36.4
Greatest dorsoventral diameter (dorsoplantar) of proximal end	34.7	33.9	27.1	25.8	27.0	25.6	33.2	26.9	34.2	28.2
Transverse diameter at mid shaft	19.4	20.0	17.9	17.7	18.2	15.8	23.3	19.3	23.4	20.3
Dorsoventral diameter at mid shaft	16.1	15.4	16.3	15.1	16.4	12.8	20.5	16.8	20.0	18.4
Greatest transverse diameter at distal end of shaft	25.0	24.8	26.4	23.8	26.7	22.7	29.7	24.3	30.0	24.9
Greatest dorsoventral dimension of distal end	23.6	22.8	—	—	—	—	—	—	—	—

Table 4.32. Right and left metatarsal IV measurements (mm): *Xenosmilus hodsonae.* Compared specimens (observed range), *Smilodon fatalis, Panthera atrox* (Merriam and Stock, 1932).

	X. hodsonae BIOPSI 101		Smilodon fatalis 2021		Smilodon fatalis 2021		Panthera atrox 2919		Panthera atrox 2919	
			R-1 Max.	R-3 Min.	L-1 Max.	L-3 Min.	R-1 Max.	R-3 Min.	L-1 Max.	L-3 Min.
	R	L								
Greatest length	98.3	94.1	113.3	98.9	113.8	83.6	167.3	135.8	166.9	135.1
Greatest transverse diameter (mediolateral) of proximal end	23.2	22.5	30.9	26.3	30.0	25.8	35.0	28.7	33.4	28.8
Greatest dorsoventral diameter (dorsoplantar) of proximal end	28.4	31.5	18.6	17.5	19.9	16.1	23.0	18.7	25.8	20.6
Transverse diameter at mid shaft	16.0	16.7	16.0	15.1	17.0	13.4	19.7	17.0	20.4	17.4
Dorsoventral diameter at mid shaft	16.9	17.2	17.0	14.6	14.9	13.0	21.6	17.2	21.3	18.8
Greatest transverse diameter at distal end of shaft	23.5	24.6	25.2	21.9	22.3	18.5	26.6	22.8	27.7	21.7
Greatest dorsoventral dimension of distal end	22.0	21.0	—	—	—	—	—	—	—	—

medial prominence on the distal end that extends onto the plantar surface. *Smilodon* shows a much smaller development in this same region. As with the other metatarsals, metatarsal V in the Idaho homothere is significantly longer than the metacarpal V, but of equal robustness as is also the case for *Panthera.*

PHALANGES

Proximal phalanges II, III, IV, and V from the left hind limb are present in *Xenosmilus* (BIOPSI 101, fig. 4.25A–U, table 4.34). Compared to *Panthera,* the proximal phalanges are shorter, broader, and more curved. The scars for the flexor tendons

Table 4.33. Right metatarsal V measurements (mm): *Xenosmilus hodsonae.* Compared specimens (observed range), *Smilodon fatalis, Panthera atrox* (Merriam and Stock, 1932).

	X. hodsonae UF 60,000	*Smilodon fatalis* 2022		*Smilodon fatalis* 2022		*Panthera atrox* 2920		*Panthera atrox* 2920	
	R	R-1 Max.	R-3 Min.	L-1 Max.	L-3 Min.	R-1 Max.	R-3 Min.	L-1 Max.	L-3 Min.
Greatest length	84.3	91.9	70.8	94.8	72.7	155	125.4	153.3	123.6
Greatest transverse diameter (mediolateral) of proximal end	22.0	—	—	—	—	—	—	—	—
Greatest dorsoventral diameter (dorsoplantar) of proximal end	28.3	28.6	24.4	28.2	24.9	27.7	23.7	31.6	23.3
Transverse diameter at mid shaft	13.2	13.6	10.5	13.5	13.8	14.9	12.1	16.6	13.0
Dorsoventral diameter at mid shaft	11.9	16.0	12.1	14.9	13.9	15.0	12.7	15.8	13.0
Greatest transverse diameter at distal end of shaft	22.9	21.6	17.1	21.6	19.0	23.7	20.0	25.7	20.2
Greatest dorsoventral dimension of distal end	18.6	—	—	—	—	—	—	—	—

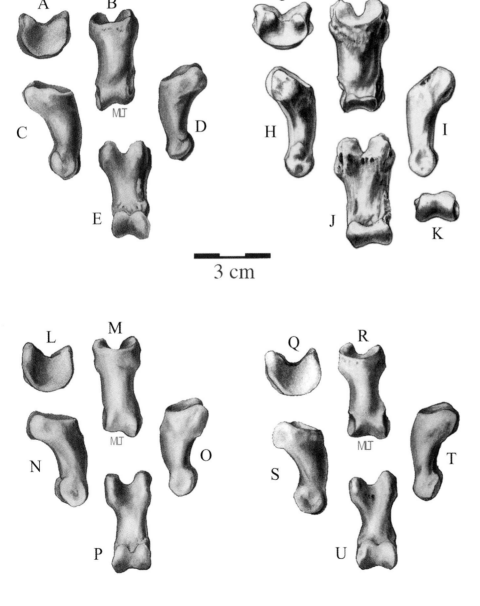

3 cm

Figure 4.25. The first phalanx of the right digit II (pes) of *Xenosmilus hodsonae*, BIOPSI 101: *A*, proximal end view; *B*, dorsal view; *C*, lateral view; *D*, medial view; *E*, ventral view. The first phalanx of digit III (pes) of *X. hodsonae*, UF 60,000: *F*, proximal end view; *G*, dorsal view; *H*, lateral view; *I*, medial view; *J*, ventral view; *K*, distal end view. The first phalanx of the right digit IV (pes) of *X. hodsonae*, BIOPSI 101: *L*, proximal end view; *M*, dorsal view; *N*, lateral view; *O*, medial view; *P*, ventral view. The first phalanx of the right digit V (pes) of *X. hodsonae*, UF 60,000: *Q*, proximal end view; *R*, dorsal view; *S*, lateral view; *T*, medial view; *U*, ventral view. (drawing Mary Tanner)

Table 4.34. Pes phalanges digits II, III, IV, V measurements (mm): *Xenosmilus hodsonae.* (suggested numbers)

| | *Xenosmilus hodsonae* BIOPSI 101 | | | |
	II	III	IV	V
Greatest length	38.7	45.4	38.4	38.0
Greatest dorsoventral (dorsoplantar) diameter of proximal end	17.6	18.6	17.0	17.8
Greatest transverse diameter (mediolateral) of proximal end	19.2	24.0	19.0	19.2
Dorsoventral diameter at mid shaft	11.2	11.7	12.4	11.9
Transverse diameter at mid shaft	13.9	16.9	11.6	12.1
Greatest transverse diameter at distal end of shaft	15.3	16.7	15.6	15.9
Greatest dorsoventral diameter of distal end	10.1	11.3	10.5	11.2

are more prominent and distal. The largest proximal phalanx is the one for digit III. Digits II and V are smaller and narrower. The proximal phalanx for digit II has a lateral facet that fits against a facet for the proximal phalanx for digit III, locking the two phalanges together at their base while spreading them. The ventral surface of the proximal facet on the phalanges is deeply notched for the posterior keel on the metatarsals, so that the toe can be strongly flexed. The distal ends of the phalanges are comparatively broad in general aspect. The phalanges indicate a foot with short toes that could be strongly flexed.

CONCLUSIONS

Xenosmilus remains isolated in terms of its origin and world distribution, and until recently, it was only known from two specimens from a single Florida locality. Now, there is a record from Uruguay, South America (Mones and Rinderknecht, 2004), consisting of a partial ramus, and there is a radius, possibly belonging to *Xenosmilus,* from the Blancan of Arizona (White, Morgan, and White, 2005). *Xenosmilus* is clearly a homothere but deviates sharply from other members of that group. It is likely that the broad forehead characteristic of *Homotherium,* along with elongated legs, are derived features for that genus. We do not think that the common ancestor of *Xenosmilus* and *Homotherium* had legs more elongate than those of a lion (*Panthera*). *Xenosmilus* is the result of an evolutionary trend toward shorter legs and *Homotherium* an evolutionary trend toward longer ones. Pliocene examples of homotheres already show the broad forehead and elongated limbs of *Homotherium,* and we suspect that *Xenosmilus* and *Homotherium* were already separate lineages in the Pliocene.

Xenosmilus, when it was first discovered, was thought to be a new occurrence of *H. serum,* because that species was already known from fragmentary material in Florida. Because the Texas material had been described twice in detail (Meade, 1961; Rawn-Schatzinger, 1992), there was little incentive to describe the Florida material, and it remained

unstudied. The skull in the University of Florida collection was incomplete and restored with a broad forehead like that of the Texas specimen. Study of the second skull (BIOPSI 101), in which the forehead was preserved, showed this reconstruction to be incorrect. The first clue to the uniqueness of *Xenosmilus* came from the shortened limb skeleton.

The locality that produced *Xenosmilus* contained fragmentary remains of another kind of saber-tooth cat, *Smilodon gracilis.* Although *Smilodon* eventually evolved to the size of *Xenosmilus, S. gracilis* is about the size of a jaguar and only two-thirds the size of *Xenosmilus. Homotherium* was not found at the *Xenosmilus* locality and probably should not be expected.

Homotheres and smilodontins must have had profoundly different ecological requirements, and a good locality for one is likely to provide scant evidence of the other. *Xenosmilus* and *Smilodon,* with their short stout legs, were likely ambush predators that shared the need for lots of cover. During the Pleistocene, Florida had local areas of tall grass savanna. Phillip V. Wells produced a map of North American Late Pleistocene vegetation for Martin, Rogers, and Neuner (1985), who related his plant distributions to those of distinctive combinations of mammals (fig. 4.26). His interpretation of the vegetation in central Florida ranged from deciduous woodlands to savannah. The *Xenosmilus* locality was located in a topographically featureless plain where savannah was likely. This is similar to the type of habitat where tigers are normally found and may have contributed to the evolution of the tiger's stripes. Because of this comparison, we have striped our reconstruction of *Xenosmilus* (fig. 4.1B). We think that short-legged sabercats wrestled with their prey while biting it, and that this required them to be able to stabilize their hind feet through a plantigrade stance, as is seen in bears. The plantigrade posture in *Xenosmilus* can be confirmed through the articulations of the ankle with the tibia and accounts for the exceptionally robust hind foot of *Xenosmilus.* As in other homotheres, the upper and lower incisors are enlarged, serrated, and arranged in a semicircle. The upper I[3] is especially large and has a serrated posterolateral

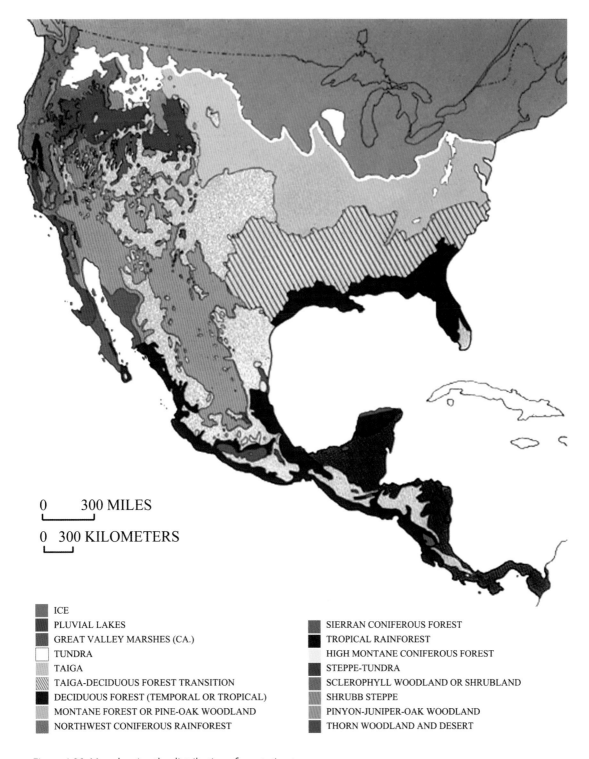

ICE
PLUVIAL LAKES
GREAT VALLEY MARSHES (CA.)
TUNDRA
TAIGA
TAIGA-DECIDUOUS FOREST TRANSITION
DECIDUOUS FOREST (TEMPORAL OR TROPICAL)
MONTANE FOREST OR PINE-OAK WOODLAND
NORTHWEST CONIFEROUS RAINFOREST

SIERRAN CONIFEROUS FOREST
TROPICAL RAINFOREST
HIGH MONTANE CONIFEROUS FOREST
STEPPE-TUNDRA
SCLEROPHYLL WOODLAND OR SHRUBLAND
SHRUBB STEPPE
PINYON-JUNIPER-OAK WOODLAND
THORN WOODLAND AND DESERT

Figure 4.26. Map showing the distribution of vegetation types in the Late Pleistocene of North America. (drawing by Mary Ann Brooks under the direction of Philip V. Wells)

cutting edge. The upper canines are of a length that they would have been inserted to about their mid portion at the time the incisors were engaged. This results in the incisors and the canines forming a continuous cutting surface. As the jaws closed, a large bolus of meat was extracted. This would be useful both for feeding and killing. Because both the upper and lower canines and incisors are engaged before the carnassials, the chief bite force occurs earlier in jaw closure than in other cats. The low inclination of the occiput in *Xenosmilus*, and the resultant elongation of the temporalis muscle, is unusual for saber-tooth cats, in which the occiputs are usually more vertical than in modern felines and the temporalis muscle shorter. We think that in *Xenosmilus,* the elongation of the temporal fossa compensates for the change in position of the bite force away from the mandibular joint. We suspect that *Xenosmilus* killed its prey by creating large, open, bleeding wounds. Although the bite of all scimitar-tooth cats probably fit some aspect of this description, the combination of features in *Xenosmilus* is so unique that we have suggested a different terminology for this combination of temporalis and bite structures, referring to animals that exhibit these features as cookie-cutter cats. Given the opportunity, *Xenosmilus* would have bitten the throats of prey, as do other cats, but we suspect that its potential anatomical targets were broader than those of *Smilodon* and that it may have attacked the abdomen, especially when dealing with smaller animals such as peccaries.

The general appearance of *Xenosmilus* would not have been especially cat-like. Its limbs and body proportions resemble those of the giant panda as described by Davis (1964; Naples et al., 2003). This suggests an animal with powerful forelimbs that could grapple with prey and possibly climb trees, at least to the degree that bears do today. The curvature of the radius and the short curved fingers indicates that the hands could be retracted toward the body with great force and that they could reach around a large object, pulling it toward the mouth. That movement pattern would be especially useful for positioning live prey prior to making a bite. We propose that the canines worked in concert with the incisors, removing a large semicircular portion of flesh. While the prey may have been killed by some specialized directed bite, such as a bite to the throat, it might also have been bitten to death by multiple bites, resulting in extensive bleeding and shock. In *Xenosmilus,* the enlargement of the upper and lower incisors, coupled with the shortening of the sabers, makes this kind of bite possible. We have proposed (Martin et al., 2000) that this presents a fourth morphotype for cats (cookie-cutter cats). If this is true, we should expect to be able to identify other examples in the fossil record. We suggest that the late Oligocene nimravid genus, *Pogonodon Cope* (Cope,

1887), is such an example, being characterized by coarsely serrated short sabers, an elongated temporalis, and shortened legs (based on nearly complete skeletal material in the collection of BIOPSI).

We examined North American homothere specimens from Alaska, California, Florida, Idaho, Kansas, Oklahoma, Nebraska, South Dakota, and Texas; European specimens from England, France, Holland, Netherlands, and Spain; and Asian specimens from Russia and China, in hopes of finding *Xenosmilus,* but we have discovered no additional specimens beyond those recently reported. Our knowledge of *Xenosmilus* is still almost entirely based on the Irvingtonian of Florida. We don't know if it reached the Rancholabrean, but we wonder if *Smilodon* would have gotten so large if it had. There are a few other examples of large mammals with a last known appearance in the early Irvingtonian, including a giraffe-like camel *Titanotylopus* and a hunting dog *Xenocyon texicanus.* Early Irvingtonian sites are rare and have not produced a rich sample of carnivores. Perhaps when more such localities are described, additional specimens of *Xenosmilus* will be identified.

Appendix: Comparison Measurement Table

The tables in this appendix are comparison measurement tables of the skull and mandible of *Xenosmilus hodsonae, Homotherium crenatidens, Smilodon fatalis, Panthera atrox,* and *Homotherium* cf. *serum,* used in this study. The provenience identifications are listed under the respective table headings. All measurements were taken, using dial or digital calipers, from the actual specimens whenever possible; otherwise they were taken directly from the literature as noted in the caption heading. Measurements act as a reference to distinguish multiple variations among taxa, not only between genera, but even at the species level. These are needed whenever new genera or species are described. Note: The *Smilodon fatalis* table caption numbers represent current museum collection numbers and acronyms. The tables represent their earlier identifications from Merriam and Stock (1932).

Table 4.A.1. Skull measurements of *X. hodsonae*, BIOPSI 101 and UF 60,000; *H. crenatidens*, PIN 3120-610; *S. fatalis*, LACMHC 2001-24 and LACMHC 2001-76; *S. fatalis*, KUVP 98124; *P. atrox*, KUVP 44409. (*a* = approximate, *e* = estimate, + = damaged element, — = dimension not available)

No.	Measurement Points	Measurement description	*X. hodsonae* Martin, Naples, Babiarz, Hearst, 2000 BIOPSI 101 Holotype	*X. hodsonae* Martin, Naples, Babiarz, Hearst, 2000 UF 60,000 Paratype	*H. crenatidens* Kondrashov, Martin, this vol. PIN 3120-610	*S. fatalis* Merriam, Stock, 1932 2001-24	*S. fatalis* Merriam, Stock, 1932 2001-76	*S. fatalis* Martin, Naples, Babiarz, this vol. KUVP 98124	*P. atrox* Martin, Naples, Babiarz, this vol. KUVP 44409
1	1 & 2	Length from anterior end of alveolus of I^1 to posterior end of condyles	294.0	304.0	263.0	344.1	271.4	327.0	305.0
2	1 & 3	Basal length from anterior end of alveolus of I^1 to inferior notch between condyles	274.0	281.0	245.0	324.9	260.0	306.0	281.0
3	1 & 4	Length from anterior end of alveolus of I^1 to inion (apex of occipital crest)	324.0	—	271.0	377.9	304.8	355.0	338.0
4	1 & 5	Length from anterior end of alveolus of I^1 to anterior end of posterior nasal opening	145.0	—	135.0*e*	174.9	137.6	160.0	151.4
5	1 & 6	Length of palate from anterior end of alveolus of I^1 to posterior border of palatine (anterior border of pterygoid fossa)	171.0	—	164.0*e*	—	—	187.0	177.0
6	7 & 2	Length from posterior end of glenoid fossa to posterior end of condyles	97.0	103.0	82.0	109.0	83.4	113.0	79.0
7	8 & 8.1	Maximum length of nasal bone (anterior posterior distance medial edge)	70.1	—	65.6	85.9	69.7	R 98.1 L 95.4	—
8	8 & 9	Length of anterior nares (anterior nasal aperture)	52.7	—	28.0*e*	—	—	57.4	55.4
9	10 & 11	Width of anterior nares (anterior nasal aperture)	36.7	—	24.0*e*	63.4*a*	46.9	53.0	101.0
10	12 & 13	Maximum width across the muzzle at canines	83.6	100.0*e*	56.0*e*	114.5	93.6	98.7	76.6
11	14 & 15	Minimum width between superior borders of orbits	67.0	67.0	72.0	101.0	81.3	90.1	101.1
12	16 & 17	Width across postorbital processes	78.0	—	88.0*e*	124.8	109.1	116.6	101.1
13	18 & 19	Minimum width of postorbital constriction	56.1	—	48.0*e*	63.9	60.0	60.5	74.3
14	20 & 21	Maximum width across zygomatic arches	176.5	—	150.0*e* +	232.4	186.3	225.0	214.0
15	22 & 23	Minimum anterior palatal width between superior canines	48.2	—	32.0	63.0	48.9	53.8	53.3
16	24 & 25	Width across palate between posterior ends of alveoli for superior carnassials	102.0	—	69.0	129.0	115.8	124.0	118.6
17	26 & 27	Length of auditory bulla from posterior lacerate foramen to external auditory meatus	38.1	—	38.0*e*	56.9	47.9	60.0	28.0
18	28 & 29	Maximum width across mastoid processes	119.5	—	101.0*e*	154.3	122.8	133.3	133.8
19	30 & 31	Maximum width across occipital condyles	68.0	70.0	56.0	70.4	61.0	72.0	66.9
20	32 & 33	Length from anterior end of alveolus of I^3 to posterior end of premaxillary process	108.0	—	99.0*e*	—	—	116.0	95.4
21	6 & 3	Length from posterior border surfaces of palantine to inferior condylar notch	102.6	—	80.8	—	—	126.0	104.5+

(continued)

Table 4.A.1. (continued)

No.	Measurement Points	Genus and species: Publication: ID Number	X. hodsonae Martin, Naples, Babiarz, Hearst, 2000 BIOPSI 101 Holotype	X. hodsonae Martin, Naples, Babiarz, Hearst, 2000 UF 60,000 Paratype	H. crenatidens Kondrashov, Martin, this vol. PIN 3120-610	S. fatalis Merriam, Stock, 1932 2001-24	S. fatalis Merriam, Stock, 1932 2001-76	S. fatalis Martin, Naples, Babiarz, this vol. KUVP 98124	P. atrox Martin, Naples, Babiarz, this vol. KUVP 44409
22	34 & 38	Perpendicular height from base of condyles to top of sagittal crest	119.6	—	82.0	—	—	104.0	103.8
23	34 & 35	Perpendicular height of skull from apex of occipital crest (inion) to dorsal border of foramen magnum	85.0	—	50.8	—	—	111.7	70.8
24	36 & 37	Maximum width of foramen magnum	32.7	—	25.0	—	—	35.0	31.1
25	35 & 38	Height of foramen magnum	28.0	30.2	—	—	—	34.0	29.3
26	39 & 40	Width of occiput just above level of condyles	90.3	—	64.5	—	—	106.0	96.5
27	66 & 67	Width between glenoid processes measured from medial edges of glenoid fossae	62.7	—	65.0e	—	—	94.3	79.1

Table 4.A.2. Maxillary dentition measurements of X. hodsonae, BIOPSI 101 and UF 60,000; H. crenatidens, PIN 3120-610; S. fatalis, LACMHC 2001-24, and LACMHC 2001-76; S. fatalis KUVP 98124; P. atrox, KUVP 44409. (a = approximate, e = estimate, — = dimension not available, N/A = not available/not developed)

No.	Measurement Points		X. hodsonae Martin, Babiarz, Naples, Hearst, 2000 BIOPSI 101 Holotype		X. hodsonae Martin, Babiarz, Naples, Hearst, 2000 UF 60,000 Paratype		H. crenatidens Kondrashov, Martin, this vol. PIN 3120-610		S. fatalis Merriam, Stock, 1932 2001-24	S. fatalis Merriam, Stock, 1932 2001-76	S. fatalis Martin, Naples, Babiarz, this vol. KUVP 98124		P. atrox Martin, Naples, Babiarz, this vol. KUVP 4440	
		ID Number	R	L	R	L	R	L	R		R	L	R	L
1	41 & 42	Length from anterior end of alveolus of C to posterior end of alveolus of P^4	96.9	96.9	102.0e	104.0e	97.4	97.1	126.3	102.0a	120.6	120.5	102.5	102.6
2	42 & 43	Length from anterior end of alveolus of P^3 to posterior end of alveolus of P^4	51.2	52.2	53.0	53.0	—	48.0	63.7	48.8a	57.4	58.1	58.4	59.3
3	43 & 44	Length of diastema from posterior end of alveolus of C to anterior end of alveolus of P^3	10.1	10.3	13.0e	13.0e	N/A	16.0	18.0	15.0	20.0	20.2	18.9	19.2
4	45 & 46	Width of incisor series measured between lateral borders of alveoli of right and left I^3	67.0		—		46.0e	—	63.5	50.5	58.1		43.4	
5	47.1 & 48.1	Maximum anterior posterior diameter of I^1 at alveolus (mesiodistal)	12.1e	12.1e	—	—	10.3e	10.0e	—	—	12.6	12.6	8.8	8.5
6	47 & 48	Maximum width of I^1 at alveolus (buccolingual/transverse diameter)	7.8e	7.8e	—	11.0	8.0e	8.0e	7.6	6.3	6.3	6.3	4.0	3.8
7	49.2 & 50.2	Maximum anterior posterior diameter of I^2 at alveolus	12.4e	12.0e	—	—	11.3e	11.9e	—	—	13.7	13.5	4.4	4.4
8	49 & 50	Maximum width of I^2 at alveolus	13.0e	12.0e	—	—	10.5e	10.8e	9.5	7.9	7.1	7.1	10.5	10.1
9	46.3 & 51.3	Maximum anterior posterior diameter of I^3 at alveolus	20.0e	20.0e	—	—	15.3e	14.8e	—	—	16.8	16.8	12.3	13.1
10	46 & 51	Maximum width of I^3 at alveolus	13.4	13.3	—	—	11.4e	10.9	13.0	12.0	11.6	11.6	11.6	10.8
11	41 & 44	Anteroposterior length of C at alveolus (mesiodistal diameter)	35.0	37.0	40.0	42.0	33.2	32.6	46.1	39.0	40.8	40.8	23.9	23.2
12	23 & 52	Width of C at alveolus	17.3	17.3	21.0	—	13.4	13.1	22.9	18.0	19.0	20.6	18.0	18.1
13	43 & 53	Anteroposterior length of P^3 at alveolus	11.9e	11.9e	12.0	11.0	N/A	10.5	18.5	—	17.4	17.4	24.2	23.7
14	54 & 55	Width of P^3 alveolus	7.7	7.7	10.0	10.0	N/A	6.6	10.6	7.9	8.6	7.9	15.3	14.1
15	56 & 57	Anteroposterior length of P^4 at base of crown	41.5	41.4	44.0	44.0	39.5	38.6	46.0	33.4a	41.1	43.2	35.8	35.1
16	58 & 59	Maximum width of P^4 across protocone	13.0e	13.0e	16.0	—	12.2	11.8	19.3	14.2	19.7	20.7	19.0	18.4
17	60 & 61	Anteroposterior length of P^4 paracone at base of crown	13.2	14.0	23.0	—	14.0	13.0	13.7	—	14.7	15.4	14.0	13.6
18	42 & 53	Anteroposterior length of P^4 at alveolus	41.1	41.1	—	—	38.9	38.3	—	—	38.2	38.6	33.1	33.6
19	60 & 56	Length of P^4 parastyle from anterior base of paracone to anterior end of tooth	10.4	10.4	19.0	—	8.0	8.0	10.1	—	11.7	11.7	8.6	8.4
20	61 & 57	Length of P^4 metacone blade from base of crown above carnassial notch to most posterior point on base of metacone	16.9	16.3	—	—	16.5	16.6	15.2	—	19.7	19.8	13.1	12.9
21	62 & 63	Anteroposterior length of M^1 at alveolus	N/A	N/A	N/A	N/A	N/A	N/A	—	—	12.7	13.0	9.6	9.6
22	64 & 65	Width of M^1 at alveolus	N/A	N/A	N/A	N/A	N/A	N/A	—	—	4.2	5.0	4.1	4.6

Table 4.A.3. Mandible measurements of X. hodsonae, BIOPSI 101 and UF 60,000; S. fatalis, LACMHC 2002-72 and LACMHC 2002-104; H. cf. serum, BIOPSI 112; P. atrox, KUVP 44409. (— = dimension not available)

No.	Measurement Points	Measurement	X. hodsonae Martin, Naples, Babiarz, Hearst, 2000 BIOPSI 101 Holotype	X. hodsonae Martin, Naples, Babiarz, Hearst, 2000 UF 60,000 Paratype	S. fatalis Merriam, Stock, 1932 2002-72	S. fatalis Merriam, Stock, 1932 2002-104	H. serum Martin, Naples, Babiarz, this vol BIOPSI 112 R	H. serum L	P. atrox Martin, Naples, Babiarz, this vol KUVP 44409 L
1	1 & 2	Length of ramus from anterior border of I_1 alveolus at symphysis to posterior end of condyle at center	202.0	204.0	230.0	178.3	219.0	221.5	230.2
2	2 & 3	Length of ramus from anterior end of outer flange to posterior end of condyle	171.0	184.0	218.8	166.1	188.0	192.0	—
3	6 & 7	Distance from alveolus of C to ventral border of flange	57.0	—	—	—	70.5	71.0	—
4		Length of symphysis measured along anterior border	71.0	73.0	72.7	48.7	73.3	73.3	66.2
5	32 & 33	Minimum depth of ramus below diastema	42.0	46.0	38.7	29.6	46.1	45.0	43.1
6	5 & 9	Depth of ramus below posterior border of M_1 alveolus	48.0	49.0	45.6	36.0	45.6	44.4	47.3
7	10 & 11	Depth of ramus below anterior border of P_4 alveolus	43.0	48.0	—	—	46.2	44.6	42.1
8	12 & 13	Thickness of ramus below M_1 (buccolingual diameter)	20.0	21.0	22.5	19.2	18.3	18.3	19.1
9	14 & 15	Height from inferior border of angular process to summit of condyle	42.0	45.0	37.4	32.3	45.5	46.4	46.8
10	14 & 16	Height from inferior border of angular process to summit of coronoid process	82.0	84.0	75.2	58.0	79.2	80.8	108.2
11	17 & 18	Transverse width of condyle	45.0	45.0	51.4	38.7	47.8	48.4	50.5
12	15 & 19	Maximum depth of condyle	19.0	21.0	18.9	15.8	20.0	21.4	18.8

Table 4.A.4. Mandibular dentition measurements of X. hodsonae, BIOPSI 101 and UF 60,000; S. fatalis, LACMHC 2002-72 and LACMHC 2002-104; H. cf. serum, BIOPSI 112; P. atrox, KUVP 44409. (— = dimension not available)

No.	Measurement Points	Measurement	X. hodsonae — Martin, Naples, Babiarz, Hearst, 2000 — BIOPSI 101 Holotype	X. hodsonae — Martin, Naples, Babiarz, Hearst, 2000 — UF 60,000 Paratype	S. fatalis — Merriam, Stock, 1932 — 2002-72	S. fatalis — Merriam, Stock, 1932 — 2002-104	H. serum — Martin, Naples, Babiarz, this vol. — BIOPSI 112 R	H. serum — Martin, Naples, Babiarz, this vol. — BIOPSI 112 L	P. atrox — Martin, Naples, Babiarz, this vol. — KUVP 44409 L
1	21 & 5	Length from anterior border of C alveolus to posterior border of M_1 alveolus	112.4	121.7	141.8	115.7	119.3	119.5	124.4
2	4 & 8	Length of diastema from posterior border of C alveolus to anterior border of P_3 alveolus	N/A	N/A	—	—	37.1	36.8	26.2
3	4 & 10	Length of diastema from posterior border of C alveolus to anterior border of P_4 alveolus	38.5	41.6	65.5	46.3	51.0	52.5	45.6
4	8.1 & 10	Length of diastema from posterior border of P_3 alveolus to anterior border of P_4 alveolus	N/A	N/A	—	—	6.3	7.3	1.8
5	8 & 5	Length from anterior border of P_3 alveolus to posterior border of M_1 alveolus	N/A	N/A	—	—	66.4	66.6	72.7
6	10 & 5	Length from anterior border of P_4 alveolus to posterior border of M_1 alveolus	55.0	56.0	57.8	52.8	50.3	50.3	56.5
7	1 & 20	I_1 maximum anteroposterior diameter at alveolus	10.0	—	—	—	9.0	9.0	6.3
8	1.1 & 27	I_1 maximum transverse diameter at alveolus	5.0	—	5.0	—	5.0	5.0	2.9
9	3.2 & 20.2	I_2 maximum anteroposterior diameter at alveolus	14.0	14.3	7.1	—	11.5	12.0	7.0
10	27.2 & 28	I_2 maximum transverse diameter at alveolus	8.0	9.2	—	—	6.9	6.9	4.0
11	3.3 & 20.3	I_3 maximum anteroposterior diameter at alveolus	15.1	15.8	8.6	—	11.7	11.0	8.2
12	28.3 & 29	I_3 maximum transverse diameter at alveolus	11.5	13.0	—	8.0	9.3	9.0	5.7
13	30 & 31	C maximum anteroposterior diameter at base of enamel	20.5	21.0	16.0	14.7	16.5	16.2	22.0
14	30.1 & 31.1	C maximum transverse diameter at base of enamel	13.5	15.2	10.5	10.4	11.8	11.7	15.0
15	21 & 4	C maximum anteroposterior diameter at alveolus	21.3	22.6	—	—	12.0	11.6	25.5
16	34 & 35	C maximum transverse diameter at alveolus	13.2	15.0	—	—	11.0	11.4	15.0
17		P_2 maximum anteroposterior diameter	N/A	N/A	—	—	—	—	—
18		P_2 maximum transverse diameter	N/A	N/A	—	—	—	—	—
19	8 & 8.1	P_3 maximum anteroposterior diameter	N/A	N/A	—	—	8.3	8.3	17.0
20	36 & 37	P_3 maximum transverse diameter	N/A	N/A	—	—	5.4	5.4	9.9
21	10 & 23	P_4 maximum anteroposterior diameter	23.7	23.1	26.0	23.2	18.6	18.4	24.1
22	38 & 39	P_4 maximum transverse diameter	11.3	11.8	12.7	11.3	9.0	8.8	13.1
23	25 & 23.1	M_1 maximum anteroposterior diameter	31.0	32.8	32.1	27.4	29.0	30.0	27.2
24	40 & 41	M_1 maximum transverse diameter	14.4	—	16.1	14.0	10.7	11.0	13.2
25	23.1 & 24	M_1 length of paraconid blade at base of crown	15.0	—	—	—	14.0	14.9	14.1
26	24 & 25	M_1 length of protoconid blade at base of crown	16.0	17.5	18.0	16.0	15.0	14.7	15.2
27		M_2 maximum anteroposterior diameter at alveolus	N/A	N/A	—	—	—	—	—
28		M_2 maximum transverse diameter at alveolus	N/A	N/A	—	—	—	—	—

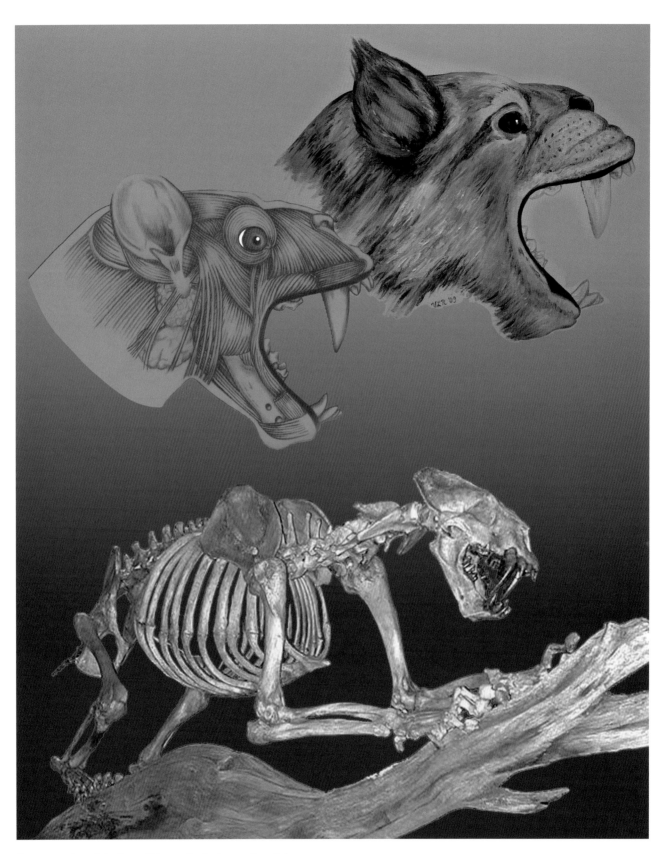

Reconstruction of the face of *Xenosmilus hodsonae* at partial gape
(*top*). Reconstruction of the craniofacial musculature at the same
degree of gape (*middle*). Skeleton reconstructed in a life position
(*bottom*).

5

The Musculature of *Xenosmilus,* and the Reconstruction of Its Appearance

VIRGINIA L. NAPLES

PREDICTING THE APPEARANCE of an extinct animal depends on the quality of the reconstruction of both internal anatomical structures and the external features of skin and pelage. When the skeleton of the fossil is complete, it is easier to provide an accurate picture of the animal, because the missing soft tissues can be reconstructed with a degree of accuracy. When skeletal as well as soft tissues must be reconstructed, each researcher is likely to infuse his or her interpretation of these body regions and systems. As each paleoartist has a personal style, the reconstructed appearance reflects these nuances. Such factors have resulted in a wide range of variation in the reconstruction of fossil animals. For most fossil animals, no living analogues exist; therefore the best models are the most closely related living species. Unfortunately, detailed anatomy of even well-known living species is often unavailable. Consequently, reconstructions can be inaccurate.

There are no living saber-tooth cats; therefore, visualizing a living felid, and adding to it the features that typify a saber-tooth, is likely to result in an animal that mimics the living species, but does not accurately resemble the extinct form. The best method is to assemble the available skeletal elements feature by feature to provide an estimate of the entire saber-tooth, including the overlying skin and pelage. As the complete skeleton of *Xenosmilus hodsonae* is unavailable, reconstruction of the musculature will be limited to the body regions for which bones are available. As with many fossils, the smaller elements were not collected, and many elements are incomplete. For *Xenosmilus,* some axial elements, including some vertebrae, were missing. None of the ribs, sternum, or cervical vertebrae was located. Appendicular skeletal elements (limbs and supporting girdles) that are incomplete, damaged, or unavailable, include parts of the forelimb and hindlimb girdles, scapula, and pelvis, as well as some smaller bones of the forelimb including the wrist and hand (carpals, metacarpals, and phalanges), some bones of the ankle (tarsals), and foot (metatarsals and phalanges). Accordingly, these areas are not included in the reconstruction of the musculature of *Xenosmilus* presented here.

Texture, length, and distribution of the hair coat, as well as color and pattern, must be reconstructed using reasonable estimates based on closely related living forms. Living felid species range in color from light colored (often tawny) to solid dark or black (or mostly solid) with a wide variety of spotted and striped patterns (Nowak, 1999). Usually the hair coat of each species is adapted so that the animal blends into its environment. All living cats capture prey by stealth, carefully stalking unwary or careless animals until the predator is close enough to make a final rush. This stalk allows the predator to get close enough to grasp, bite, and kill the victim. Cheetahs differ from other cats because of their great speed in pursuit of nearly equally fleet prey, but their final rush only occurs after getting close enough by stalking to increase the likelihood of capture. Unlike canids, cats do not pursue prey over long distances, and even the most robust among felids, making the final rush toward the prey, become exhausted after running relatively short distances at top speeds (Caro, 1994; Schaller, 1972).

Coat texture and length also differ among living cat species. As with color and pattern, these differences are related to the environment and vary from short, smooth coats in cats that live in tropical or warm environments to long, luxuriant heavy outer coats that overlie fine, dense, insulating undercoats in species that live in colder climates, including mountainous areas (Kitchener, 1991; Nowak, 1999). Therefore, in addition to mounting a skeleton and making muscular reconstructions, it is essential to know about the environment in which a species lived to make the reconstruction as accurate as possible.

MUSCULOSKELETAL RECONSTRUCTION OF THE CRANIAL REGION

Xenosmilus has a very small and narrow skull when compared to most living cat species of similar body size and to another short-legged extinct ambush predator, *Smilodon*, with its longer, wider skull. In life, *Xenosmilus* probably had a smaller head, giving it different body proportions from contemporary felids. The shape of the body is based on the musculoskeletal structures underlying the superficial tissues of the skin and pelage. In *Xenosmilus*, two groups of muscles determine the shape and size of the head: the larger and deeper muscles, and those most responsible for affecting head shape, the muscles of mastication. This muscle group includes, most prominently, the temporalis and masseter muscles. The other masticatory muscles, the medial and lateral pterygoid muscles, are deep to the masseter and temporalis muscles, so they only minimally affect the appearance of the head. Masticatory muscles move the lower jaw and are adapted to allow the animal to bite and kill prey as well as to remove the flesh efficiently from the carcasses of prey species. A second and more superficial group of muscles that also have an influence on head shape, specifically facial shape, are the muscles of facial expression. These control the movement of facial features such as the eyes, ears, nose, and mouth. Smaller than the masticatory muscles, in general, they leave less prominent muscle scars on the bones of the skull. Nevertheless, some of these muscles are important in making a reconstruction, either because they leave characteristic markings on skull bones or because their functions are well enough discerned that they can be estimated (see

figure on page 98). Of particular importance in restoring sabercats is the shape and size of the mouth at all degrees of gape, the likely "droop" of the lips, and the length of the exposed, elongate upper canines.

The Muscles of Mastication

The M. temporalis

The M. temporalis is large and thick, and arises in the temporal fossa on the posterior region of the skull. It is largely responsible for elevation of the mandible. The narrowness of the *Xenosmilus* skull restricts the space for origin of this muscle, making the temporal fossa shallower than in other saber-tooths. If all other skull proportions had remained the same, this muscle in *Xenosmilus* would have had a greatly decreased volume and thus a reduced ability to generate force for closure of the mandible. A small M. temporalis would be a serious disadvantage for an animal that depends on the force its bite can generate to survive. In *Xenosmilus*, the reduction of volume in the temporal fossa in a mediolateral direction is compensated by an increase in anteroposterior length, as shown by the posterior extension of the occiput. It is unusual for this region of the skull of a saber-tooth to extend over and posterior to the occipital condyles to the extent seen in this animal, because this feature decreases the ability of the cat to elevate the head, a movement that contributes greatly to increasing gape in other saber-tooths. In *Xenosmilus*, the required gape to clear the lower canines is less than in a dirk-tooth saber-tooth of comparable cranial size, and this change in skull shape allows the M. temporalis to maintain a relative volume similar to that of other felids (fig. 5.1). The insertion of the muscle in *Xenosmilus* is similar in placement to that in other saber-tooths, although the coronoid process of *Xenosmilus* is smaller and more vertically oriented than that of conical-tooth cats of the same mandible size. Reduction in coronoid process height serves to increase the excursion of the M. temporalis, an attribute shared with *Smilodon*, among other saber-tooths.

The M. masseter superficialis

The masseter musculature in mammals consists of several divisions, all of which assist the temporalis musculature with elevation of the mandible. However, only the superficial components of the muscle have a major influence on the appearance of the head, as the other portions lie deep to it or to the zygomatic arch. The M. temporalis passes deep to the zygomatic arch from above, and the superficial portion of the masseter muscle arises from its lateral and ventral surfaces. From these origins, the divisions of this muscle (which cannot be individually identified for this study) pass to an insertion on the lateral and ventral surface of the mandible, covering the angular process posteriorly (figs. 5.1, 5.2). This muscle forms the bulge of the cheek. The most anterior portion limits the extent to which the mandible can be de-

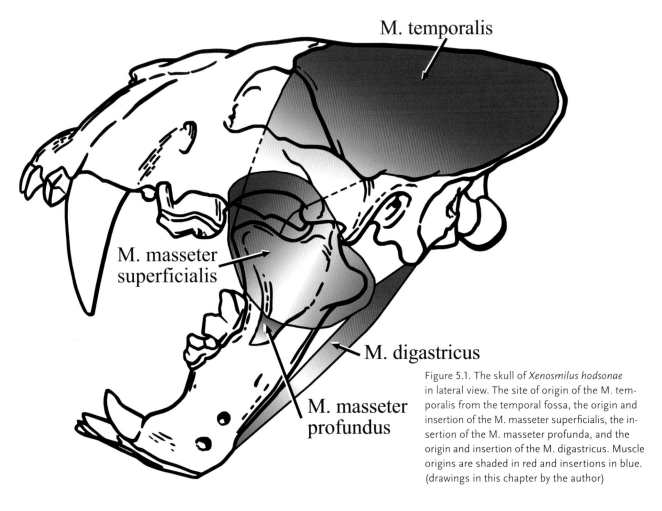

M. temporalis

M. masseter
superficialis

M. digastricus

M. masseter
profundus

Figure 5.1. The skull of *Xenosmilus hodsonae*
in lateral view. The site of origin of the M. temporalis from the temporal fossa, the origin and
insertion of the M. masseter superficialis, the insertion of the M. masseter profunda, and the
origin and insertion of the M. digastricus. Muscle
origins are shaded in red and insertions in blue.
(drawings in this chapter by the author)

pressed, thus limiting the degree to which an animal can gape (Naples, 1985; fig. 5.2D). Because saber-tooths have greatly elongated upper canine teeth that must be cleared prior to biting, this distance is of major importance (fig. 5.2).

The M. masseter profundus and the M. zygomaticomandibularis

The M. masseter profundus and the M. zygomaticomandibularis assist with the elevation of the mandible (fig. 5.3B, 5.3E). The M. masseter profundus is a deep portion of the masseter muscle complex, but the M. zygomaticomandibularis is a deep portion of the temporalis musculature. Both muscle divisions originate from the deep surface of the zygomatic arch of the skull and so are not depicted with the reconstructions of the superficial muscles (see figure on page 98; fig. 5.2). At their origins, the fibers of these muscles intermingle, and individual muscles are often difficult to distinguish. They course ventrally to insert into depressions on the lateral surface of the mandibular body. In living felids, these muscles and their respective scars can be distinguished. However, in *Xenosmilus hodsonae,* the limits of the origins and insertions of these muscles cannot be easily differentiated, and therefore, their insertion scars have been depicted together (fig. 5.2B). The areas of their collective origins and insertions

are relatively large, suggesting that they supported muscles of large mass, which could have increased the mediolateral thickness of the muscles in these areas. Therefore, they do contribute, albeit indirectly, to the shape of the cheek region of *Xenosmilus.*

The pterygoideus musculature

The pterygoideus group of masticatory muscles consists of the large M. pterygoideus medius, a muscle that not only assists with mandibular elevation, but also is the primary effector of the mediolateral mandibular movements that allow the saber-tooth to slice meat using the carnassial teeth. In *Xenosmilus* and *Smilodon,* the attachments of these muscles are small, even when compared to those of conical-tooth cats, which in turn are relatively smaller than the attachments of those muscles in canids and other carnivores. Reduction in the relative sizes of these muscles probably reflects the reduction of mediolateral mandibular movements in felids generally, but the additional reduction in the relative size of these muscles in the saber-tooths likely reflects the requirement for a stricter propralinal (anteroposterior) mandibular movement pattern that ensures correct alignment of the elongate canine teeth. In comparison to the other muscle of this group, the M. pterygoideus lateralis consists of two

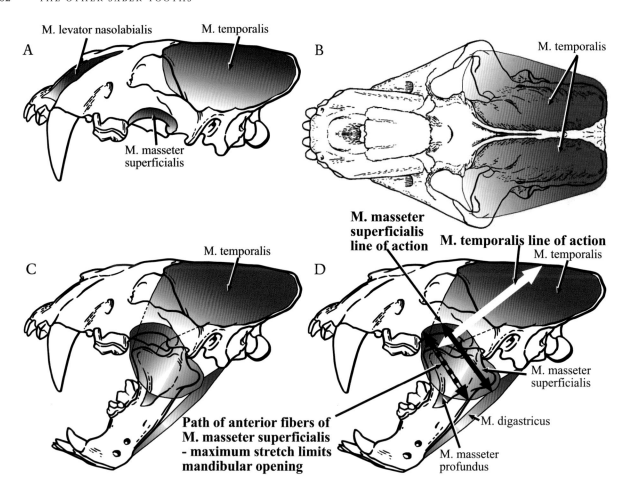

Figure 5.2. *A*, The origins of the masticatory muscles, M. temporalis and M. masseter superficialis, and the facial muscle, M. levator nasolabialis, of *Xenosmilus hodsonae*, lateral view, to demonstrate their relative sizes, shapes, and locations. *B*, The origin of the M. temporalis, dorsal view. *C*, Reconstruction of the path of the muscle belly of the M. temporalis from origin (red shading) to insertion (blue shading). *D*, A white arrow shows the line of action (direction in which the main force of a muscle pulls during contraction) of the M. temporalis, and a heavy black arrow shows the line of action of the M. masseter superficialis. The heavy, black-striped arrow indicates the limitation of the ability of the anterior portion of the M. masseter superficialis to stretch to accommodate maximum mandibular opening.

small bellies (fig. 5.3E). These assist with control of the position of the mandibular condyle during both elevation and depression. However, as these are both deep to the larger masticatory muscles they play no role in the appearance of the head of *Xenosmilus*.

M. digastricus
Mandibular depression in most mammals is partially achieved by gravity, as well as by a relatively small muscle, the M. digastricus. In saber-tooth cats, the muscle origin and insertion are both large, because at gapes of more than 90°, gravity will no longer assist with opening (figs. 5.1, 5.3B–E).

In this study the muscle has been restored as relatively thick and with an excursion that would permit mandibular opening to at least 110°, facilitated by the posterior lateral flare of the mandibular angular processes that pass external to the mastoid processes.

The Muscles of Facial Expression

Muscles of the Oral Region
Several facial muscles are associated with the positioning, size, and shape of the nose and mouth in mammals. Only those that have left evidence of their attachments on cranial

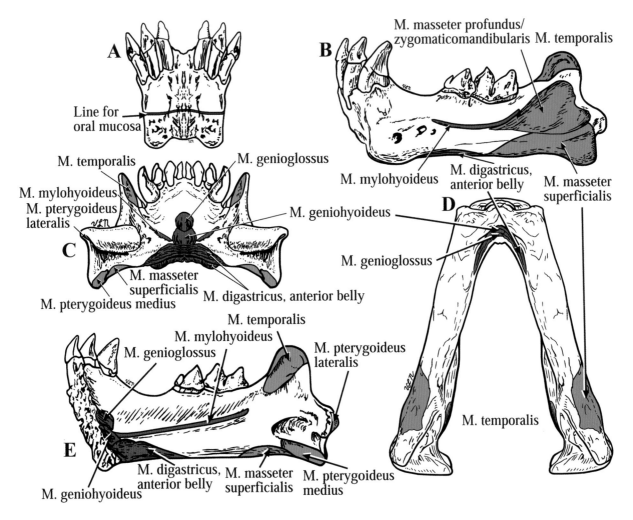

Figure 5.3. *A*, The mandible in *Xenosmilus hodsonae* in anterior view where the oral mucosa attaches tissues of the lip to the mandible. *B*, The mandible in lateral view with origins of the M. mylohyoideus and M. digastricus and insertions of the masticatory muscles M. temporalis, M. masseter superficialis, and M. masseter profundus/zygomaticomandibularis. *C*, A posterior view of the mandible with the origins of tongue and throat muscles M. genioglossus, M. geniohyoideus, M. mylohyoideus, and M. digastricus anterior belly, and the insertions of masticatory muscles M. temporalis, M. masseter superficialis, M. pterygoideus medius, and M. pterygoideus lateralis. *D*, The mandible in ventral view with the origins of throat and tongue muscles M. genioglossus and M. digastricus anterior belly and the masticatory muscles M. masseter superficialis and M. temporalis. *E*, The mandible in medial view with the scars of insertion of the medial slips of the M. temporalis, the M. masseter superficialis, the M. pterygoideus medius, and the M. pterygoideus lateralis. Separate scars for the typical two bellies of the M. pterygoideus lateralis occur in living species, but were not distinguishable in *Xenosmilus hodsonae*. Therefore, only a single scar is visible as well as scars of origin of M. mylohyoideus, M. genioglossus, M. geniohyoideus, and M. digastricus anterior belly. Origins are red and insertions blue.

bones or that can be reconstructed from the shape of these features will be considered here.

The M. levator nasolabialis

The scar for the M. levator nasolabialis in *Xenosmilus* is unusually large and fills a deep ovate impression, with the longer axis anteroposteriorly oriented on the dorsal border of the maxilla. The shape of the anterior skull would allow passage of a wide belly of this muscle lateral to the nasal openings. The insertion of the muscle is into the dorsal border of the upper lip above the attachment of the oral mucosa (fig. 5.4A). This muscle would function to elevate the upper lip and is large in *Xenosmilus hodsonae*, because the soft tissues of the upper lip and the more dorsal nasal region must be strongly retracted and elevated when the cat has opened the mouth to make a bite. The position and size of this muscle suggests that *Xenosmilus* had a more retracted nasal region than do the living large conical-tooth cats such as *Panthera leo* and the extinct dirk-tooth *Smilodon*. The shortened nasal bones as well as the broad muzzle required to accommodate this musculature give *Xenosmilus* a relatively shorter, anteriorly broader muzzle with a facial profile that is slanted pos-

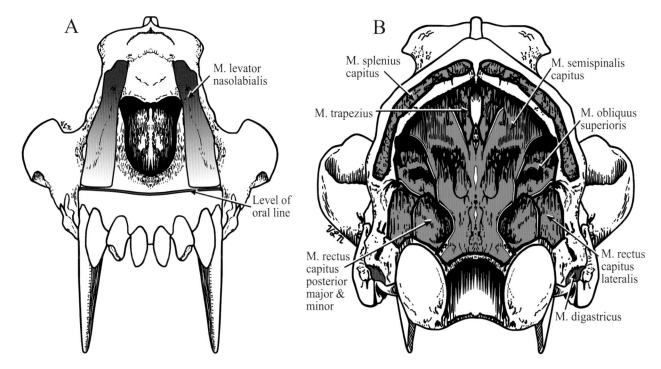

Figure 5.4. *A,* The skull of *Xenosmilus hodsonae* in anterior view with the origin and muscle belly of the M. levator nasolabialis (red shading) and the oral line on the mucosa. *B,* The posterior view of the skull with the insertions (blue) of muscles that insert on the neck and control head position and the origin of the M. digastricus posterior belly (red).

teriorly rather than being vertical or canted anteriorly, as in most conical-tooth cats. A shorter nasal profile also occurs in *Smilodon,* although this feature is more prominent in *Xenosmilus* because of the large, procumbent incisors, which in *Xenosmilus* are larger than those of other saber-tooths, including *Smilodon* (fig. 5.1).

The M. orbicularis oris

The M. orbicularis oris surrounds the oral opening and controls the tension of the lips, whether fully pursed or stretched to maximum gape. Maximum gape in saber-tooths is controlled by this muscle as well as the anterior part of the M. masseter superficialis. The M. orbicularis oris leaves only slight traces of its attachments on the maxilla and mandible in *Xenosmilus,* but those lines extend across the premaxilla and the maxilla, ending posterior to the location of the rear of the upper and lower carnassial teeth (fig. 5.3A, 5.4A).

Muscles of the Occipital Region

The occipital region in *Xenosmilus* overhangs the neck, an unusual feature in saber-tooth cats that is not reflected in the skull of *Smilodon.* The posterior surface of the *Xenosmilus* skull shows a series of depressions that were for the insertions of the portions of the cervical muscles that attach to the head (fig. 5.4B). As the origins of the muscles are from the vertebral

column, they are not available to describe here. Collectively, these muscles draw the head backward, when acting simultaneously on both sides, and turn the face toward the side of the muscle that is contracting when acting unilaterally (Reighardt and Jennings, 1930; Williams et al., 1989; Gilroy et al., 2008). Generally these muscles are of particular importance in cats because these animals elevate the head as a contribution to increasing gape, an action not commonly used among other carnivorans (Gorniak and Gans, 1980). The most dorsal scar is for the insertion of the M. splenius capitus (fig. 5.4B). Ventrally and medially, there are deeply depressed, dorsoventrally elongated scars for the insertion of the M. trapezius. Ventral and lateral to these insertions are the scars of insertion of the M. semispinalis capitus. Deep to these large, superficial muscles is a group that acts to control head posture, consisting of the M. obliquus superioris and the Mm. rectus capitus posterior major and minor. An additional muscle that is inserted on the posterior face of the skull, ventral and lateral to the previous muscle group, is the M. rectus capitus lateralis.

BIOMECHANICS OF CRANIAL MUSCULATURE

The shape and relative positions of the masticatory muscles as shown in previous reconstructions not only reveal the shape of the head, but also the amount and direction of forces the cat can employ in closing the mandible. The elongated posterior portion of the skull in *Xenosmilus* correlates with the anteroposteriorly elongated M. temporalis, and reflects that the main orientation in which this muscle can apply

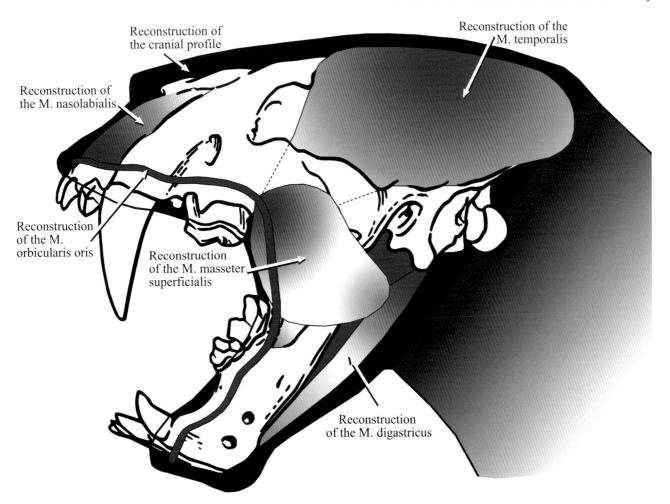

Reconstruction of
the cranial profile

Reconstruction of the
M. temporalis

Reconstruction of
the M. nasolabialis

Reconstruction
of the M.
orbicularis oris

Reconstruction
of the M. masseter
superficialis

Reconstruction
of the M. digastricus

Figure 5.5. The skull of *Xenosmilus hodsonae* in lateral view, with reconstructions of the relative size of cranial muscles. The cranial profile, including an estimate of the volume of other cranial soft tissues, is indicated in black and grey shading surrounding the skull and restored musculature. The nasal profile is more retracted than in other saber-tooths.

force to close the jaw is more horizontal than in other saber-tooths such as *Smilodon* (fig. 5.2D). Likewise, orientation of the M. masseter superficialis in *Xenosmilus* also shows a strong horizontal component of motion. This difference is important in this cat because the relatively more horizontally oriented line of these muscles' action changes the location of the part of the arced path of the mandible's closure, where maximum forces can be exerted. The arrangement in *Xenosmilus* allows greater force to be generated at the canines during the early part of the bite, when the canines have just begun to engage in the flesh of the prey. In *Xenosmilus,* the relatively larger insertion scars of the muscles attaching on the posterior surface of the skull also act with the overhanging occipital region to assist with the increasing gape, allowing this cat to elevate the head as a integral part of making a bite with the elongate canine teeth. Together, these features contribute to the unique cranial shape, relative size, and ori-

entations of muscles of facial expression and mastication in *Xenosmilus* (fig. 5.5).

BONES AND MUSCULATURE OF THE FORELIMB

The forelimb of *Xenosmilus* is represented by a fragment of the pectoral girdle (the distal part of the scapula) and limb bones, including the humerus, radius, ulna, and a few elements of the manus. Based on muscle scars identifiable on these bones, the origins and insertions of many of the larger muscles of this body region can be reconstructed.

The Scapula
The distal portion of the scapula recovered is missing most of the processes. However, a broad depression, the infraglenoid tubercle, representing the origin of the M. triceps, long head, is preserved (fig. 5.6A).

The Humerus
The humeral head in *Xenosmilus* is relatively larger than that of other saber-tooths, including *Smilodon* (fig. 5.6B). The articular surface is not only relatively larger, broader, and more

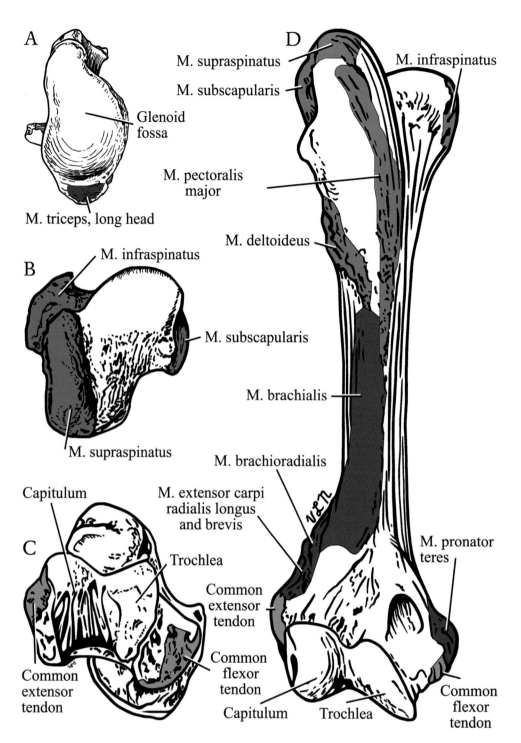

A
Glenoid
fossa

M. triceps, long head

M. supraspinatus
M. subscapularis

M. pectoralis
major

M. infraspinatus

B
M. infraspinatus

M. subscapularis

M. supraspinatus

M. deltoideus

M. brachialis

Capitulum

M. extensor carpi
radialis longus
and brevis

Trochlea

M. brachioradialis

C

Common
extensor
tendon

Common
flexor
tendon

Common
extensor
tendon

Capitulum   Trochlea

M. pronator
teres

Common
flexor
tendon

Figure 5.6. *A*, The distal end of the right scapula of *Xenosmilus hodsonae* with the origin of the long head of the M. triceps from the depression of the infraglenoid tubercle. *B*, The proximal end of the humerus, showing the shape of the humeral head and the scars of insertion of three of the four muscles of the rotator cuff. *C*, The distal view of the humerus, identifying features of the radioulnar articulation and tendinous attachment sites for common flexor and extensor muscles of the forearm. *D*, The anterior view of the humerus, showing insertions of scapulohumeral muscles, origins of humeroradial and humeroulnar muscles, and tendinous origins of forearm muscles. Origins are red, insertions blue, and tendinous attachments green.

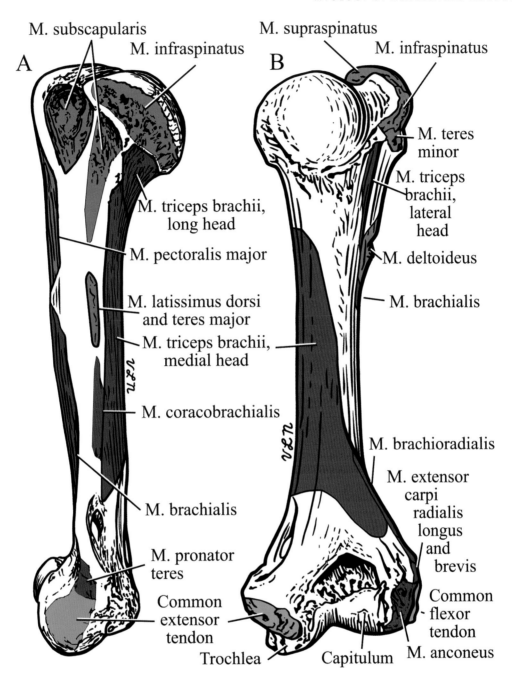

Figure 5.7. The humerus of *Xenosmilus hodsonae: A,* medial view; *B,* posterior view. Both views show insertions of scapulohumeral and thoracic muscles and origins of muscles attaching to the forearm. Insertions are blue, origins red, and tendinous attachments green.

circular, but also continues farther distally onto the posterior aspect of the bone, allowing a greater angle of rotation in the glenoid fossa (figs. 5.6A, 5.7). In contrast, in *Smilodon,* the posterior 20% of the head narrows, indicating that at maximum flexion (forward movement of the forelimb), *Xenosmilus* retains a greater capacity for mediolateral movement. These structural relationships suggest that *Xenosmilus* is able both

to bring the forelimb farther forward than can *Smilodon* and to abduct and adduct it to a greater extent. For a predator that needs to grasp large, active prey, increasing the ability to reach anteriorly as well as mediolaterally at the limit of limb flexion would be especially beneficial. The greater and lesser tuberosities are large in both *Xenosmilus* and *Smilodon,* but with subtle differences that reflect slight differences in position and proportion of muscles of the shoulder region. The greater tuberosity in *Xenosmilus* is more rounded than that of *Smilodon,* but projects farther dorsally and shows a larger lateral muscle scar excavation than does *Smilodon.* The insertion scars of the rotator cuff muscles are larger than in *Smilodon* (figs. 5.6B, 5.6D, 5.7). The rounded humeral head allows for a large range of motion as well as a significant capability for circumduction. The humeral shaft in both saber-tooths is rounded in cross section, although there is a sharp anterior ridge in *Smilodon,* making the shape somewhat triangular. The same feature in *Xenosmilus* is rounded, although prominent and rugose. The scars for some of the muscles originating and inserting along the humeral shaft are also relatively larger than in *Smilodon* (figs. 5.6D, 5.7).

The shape of the distal end of the humerus is asymmetrical, with the medial side of the trochlea projecting more distally in both *Xenosmilus* and *Smilodon.* The capitulum is rounded, allowing for a great range of motion at the elbow, which is also reflected in the deeply pocketed olecranon fossa on the posterior aspect of the humerus (fig. 5.6C–D). Although this fossa is deep in both species, in *Xenosmilus* it is approximately one-third shorter in the dorsoventral dimension than in *Smilodon,* suggesting that the ulnar olecranon process tip is either smaller or tapers more dramatically toward the tip. In distal view, the articular surfaces are significantly deeper anteroposteriorly in *Xenosmilus* when compared to those of *Smilodon.*

Muscles with Attachments to the Humerus

In the human forelimb, four relatively short shoulder muscles (Mm. supraspinatus, infraspinatus, subscapularis, and teres minor) assist with controlling the position of the large, globular humeral head. In humans, the shape of the humeral head and glenoid fossa permit the shoulder joint to have the greatest freedom of movement of any joint, allowing movement in all planes, which is described as circumduction. These are the muscles of the rotator cuff and, although this term is not generally used for describing the anatomy of other species, here that term is extended to saber-tooth cats because these muscles clearly serve the same function.

The M. supraspinatus

In *Xenosmilus,* the origin of the M. supraspinatus is unavailable, but it inserts onto an ovate rugosity atop the most dorsal part of the greater tuberosity of the humerus (figs. 5.6B, 5.6D, 5.7B). The insertion scar is similarly positioned in *Smilodon,* in which it is more rectangular. Occupying the most dorsal position of the group, the M. supraspinatus is responsible for preventing direct ventral dislocation, or subluxation, of the humeral head.

The M. infraspinatus

The origin of the M. supraspinatus is unavailable for *Xenosmilus.* The insertion of the muscle is into an ovate depression on the lateral aspect of the greater tubercle of the humerus (figs. 5.6B, 5.6D, 5.7). This insertion is about twice as large as the corresponding depression in *Smilodon,* suggesting that in *Xenosmilus,* the requirement for preventing the humeral head from slipping medially is of particular importance.

The M. subscapularis

As for the previous two muscles, the origin of the M. subscapularis is unavailable. The insertion in *Xenosmilus* is into an elongated, rugose area on the medial aspect of the lesser tuberosity of the humeral head (figs. 5.6B, 5.6D, 5.7A). This insertion is approximately twice the size of that of *Smilodon,* again suggesting that control of mediolateral slippage of the humeral head is a more significant issue in *Xenosmilus.*

The M. teres minor

The scapular origin of the M. teres minor is unavailable, but the insertion in *Xenosmilus* and *Smilodon* is similarly located, on the lateral aspect of the humerus, distal to the insertion of the M. infraspinatus and proximal to that of the M. deltoideus (fig. 5.7B). As with other muscles of the rotator cuff group, the scar of insertion is more prominently marked by a rugose depression in *Xenosmilus.* This muscle would have acted with the M. infraspinatus to restrict medial sliding of the humeral head, although its more distal insertion would give this muscle a greater mechanical advantage in steadying the humeral head, particularly toward the anterior limits of flexion.

Collectively, the actions of these four muscles (Mm. supraspinatus, infraspinatus, teres minor, and subscapularis) enhance the ability of these cats to maintain control of the position of the forelimb at the shoulder, when reaching anteriorly toward the maximum extent of its range of motion. However, as all the insertion scars in *Xenosmilus* may be as much as twice as large as those in *Smilodon,* it is likely that maintaining control over the head of the humerus, particularly when the cat was reaching anteriorly to the fullest extent, was of even more importance in the former.

The M. pectoralis major

The origin of the M. pectoralis major cannot be described, but in *Xenosmilus* the insertion is from a large, rounded ridge that occupies a ridge on the anterior surface of the humerus (figs. 5.6D, 5.7A). The feature is strikingly different from the condition in *Smilodon,* in which this feature is marked by an extremely prominent sharp-edged ridge that has a medial orientation. This muscle acts as both an adductor and medial rotator of the forelimb, especially when this movement must overcome resistance. It acts in conjunction with the Mm. latissimus dorsi and teres major. The muscle also draws the forelimb forward and medially from a posterior and lateral orientation in both saber-tooths and may act with the Mm. coracobrachialis and the deltoideus (anterior portion only). Because of the relatively greater prominence of the medially oriented ridge in *Smilodon,* as well as the relatively smaller muscles controlling the humeral head, I suggest that this muscle provides a greater component of the force to move the forelimb medially than in *Xenosmilus;* although both saber-tooths require similar forelimb movement patterns, I suggest they achieve these movements by emphasizing an entirely different group of muscles.

The M. deltoideus

The origin of the M. deltoideus is from the scapula and, therefore, is not available for this study. The insertion, however, is a marked rugosity on the lateral face of the humerus (figs. 5.6D, 5.7) as a robust ovate scar that is a continuation of the raised ridge that forms the origin of the M. triceps lateral head. The deltoid insertion is more robust than that of *Smilodon,* although the relative positions are similar in both animals. This muscle is fan-shaped in most mammals, with multiple functions that use some parts as well as movements that result from action of the entire muscle. The anterior part would have assisted in drawing the forelimb forward, acting with the M. pectoralis major, to contribute to medial rotation, while the fibers of the posterior portion would assist in retraction of the limb and lateral rotation, especially in conjunction with the M. latissimus dorsi and M. teres major. The midportion of the muscle would effect most of the abduction of the limb, with stretching of the anterior and posterior portions helping to steady the raised forelimb. The rotator cuff muscles would also assist the action of this part of the muscle in steadying the humeral head.

The M. triceps

The M. triceps is typically divided into three heads in mammals: long, lateral, and medial. The scars for these muscle divisions include a rugose area along the lateral humeral shaft, proximal to the scar of insertion of the M. deltoideus for the origin of the lateral head (fig. 5.7A) and an elongated rugosity for the medial head on the medial face of the bone (fig. 5.7B). The M. triceps long head originates from the infraglenoid tubercle on the ventral (posterior) border of the scapula in most mammals, and was discussed with the scapula (fig. 5.6A). The common insertion of the slips of this muscle is onto the olecranon process of the ulna and will be discussed with other features of that bone.

The M. teres major and M. latissimus dorsi

Elongate, rugose features on the anteromedial aspect of the humeral shaft in *Xenosmilus* mark the insertions of the M. teres major and the M. latissimus dorsi (fig. 5.7A). These scars are probably separate, but their respective boundaries cannot be identified on *Xenosmilus* or *Smilodon*. Although many features of the appendicular skeleton are more robust and suggest an increased mechanical advantage in *Xenosmilus* as compared to *Smilodon*, this scar in *Xenosmilus* is slightly smaller and somewhat more proximal, reducing the mechanical advantage of these two muscles. This arrangement sug-

gests that *Xenosmilus* depended somewhat less on these muscles for medial rotation, adduction, and extension of the forelimb than did *Smilodon*. However, in both animals, the relative size, robustness, and location of these insertion scars suggest that the muscles had excellent mechanical advantage for medial rotation of the forelimb and were large relative to those of saber-tooths that were not ambush predators. Such movement capability would enhance the ability of *Xenosmilus* and *Smilodon* to bring the forelimb toward the body and would be important in grasping with the forepaws.

The M. coracobrachialis

The origin of the M. coracobrachialis is not available, but the scar of insertion in both saber-tooths covers a longitudinally ovate region of the anteromedial aspect of the humeral shaft, distal and medial to the scars of the Mm. teres major and latissimus dorsi (fig. 5.7A). The muscle acts to assist in drawing the forelimb anteriorly, especially from toward the posterior limits of its excursion. When the forelimb is abducted, this muscle assists the anterior portion of the M. deltoideus in preventing side sway of the limb. The collective insertion scars of this muscle, with the Mm. teres major and latissimus dorsi, are less prominent in *Xenosmilus* than in *Smilodon,* again suggesting that the former species is more dependant upon the action of the relatively larger rotator cuff to control the position and steadying of the humeral head in the glenoid fossa.

The M. brachialis

Although the scar of origin of the M. brachialis (figs. 5.6D, 5.7B) is equally robust in *Xenosmilus* and *Smilodon,* the muscle originates more distally in *Xenosmilus;* this results in a shorter muscle with a more direct line of action for flexion of the forearm at the elbow and, therefore, a greater mechanical advantage than is the case for the same muscle in *Smilodon*.

The M. pronator teres

In *Xenosmilus,* the M. pronator teres arises from an oval rugosity on the medial epicondyle of the distal part of the humerus (figs. 5.6D, 5.7A). It is distinct from the common flexor tendon and more distal to it in both *Xenosmilus* and *Smilodon*. Relatively speaking, this muscle origin is located similarly in *Smilodon,* but is approximately one-third larger. In many species there is also an ulnar head of this muscle; however, for the saber-tooths such a muscle scar was not discernible. The muscle acts to rotate the radius on the ulna, an action that would turn the paw pad surface of the manus ventrally.

The Common Flexor Tendon

In *Xenosmilus,* the common flexor tendon is the site of origin of the superficial group of forearm flexor muscles (figs. 5.6C–D, 5.7B). These are the Mm. palmaris longus, flexors carpi radialis and ulnaris, and the flexor digitorum superficialis. The origin is a rugose tuberosity that extends from the medial to the posterior surface of the distal half of the medial epicondyle of the humerus. In *Xenosmilus* and *Smilodon,* this feature is similar, being relatively large in size, shape, and position, suggesting that the actions of these muscles are for flexion of the wrist and abduction of the manus. With its belly crossing the flexor surface of the forearm, the M. flexor carpi ulnaris is an antagonist to the rest of the muscles of the group, and adducts the manus. All of these actions would be of particular importance in cats generally, and especially in sabercats for controlling and manipulating large prey.

The M. brachioradialis

The M. brachioradialis arises from a longitudinally oriented lunate rugosity on the lateral aspect of the distal humerus immediately proximal to the medial epicondyle (figs. 5.6D, 5.7B). In *Xenosmilus* and *Smilodon,* this scar is of relatively large size and similarly positioned. This muscle is probably equally important in these two saber-tooths because it is most active in rapid flexion and extension of the elbow joint and acts to steady the joint in either movement.

The M. anconeus

In both *Xenosmilus* and *Smilodon,* the scar of origin of the M. anconeus arises from a flattened ovate ridge-bounded surface from the most distal part of the lateral epicondyle of the humerus (fig. 5.7B). This muscle assists the triceps in extending the elbow joint and is probably especially important in rotating and slightly abducting the ulna during pronation of the forearm.

The Common Extensor Tendon

The muscles arising from the common extensor tendon act on the elbow joint and the more distal joints of the forearm (radiocarpal, carpal, and digital joints) as extensors (figs. 5.6C–D, 5.7). These five muscles are the Mm. extensor carpi radialis longus and brevis, extensor digitorum, extensor digiti minimi, and extensor carpi ulnaris. All of these muscles insert onto bones of the wrist and manus not available to this study; therefore, any specializations cannot be discussed, although the size of their scars of origin suggest that they are similarly important to all saber cats, which need precise control and flexibility at the wrist.

The Bones of the Forearm
The Radius and Ulna

In *Xenosmilus,* the radius and ulna are relatively short and robust, with prominent scars of origin and insertion of the flexor and extensor muscles (fig. 5.8). The proximal articulation with the humerus is well-defined, with tightly apposed reciprocal concave and convex articulations. It appears that

the elbow in *Xenosmilus* can be extended to a slightly greater degree than can that of *Smilodon* and also that the olecranon process fits more compactly into the olecranon fossa of the humerus.

The Ulna

The olecranon process is robust and thick mediolaterally, although with less anteroposterior depth than in that of *Smilodon.* In both species it shows a large rugosity on the proximal surface. It is slightly shorter in *Xenosmilus* than in *Smilodon.* The tip of the olecranon surface in *Xenosmilus* has a knob-like projection at the anterior aspect that aligns with the orientation of the anterior limit of the radial notch distal to it. In *Smilodon,* this projection has a more medial orientation and is more prominent and pointed, projecting farther dorsally. In *Xenosmilus,* the trochlear notch is large and asymmetrical mediolaterally, with the medial aspect extending more distally to a slightly greater extent than in *Smilodon.* Both the articular surfaces are slightly broader in *Xenosmilus* than in *Smilodon.* The radial notch of the proximal radioulnar joint in *Xenosmilus* is more sharply concave, covering approximately 120°; it is more sharply defined but slightly shorter dorsoventrally than in *Smilodon.* The shaft of the ulna is slightly more gracile in *Xenosmilus* than in *Smilodon.* The medial aspects of both forearm bones show a large rugosity that serves as the attachment for an interosseus membrane that extends the entire length of the shafts. This ligament would have provided an enlargement of the surface area between the shafts of these bones, especially for the attachments of the deep flexor muscles of the manus digits. The lateral aspect of the trochlear notch in *Xenosmilus* is flatter, flared more laterally, and has a larger surface area than in *Smilodon.*

The Radius

The radial head in *Xenosmilus* is relatively smaller than in *Smilodon,* and is a rounder oval, lacking the depression on the posterior face that continues as a groove distally. The radial shaft is slightly more gracile in *Xenosmilus* and is bowed medially along the distal half in comparison to *Smilodon.* The shapes of both the anterior and posterior faces of the bone show similar indications of muscle attachment and the passage of tendons. The proportions of these features are generally alike in both species.

The Muscles that Attach to the Forelimb Bones

The muscles that attach to the forelimb bones serve to control the extent and speed of flexion and extension of the forelimb as well as the medial and lateral rotational capabilities. They also effect the adduction and abduction of the manus. In some cases, in addition to the scars of attachment, the tendons of the extrinsic muscles of the forearm pass through identifiable grooves on the anterior and posterior surfaces of the radius and ulna. Many of the muscles that

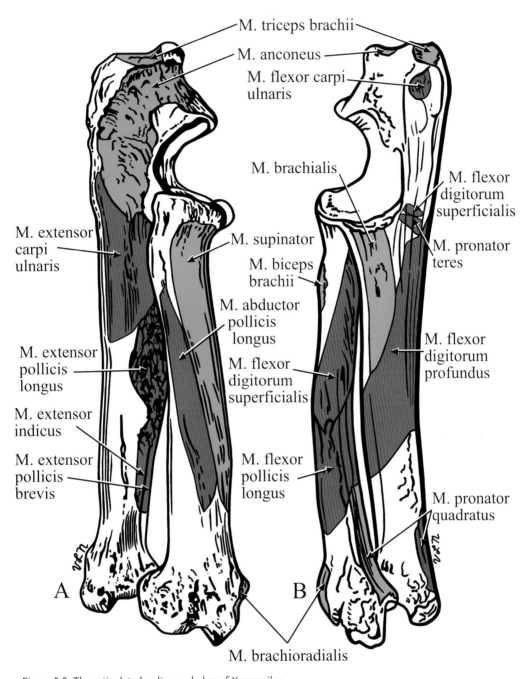

Figure 5.8. The articulated radius and ulna of *Xenosmilus hodsonae: A,* anterior view; *B,* posterior view. The areas of attachment of arm and forearm flexor and extensor muscles and the digital flexor and extensor muscles. Origins are red and insertions blue.

control movement of the manus arise from common tendons and insert onto carpal bones that were not recovered for *Xenosmilus* and, therefore, have not been included in this description. As identification of digital phalanges that were collected with the *Xenosmilus* specimens is difficult, the details of forepaw function in this cat cannot be reconstructed in greater detail than what can be concluded from a few of the forearm muscles.

The Muscles on the Flexor Surface of the Forearm
The M. biceps brachii

In most mammals, the M. biceps brachii is among the largest of those of the forelimb and occupies the anterior face of the humerus. This muscle is the main flexor of the forearm. In this study, the scapular origins of the two heads are unavailable, but in *Xenosmilus,* their common insertion is onto a large ridge-bounded, teardrop-shaped facet (larger aspect ventral) on the anterior and lateral aspect of the radius (fig. 5.8A). This is also the case in *Smilodon*. The insertion is slightly more proximal in *Xenosmilus* but has relatively the same size as in the dirk-tooth cat.

The M. flexor pollicis longus

The origin of the M. flexor pollicis longus is the large longitudinally oriented rugose scar that is located mostly on the anterior face of the shaft of the radius and the adjacent medial aspect of the interosseus membrane attaching to the ulna; this muscle flexes the thumb and positions the tip of the digit to maximize the extension of the claws (fig. 5.8A). Although these scars in both cats occupy about the same relative surface area on the radial shaft, this scar is less rugose in *Xenosmilus* than in *Smilodon* and is not separated from the more medial aspect of the bone shaft by a longitudinal concavity as in *Smilodon*. Both these cats have relatively large thumbs, and a large scar of origin for this muscle would be predicted, as it would permit the cat to flex the thumb strongly in conjunction with the rest of the fingers in grasping with extended claws.

The M. flexor digitorum profundus

The anterior ulnar surface and the adjacent interosseus membrane are the origin of the M. flexor digitorum profundus, the muscle responsible for flexing the distal phalanges of digits II–V (fig. 5.8B). The relative size of the muscle scars in *Xenosmilus* indicates that the slips of this muscle were large and suggests that this cat had a great capability to flex the digits forcefully. The action of this muscle would have been important in positioning the digits to take maximal advantage of the ability to extend the claws for grasping prey. The scar for this muscle in *Smilodon* is similar in size and position to that of *Xenosmilus*; therefore the slips of this muscle probably functioned as did those of *Xenosmilus*.

The M. pronator quadratus

In *Xenosmilus*, the scar of origin of the M. pronator quadratus is from approximately the distal third of the anterior face of the radius, marked by a rugose surface (fig. 5.8B). The insertion of this muscle is onto a distinct rugose longitudinal ridge on the medial edge of the anterior face of the distal third of the ulnar shaft. The positions of the muscle scars and their relative sizes are similar in *Smilodon* and in *Xenosmilus*. In both animals, this muscle would have acted as a synergist with the M. pronator teres to assist in pronation of the supinated forearm and manus. Movement of the distal segments of the forelimb into pronation would be of major importance in allowing the cats to grasp with the digits, particularly the enlarged thumb.

The Muscles on the Radial Surface of the Forearm
The M. brachioradialis

The origin of the M. brachioradialis was discussed with the humerus (fig. 5.7B). The belly passes distally on the lateral surface of the radius (medial surface of a pronated cat forelimb) to insert on the medial side of the radius onto a prominent rounded crest; possibly it continues proximally along a rugose longitudinal ridge that runs along the edge of the bone, which possibly attaches to the most distal one-third of the shaft (fig. 5.7A). The muscle assists in flexing the forearm but also in initiating pronation, and as a pronator, would work as a synergist with the Mm. pronator teres and pronator quadratus. The insertion of the muscle is of similar size and position in both *Xenosmilus* and *Smilodon*.

The Muscles on the Extensor Surface of the Forearm
The M. extensor carpi ulnaris

Of the three superficial muscles of the posterior compartment of the forearm (Mm. extensor digitorum, extensor digiti minimi, and extensor carpi ulnaris) that originate from the humeral common extensor tendon, in *Xenosmilus*, only the M. extensor carpi ulnaris also has an origin from a longitudinally oriented rugose ridge on the posteromedial edge of the ulna (fig. 5.7). In contrast, this ridge is much less distinct in *Smilodon*. This muscle extends the manus and adducts the wrist.

The M. supinator

In *Xenosmilus*, the M. supinator arises from a sharp ridge on the center of the posterior proximal aspect of the ulna, from the distal aspect of the olecranon expansion as the distalmost lip of the trochlear fossa (fig. 5.8A). In *Smilodon*, the origin of the muscle is from approximately the same area, at the lateral edge of the posterior face of the ulna. The insertion of the muscle in both cats is onto the posterior face of the radius, from just distal to the radial head, and continuing along approximately half of the posterior face of the bone. The origin of the muscle in *Xenosmilus*, from the middle of the posterior face of the ulna, gives the muscle a slightly longer excursion than is possible for that in *Smilodon*. Although a subtle difference, a longer excursion of the muscle suggests that *Xenosmilus* had a greater capacity to supinate the forearm than did *Smilodon*. This advantage could either work as an antagonist to the large pronator muscles of the forearm, or it could assist the muscles involved with medial rotation in grasping with the digits.

The M. abductor pollicis longus

In *Xenosmilus*, the M. abductor pollicis longus arises from a longitudinally oriented depression on the posterior face of the ulna facing the radius, the interosseus membrane, and a longitudinal strip of smooth bone on the posterior face of the radius on the side toward the ulna (fig. 5.8A). This origin continues distally for approximately one-third of the shaft of the bone. In *Smilodon*, this muscle scar is less clearly marked but can be estimated to occupy a similar position. In both saber-tooths, this muscle functions to abduct the thumb from the palm of the hand; it would be important in assisting the cat in releasing a grip it might have obtained on prey and would assist with disengaging the extended claws.

The M. extensor pollicis brevis

In *Xenosmilus*, the M. extensor pollicis brevis arises from the posterior ulnar crest, starting slightly distal to the level of the distalmost edge of the radial notch and continuing from a sharp ridge along at least the middle third of the shaft (fig. 5.8A). A small radial origin is from a longitudinal rugosity distal to that of the M. abductor pollicis longus. In *Smilodon*, as the posterior ulnar crest is more gently rounded, it is not possible to ascertain the limits of the ulnar origin of this muscle in this cat, although the insertion from the radius is similarly positioned as in *Xenosmilus*, although more rugose. The muscle abducts the manus, an action that would also assist the cats in controlling the position of the digits for grasping.

The M. extensor pollicis longus

In *Xenosmilus*, the M. extensor pollicis longus arises more distally, from along the posterior ulnar crest, from a rugose ridge that ends slightly proximal to the dorsal tuberosity (fig. 5.8A). The position and relative size of the muscle is similar in *Smilodon*, although the muscle scar is less rugose in this saber-tooth. In both cats this muscle would assist with extension of the wrist and abduction of the hand and would be synergistic with the M. extensor pollicis brevis.

THE BONES AND MUSCLES OF THE HINDLIMB
Bones and Muscles of the Pelvis

As is the case for the pectoral girdle, only a small portion of the pelvic girdle was recovered. The long bones of the hindlimb (the femur, tibia, and fibula) are present, as well as some of the elements of the ankle and hind foot. Only the muscles that either originate or insert on these elements were considered in this study.

The Pelvis

The acetabular region of the pelvis shows a large, round acetabulum. Some of the body of the pubis is also present.

The M. rectus femoris

A prominent rugosity anterior to the acetabulum of the pelvis marks the origin of the M. rectus femoris, a muscle that forms the superficial and anterior portion of the quadriceps femoris muscle group that collectively extends the leg (fig. 5.9B).

The Femur and Associated Muscles
The Femur

The femur in *Xenosmilus* is slightly more robust in all dimensions than in *Smilodon*. The greater trochanter and proximal shaft are also more rugose. The femoral heads in both cats are similar in shape, although that in *Xenosmilus* has a slightly longer neck. The anterior femoral faces are relatively smooth in both species. In both cats the linea aspera on the posterior face of the femur is marked by a single, relatively broad raised ridge, bounded laterally by a longitudinal depression extending approximately two-thirds of the length of the bone (fig. 5.10A). Distally, the femoral condyles are large, with the medial broader than the lateral. The anterolateral surface of the lateral femoral condyle is concave in the saber-tooth cats and gently rounded in *Panthera*. In *Xenosmilus* and *homotheres*, the posterior portion of this surface is expanded laterally, resembling *Ursus*. In *Xenosmilus*, this area is roughened, indicating the presence of large extensor and flexor ligaments of the knee and scars for muscle attachments. In *Xenosmilus*, the medial surface of the condyle of the femur contains a deep pit that serves as the origin of the medial ligament of the knee. This feature is not as prominent in the other compared species, suggesting that the knee in *Xenosmilus* is better buttressed against lateral forces than are those of the other species with which it is compared here. Lateral buttressing of the knee suggests strengthening against forces that would cause lateral sliding of the lower leg and foot. This suggests a greater need to control foot placement, as would be especially critical during bipedal attacks on prey.

The Muscles of the Femur

The muscles of the femur are responsible for flexion and extension of the thigh as well as medial and lateral movements of the leg and foot. Acting with the muscles that attach to the pelvis and spinal column, these muscles assist in maintaining balance and posture, including standing bipedally.

The M. gluteus medius

The greater trochanter in *Xenosmilus* exhibits a prominent scar for the insertion of the M. gluteus medius proximally, anteriorly, and, to the greatest extent, laterally (figs. 5.9A, 5.9C–D, 5.10B). The attachment is partially divided, perhaps because of a bursa deep to the M. gluteus medius at the proximal end of the greater trochanter. The area of this muscle scar is relatively larger than in *Smilodon*, in which the greater trochanter is more gracile and slanted more anteriorly. Together with the M. gluteus minimus, this muscle abducts the thigh, and especially the anterior portion is responsible for medial thigh rotation. Together, these muscles also contribute to maintaining the trunk upright, especially when one hind foot is elevated. When a hind foot is off the ground, these muscles contract strongly enough on the contralateral side to elevate the pelvis and, therefore, prevent it from sagging on the unsupported side.

The M. gluteus minimus

Anterior to the scar for the M. gluteus medius is insertion of the M. gluteus minimus. The distal aspect of the scars for the Mm. gluteus medius and minimus converge on the anterior and posterior faces of the trochanter. In *Smilodon*, the scar for this muscle is smaller than in *Xenosmilus* (figs. 5.9A, 5.10B).

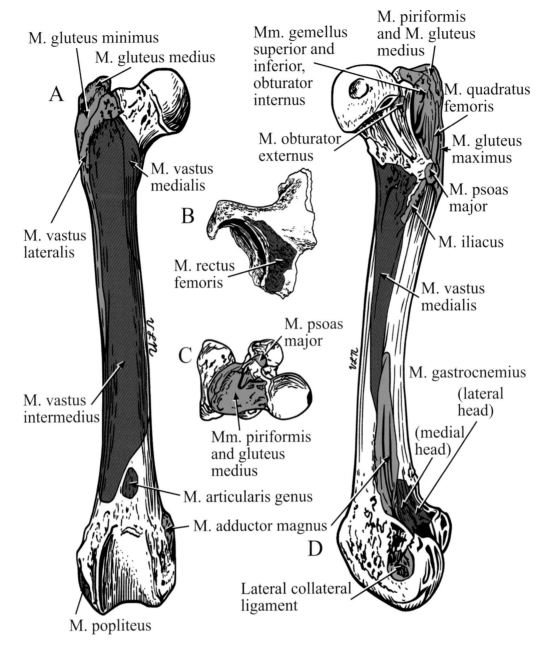

Figure 5.9. *A*, The femur of *Xenosmilus hodsonae* in anterior view, showing origins and insertions of gluteal and thigh muscles and those crossing the knee. *B*, The pelvis in lateral view, showing the rugose site of origin of the M. rectus femoris, reflected head. *C*, Proximal view of the femur, showing insertions of muscles crossing the hip joint. *D*, Medial view of the femur, showing insertion of pelvic muscles, origins of thigh muscles, and attachments of knee ligaments. Origins are red, insertions blue, and ligamentous attachments green.

The M. piriformis

The M. piriformis arises from digitations on the anterior face of the sacrum in most mammals and cannot be verified for *Xenosmilus*. The proximal end of the greater trochanter of the femur is broad and robust, with an oval depression surrounded by a raised ridge that serves as the origin at the most proximal portion of the trochanteric fossa (fig. 5.9D, 5.10B). The M. piriformis inserts in a similar position in *Smilodon;* however, the tip of the greater trochanter is more gracile and pointed and extends slightly less far proximally. The muscle is important in rotation of the thigh (lateral rotation of the extended thigh and adduction of the flexed thigh)

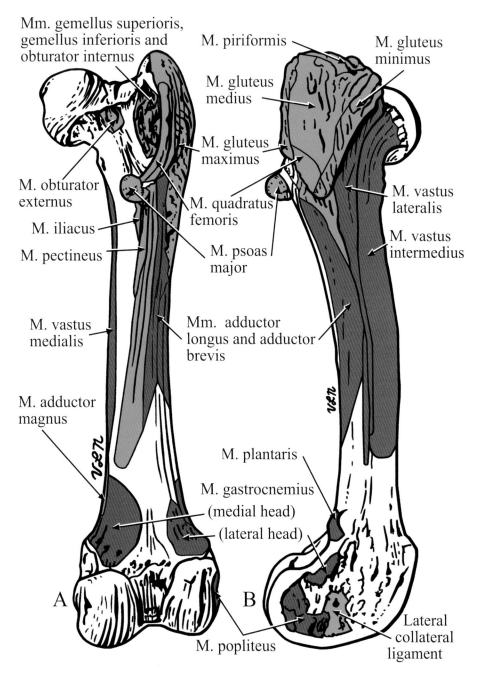

Figure 5.10. The femur of *Xenosmilus hodsonae: A*, posterior view;
B, lateral view. Insertions of pelvic muscles, origins of thigh and
lower leg muscles, and attachments of knee ligaments. Origins
are red, insertions blue, and ligamentous attachments green.

and, therefore, would be important in maintaining balance, particularly in a bipedal posture as proposed for these felids while attacking prey. This muscle acts as an antagonist of the Mm. gluteus medius and minimus when the thigh is adducted.

The quadriceps femoris

This muscle is the largest extensor of the leg in mammals and consists of four parts that cover the medial, anterior, and lateral parts of the femur. These parts include the Mm. rectus femoris, vastus medialis, vastus lateralis, and vastus intermedius. Together, these four muscles not only extend the thigh, but the knee joint, which, in *Xenosmilus*, may be able to rotate to about 140°, which is slightly greater than is usual for felids generally (Reighardt and Jennings, 1930). In contrast, in *Smilodon*, the scars of origin of the parts of this muscle are less obvious, although the knee joint may have been capable of similar extension.

The M. rectus femoris

The origin of the M. rectus femoris is from a large triangular rugose depression on the lateral surface of the pelvic fragment recovered for *Xenosmilus* (fig. 5.9B). The origin appears to be similar in size and location in *Smilodon*. In both cats the insertion of the muscle is into the patellar tendon attaching to the anterior face of the patella with the other muscles of this group.

The M. vastus medialis

In *Xenosmilus,* the M. vastus medialis occupies a rugose elongated ridge on the anteromedial aspect of the femur, starting anterior to, but at the same level with, the knob-like lesser trochanter and continuing around the femoral shaft medially to continue distally as the most medial ridge of the linea aspera (figs. 5.9D, 5.10A). The scar disappears approximately 60% of the distance distally along the posterior aspect of the femoral shaft. The distal aspect of the scar extends farther than that of the M. vastus lateralis. In *Smilodon,* the most proximal and anterior part of the scar is slightly less rugose, but the more distal aspect is less clearly marked, although, as in *Xenosmilus,* it most likely continues farther distally than that of the M. vastus lateralis.

The M. vastus lateralis

The muscle scar for the M. vastus lateralis in *Xenosmilus* is a raised rugose ridge continuing distally from the greater trochanter, distal to the insertion of the M. gluteus minimus, and as a depression across the proximal and anterior aspect of the femoral shaft (figs. 5.9A, 5.10B). It is similarly positioned in *Smilodon,* although the scar is less rugose.

The M. vastus intermedius

The anterior face of the femur in *Xenosmilus* is roughened, indicating the fleshy origin of the M. vastus intermedius (figs. 5.9A, 5.10B). This origin covers most of this region, ending distally only a short distance proximal to the patellar groove.

The M. articularis genus

The M. articularis genus, a small muscle, is a derivative of the M. vastus intermedius, making it a part of the quadriceps group (Fig. 5.9A). It arises from a rounded rugosity on the distal anterior face of the femur in *Xenosmilus*. The scar for its insertion was not obvious in *Smilodon*. This muscle slip attaches to the proximal aspect of the knee joint capsule and contracts with knee extension to prevent the patella from impinging against these soft tissues. This function would be of particular importance in *Xenosmilus,* as this cat has a greater capability for extension at the knee than is typical for felids.

The Mm. obturator internus, obturator externus, gemellus superioris, and gemellus inferioris

The posterior aspect of the greater trochanter in *Xenosmilus* shows a deep rugose depression proximally, the trochanteric fossa (figs. 5.9D, 5.10A). It serves collectively as the site of insertion for the Mm. obturator internus, obturator externus, gemellus superioris, and gemellus inferioris. The origins of these muscles are from the pelvis, from a portion that is unavailable to this study. These muscles are the small postural controllers of the hip joint and stabilize the hip in all positions, but particularly for lateral rotation, adduction, and extension. The trochanteric fossa in *Smilodon* is also particularly large and slightly wider than in *Xenosmilus* but less deep. This insertion is relatively larger in these dirk-tooth cats than in other mammals and could be related to the necessity for maintaining positional control of the pelvis.

The M. quadratus femoris

The posterior edge of the greater trochanter, distal to the scar of insertion of the M. gluteus medius in *Xenosmilus,* shows a teardrop-shaped rugosity that serves as the insertion of the M. quadratus femoris (fig. 5.10). This feature is similar in *Smilodon*. In both cats, this muscle is relatively large and serves as a synergist to the Mm. gluteus medius and gluteus minimus as well as the M. piriformis.

The M. psoas major

The lesser trochanter in *Xenosmilus* is an extremely large and rounded bony knob. It serves as the site of insertion for the M. psoas major, a muscle relatively larger in this animal than all the other felids including *Smilodon* and the bear *Ursus,* with which it has been compared (figs. 5.9D, 5.10). This muscle, along with the M. iliacus, is important in flexion and lateral rotation of the hip joint; it is also is extremely important in bending the trunk to the same side when acting unilaterally and in raising the trunk when acting bilaterally. This muscle could be inferred to be one of the main effectors in assisting the cats to raise their trunks to stand bipedally.

The M. iliacus

Distal to the lesser trochanter a depression serves as the insertion for the M. iliacus. This muscle arises from the medial face of the ilium in mammals, which is unavailable for this study, but joins the M. psoas major for a common insertion (figs. 5.9D, 5.10A). These muscles are also joined by fibers of the M. psoas minor, which arises from the lateral bodies of the lumbar vertebrae in most mammals; but such features were not discernible for *Xenosmilus* or *Smilodon*.

The M. pectineus

Along the medial edge of the linea aspera is a rugose ridge, the pectineal line, which is the insertion of the M. pectineus in *Xenosmilus* (fig. 5.10A). The ridge runs distal from the lateral aspect of the lesser tuberosity to approximately 25% of the length of the posterior femoral shaft. This feature is not as easily discernible in *Smilodon*. This muscle adducts and laterally rotates the hip and helps to stabilize the pelvis.

The Mm. adductor longus, adductor brevis, and adductor magnus

The Mm. adductor longus, adductor brevis, and adductor magnus arise from the posterior face of the femur on the raised rugose longitudinal feature, the linea aspera, but it is not possible to distinguish individual origins (fig. 5.10). The linea aspera is present on the posterior femoral shaft in *Smilodon,* and the adductor muscles likely originated similarly in this cat. The muscles of this group fuse differently among mammalian groups; therefore, identification of these is problematic even among living species. Together, these muscles stabilize the pelvis and adduct and flex the hip.

The M. biceps femoris

The M. biceps femoris has both a long and a reflected head. The origin of the long head could not be observed in *Xenosmilus.* The origin of the reflected head is along the lateral surface of the linea aspera, from a rugose ridge along the posterolateral femoral shaft and lateral to the collective scars for attachment of the adductor muscles (fig. 5.11). This feature is similar in *Smilodon.* This muscle inserts onto a raised rugosity on the lateral aspect of the fibular head. The muscle extends and stabilizes the hip and laterally flexes the knee.

The Mm. gastrocnemius

In *Xenosmilus,* the scars on the posterior surface of the femur for the origin of the lateral and medial heads of the M. gastrocnemius appear to be larger than in *Panthera, Homotherium,* and *Smilodon,* as well as *Ursus* (figs. 5.9, 5.10). Medial and proximal to the origin of the M. gastrocnemius medial head is a roughened area that serves as the continuation of the attachment of the M. adductor magnus.

The M. plantaris

In *Xenosmilus* and *Smilodon,* a slight elongated, oval depression proximal to that of the origin of the lateral head of the M. gastrocnemius marks the origin of the M. plantaris (fig. 5.10B).

The M. popliteus

In both *Xenosmilus* and *Smilodon,* distal to the origin of the lateral head of the M. gastrocnemius is an oval depression on the lateral side of the femoral condyle that marks the origin of the M. popliteus (figs. 5.9A, 5.10). The depression is slightly more prominent in *Xenosmilus.* This muscle inserts onto a flattened area of the proximal aspect of the posterior surface of the tibia.

The Bones and Muscles of the Lower Leg
The Patella

In *Xenosmilus,* the patella is large, corresponding to the large and wide patellar groove on the femur. The entire anterior surface is rugose, suggesting a large attachment for the tendon of insertion of the M. quadriceps femoris muscle group proximally and the patellar ligament distally.

The Tibia and Fibula

The proximal part of the tibia in *Xenosmilus* is extremely robust, with a triangular-shaped anterior protuberance. The bone is more robust than the tibia of *Smilodon.* The large tibial tuberosity projects anteriorly and forms the distal point of the anterior tibial prominence. This feature serves as the distal attachment of the patellar ligament, the extension of the quadriceps femoris tendon (fig. 5.11). The shaft of the tibia is bowed (convex medially) in *Xenosmilus,* leaving a wide space for the interosseus membrane, which extends the surface area for the origins of the muscles attaching to the anterior face of the bone. In posterior view, the largest feature of the tibia is a longitudinal depression, occupying more than half of the bone from medial side to middle.

The Muscles of the Anterior Face of the Tibia
The M. gluteus maximus and M. tensor fascia lata

In *Xenosmilus,* the lateral aspect of the proximal tibia is a rounded knob that serves as the distal insertion of the iliotibial tract, a band of connective tissue that is the collective insertion of the M. gluteus maximus posteriorly and the M. tensor fascia lata anteriorly (fig. 5.11).

The M. sartorius

The origin of the M. sartorius, from the anterior superior iliac spine of the pelvis, is not available for *Xenosmilus,* but the insertion of the muscle shared with the Mm. gracilis and semitendinosus into a depression bounded by a rugose area proximally on the medial aspect of the tuberosity of the tibia is distinct (fig. 5.11A). This insertion scar is similarly located in *Smilodon.* The M. sartorius laterally abducts and flexes the hip joint. It acts synergistically with the Mm. semitendinosus and gracilis in flexing and medially rotating the knee.

The M. gracilis

The origin of the M. gracilis is unavailable for *Xenosmilus,* but the insertion is in common with the Mm. sartorius and semitendinosus (fig. 5.11A). This muscle adducts and flexes the hip and medially rotates and flexes at the knee synergistically with the Mm. semitendinosus and sartorius.

The M. semitendinosus

The origin of the M. semitendinosus is not available, but the insertion is in common with the Mm. sartorius and gracilis as is also the case in *Smilodon* (fig. 5.11A). In both cats, the muscle would assist to stabilize the pelvis and can flex and medially rotate at the knee.

The M. extensor digitorum longus

In *Xenosmilus,* the M. extensor digitorum longus arises from a distinct rugose triangular scar distal to the distal lip of the

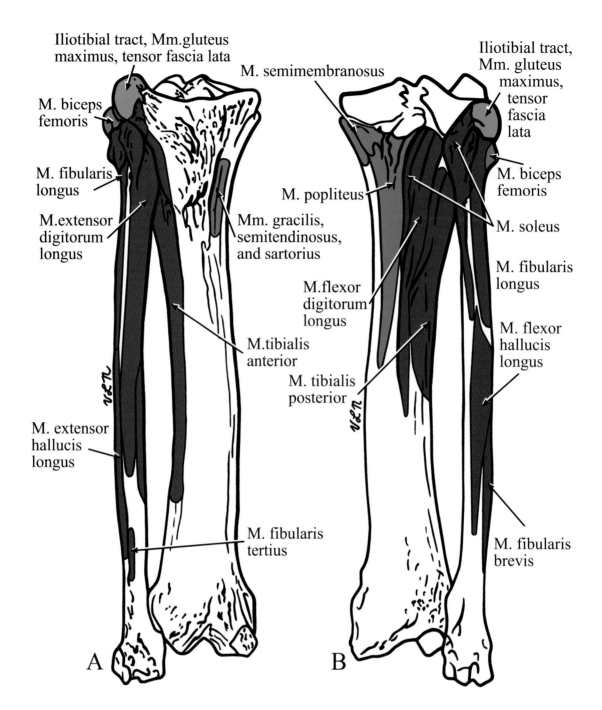

Iliotibial tract, Mm.gluteus maximus, tensor fascia lata

M. biceps femoris

M. fibularis longus

M.extensor digitorum longus

M. extensor hallucis longus

M. semimembranosus

M. popliteus

Mm. gracilis, semitendinosus, and sartorius

M.flexor digitorum longus

M.tibialis anterior

M. tibialis posterior

M. fibularis tertius

Iliotibial tract, Mm. gluteus maximus, tensor fascia lata

M. biceps femoris

M. soleus

M. fibularis longus

M. flexor hallucis longus

M. fibularis brevis

A

B

Figure 5.11. The articulated tibia and fibula of *Xenosmilus hodsonae: A,* anterior view; *B,* posterior view. The scars of insertion of hip and thigh muscles and origins of lower leg muscles. Origins are red and insertions blue.

lateral tibial condyle and also from the anterior face of the medial aspect of the fibula, arising from a sharp ridge that continues for about two-thirds of the length of the bone (fig. 5.11A). The arrangement is similar in *Smilodon*. This muscle extends the toes and dorsiflexes the foot in conjunction with the Mm. tibialis anterior and extensor hallucis longus, making it extremely important in controlling the ability of the cat to flex the claws of the hind foot.

The M. tibialis anterior
In *Xenosmilus,* there is a distinct elongate depression on the anterolateral face of the tibia that defines the origin of the M. tibialis anterior (fig. 5.11A). This feature is equally prominent and similarly placed in *Smilodon*. This muscle flexes and supinates the foot and would be an important synergist of the M. tibialis posterior.

The M. extensor hallucis longus
In *Xenosmilus,* the M. extensor hallucis longus arises from the medial surface of the fibula, marked by a long thin ridge and the interosseus membrane (fig. 5.11A). Although the membrane is not present, the medial bowing of the tibia suggests that there is a large space for the origin of this muscle. An equally prominent longitudinal ridge occurs in *Smilodon*. In both cats this muscle would be important in the extension of the first digit, a movement essential for retracting the claws.

Muscles of the Posterior Face of the Tibia and Fibula
The M. semimembranosus
The origin of the M. semimembranosus is not available, but the insertion is into the depression on the medial side of the tibial tuberosity (posterior surface), behind the insertions of the Mm. gracilis, semitendinosus, and sartorius (fig. 5.11B). This muscle acts synergistically with the M. semitendinosus to extend the hip, stabilize the pelvis, and flex and rotate the knee joint medially.

The M. tibialis posterior
In *Xenosmilus*, the M. tibialis posterior arises from elongated ovate regions on the medial edge of the fibula and the proximal half of the posterior aspect of the tibia (fig. 5.11B). The muscle scars on both bones are rugose and probably also mark the attachment of the interosseus membrane. In *Smilodon,* the ridge that marks the medial boundary of this muscle scar is farther lateral, indicating that the tibial origin of this muscle is relatively narrower. This muscle is partially responsible for plantar flexion as well as being the main inverter of the foot (Gilroy et al., 2008), suggesting that this action is of slightly lesser importance in *Smilodon* than in *Xenosmilus*. However, in both cats, this muscle would stabilize the foot against the substrate, an action important in both quadrupedal locomotion and bipedal standing.

The M. flexor digitorum longus
In *Xenosmilus*, the M. flexor digitorum longus arises from a distinct groove in the posterior face of the tibia, lateral to the soleal line, as is also the case in *Smilodon* (fig. 5.11B). This muscle is the main flexor of the toe phalanges, primarily the distal elements, and because it inserts at the distal phalanges in living species of felids, it assists with protrusion of the claws (Reighardt and Jennings, 1930). When the foot is in not in contact with the ground, the muscle also acts as a plantar flexor (fig. 5.11B). When the foot is on the ground, this muscle acts with the intrinsic muscles to maintain the paw pads in contact with the ground, enlarging the weight-bearing area of the foot and stabilizing the heads of the metatarsals.

The M. soleus
In *Xenosmilus*, the M. soleus arises from the posterior aspect of the fibular head and the tibial shaft distal to the prominent ridge that marks the soleal line (fig. 5.11B). This feature is even more prominent in *Smilodon* than in *Xenosmilus*. The muscle acts synergistically with the two heads of the M. gastrocnemius to flex the knee and plantar flex the foot on the lower leg.

The M. flexor hallucis longus
In *Xenosmilus*, a sharp, elongate ridge marks the origin of the M. flexor hallucis longus on the posterior surface medial side of the fibula (fig. 5.11). The ridge begins slightly less than halfway distally along the shaft and continues distally nearly to the distal expansion of the fibular head. This muscle is similar in *Smilodon,* and in both cats, acts synergistically with the M. flexor digitorum longus to flex the toes and protrude the claws.

Muscles Attaching to the Lateral Surface of the Tibia and Fibula
The M. fibularis longus
In *Xenosmilus*, the M. fibularis longus arises from the rugose fibular head and along approximately the first third of the lateral shaft from a distinct groove (fig. 5.11). It is also visible in the posterior view, on the lateral edge of the bone. This arrangement is similar in *Smilodon*. The muscle is a foot flexor, but would also act as an antagonist to the Mm. tibialis anterior and posterior that serve to invert the foot. It is a dorsiflexor of the foot, acting as the main antagonist of the M. tibialis posterior.

The M. fibularis brevis
In *Xenosmilus*, the origin of the M. fibularis brevis is distinct, from the lateral surface of the fibula from about midshaft nearly to the expansion of the distal fibular head. The origin scar is visible in both the anterior and posterior views of the fibula (fig. 5.11). The origin of this muscle is similar in *Smilodon*. In both cats, the muscle would act synergistically with the M.

fibularis longus, assisting with flexion of the foot and as an antagonist to the Mm. tibialis anterior and tibialis posterior.

The M. fibularis tertius

In *Xenosmilus*, the M. fibularis tertius originates from a small, longitudinal rugosity on the anterolateral face of the fibula, just proximal to the expansion of the distal fibular head (fig. 5.11A). This location is marked by a distinct depression in *Smilodon*. This muscle acts as a dorsiflexor and pronator of the foot and would have acted in part as an antagonist of the Mm. tibialis anterior and posterior, as do the other muscles of the fibularis group.

CONCLUSIONS

Cookie-cutter cats are a new fourth ecomorph of saber-tooth cat, added to the dirk-tooth, scimitar-tooth, and conical-tooth forms. The suite of characteristics noted with the initial description of *Xenosmilus* (Martin et al., 2000) includes a previously unrecognized combination of the relatively short, coarsely serrated sabers typical of a scimitar-tooth cat with the short legs and robust body form of a dirk-tooth cat. In addition to these features, this study demonstrates that in *Xenosmilus* the body form is even more robust than is that of *Smilodon*.

The main factor shaping the head in *Xenosmilus* is the requirement for the cat to achieve a gape large enough to engage prey. Also there is the need to maintain sufficient mandibular pressure throughout the arced path of the mandible during closing to complete the bite with sufficient force to remove a chunk of flesh. Cranial shape in *Xenosmilus* permits reorientation of the masticatory muscles to increase the amount of force generated throughout the entire path of mandibular closing. In contrast, *Smilodon* requires a greater amount of force toward the end of the bite to permit its even more elongate upper canine teeth to cut themselves out of the tissues of the prey in a single stroke.

However, not only does head shape reflect the adaptations of *Xenosmilus* to making a bite that leaves a large gaping wound, but the postcranial anatomy of this animal is also adapted to complement the masticatory system. The humeri in both *Xenosmilus* and *Smilodon* are very robust, but there are major proportional differences in origins and insertions of the muscles important in prey capture and control. In both species, the greater humeral tuberosity is extremely large, although in *Xenosmilus* the crest is flared more laterally, a feature that correlates with the predicted greater ability of this species to abduct the humerus. Other features that support this suggestion include the wider and more robust proximal crest of the greater tuberosity of the humerus where the M. supraspinatus and M. infraspinatus muscles insert and the larger facet on the lesser tuberosity for insertion of the M. subscapularis muscle. These muscles connect the forelimb to the pectoral girdle and, with M. teres minor, control humeral head position. In humans, they constitute the rotator cuff, a muscle group that controls the position of the humeral head to increase the stability of the glenohumeral joint in all positions (Williams et al., 1989; Gilroy et al., 2008). These muscles serve the same function in saber-tooth cats. In both humans and the saber-tooths discussed here, the glenoid fossa of the glenohumeral joint only engages about one-third of the humeral head.

These four muscles (Mm. supraspinatus, infraspinatus, subscapularis, and teres minor) are relatively larger in *Xenosmilus* than in *Smilodon*. Additionally, the muscles have shorter lever arms, particularly the most dorsal M. supraspinatus, than are seen in *Smilodon*. Shorter lever arms increase the mechanical advantage of the muscles at the expense of speed of movement. Clearly, control of the head of the humerus is extremely critical to *Xenosmilus*, supporting the suggestion that this animal is specialized as an ambush predator of large, active prey that may require much effort to subdue prior to being dispatched. A more robust distal humerus with a deeper humero-radial joint, suggesting an improved rotational capability and a greater excursion at the humero-ulnar joint, suggests that *Xenosmilus* has an even greater capability to position the forearm with precision than does *Smilodon*, an attribute that would also assist in prey capture.

The pattern of muscles acting synergistically in the forearm of *Xenosmilus* is similar to that of *Smilodon* and confirms that both cats had similar locomotor habits. However, different muscles are of greater importance in each cat, suggesting that even though they perform overall similar actions, they did not achieve the morphological details from a closely common evolutionary history. Although the differences in forelimb musculature between these animals are subtle, they suggest that *Xenosmilus* was even more likely than *Smilodon* to engage in manipulative activities with large prey and that this cat might have had a greater repertoire of killing behaviors than did *Smilodon*.

Although none of the intrinsic muscles of the manus could be reconstructed for *Xenosmilus*, the patterns of flexion and extension of the digits suggested by the muscles of the forearm support the notion that the cat was capable of a great degree of flexion and extension of the digits, as would be beneficial in grasping large-bodied prey. Even supposing that *Xenosmilus* had relatively large forepaws and long, robust claws, it is unlikely that the manus would be particularly effective in grasping large flat body surfaces, such as the abdomens of large prey. These results concur with the experimental bites of "Robocat" made on elk and bison carcasses as reported in chapter 2 of this volume.

Not only the robustness, but the patterns of synergistic and antagonistic muscles that are more robust than in other felids, can be interpreted to reflect adaptations of the hindlimbs of *Xenosmilus* for standing bipedally. The lesser trochanter of the femur is far larger in *Xenosmilus* than even in *Smilodon*, suggesting that *Xenosmilus* depended to a greater extent on the psoas major muscles that inserted on it for maintaining stability of the pelvis while standing bipedally. Second, the large and distal attachments of the sartorius, semitendinosus, and gracilis muscles of the thigh, for example, suggest that the cat was able to maintain a stable position while standing bipedally. Third, the large and prominent muscle scars of the muscles of the lower leg suggest that they played an unusually important role in stabilization of the foot in both pronation and supination. These muscles could have worked to help the cat maintain balance and stable placement of the hind foot for standing upright to capture large and active prey. Such a stance would be required of the cat for making the kind of killing bite suggested by Kurtén (1952) and discussed in the first chapter of this volume.

SUMMARY

Xenosmilus hodsonae represents a previously unrecognized saber-tooth cat morphotype. This animal shows a combination of anatomical features not seen in either dirk-tooth or typical scimitar-tooth saber-tooths (fig. 5.5). Cranial adaptations in *Xenosmilus* allow this animal to maximize the amount of pressure the mandible could exert during closing, from the beginning of the bite to full closure onto the flesh of the prey. These features include repositioning and changing of the shape of the masticatory muscles, accompanied by changes in the muscles that control the size of the oral opening. These latter muscles are those of facial expression, and they allow the cat to retract the soft tissues of the nasal region farther back than other cats can do at maximum gape. This morphology gives the face of this cat a more rounded lateral profile and a shorter muzzle. Postcranial anatomy of *Xenosmilus* reflects the need of this cat to grapple with large, active prey while attempting to make a killing bite. Comparison with the dirk-tooth *Smilodon*, an extremely robustly proportioned ambush predator, demonstrates that *Xenosmilus* was even more massive, making it one of the most fearsome of saber-tooth predators.

North American Pliocene homothere *Homotherium ischyrus*
pursuing the contemporary horse, *Equus (Plesippus) stenonis.*
(drawing Mark Hallett)

6

Osteology and Myology of *Homotherium ischyrus* from Idaho

JONENA M. HEARST

LARRY D. MARTIN

JOHN P. BABIARZ

VIRGINIA L. NAPLES

THE GENUS *HOMOTHERIUM* was originally based on material from the Villafranchian of Europe (contemporaneous with the North American Blancan), falling between 4.8 and 2.2 M.Y.A. Previously, in the fossil record of the North American Pliocene (Blancan), representatives of this genus were only known from incomplete materials. Because of this, there was confusion about the content of the genus, and, in particular, its relationship to the North American genus *Ischyrosmilus*. Opinion fluctuated between the complete synonymy of all species that were assigned first to *Ischyrosmilus* and then to *Homotherium* and the separation of *Ischyrosmilus* out of the Homotheriini, with its inclusion in the Smilodontini (Churcher, 1984). We can now describe a nearly complete skeleton from Idaho that was discovered and collected by Jonena Hearst (fig. 6.1A–B). The Idaho specimen is compared to all the various species of *Homotherium*, including the types of *Ischyrosmilus,* and to the homothere skeleton sometimes referred to as *Dinobastis* from the Late Pleistocene Friesenhahn Cave in Texas. Comparison of the nearly complete Idaho skeleton with *Dinobastis serus* shows that the two are closely related. Features that might separate *Dinobastis* from *Homotherium* include the development of an anterior pocket in the masseteric fossa and a shortened tibia. We presently concur with other workers that *Dinobastis* and the North American Blancan genus *Ischyrosmilus* can usefully be included in *Homotherium*. We also compare the long-legged Idaho *Homotherium* (fig. 6.2) with the short-legged Florida genus *Xenosmilus* and conclude that these genera share a common ancestor, probably in the late Miocene of Eurasia.

DESCRIPTION OF *HOMOTHERIUM*
THE SKULL

The skull is about the size of that of a small African lion *(Panthera leo)* and is brachycephalic with the broad frontals typical of *Homotherium* (figs. 6.3, 6.4; tables 6.A.1, 6.A.2). A cross-sectional view of the face is triangular with strong wide zygomatic arches and a squarish nasal cavity. The frontals are flat in lateral view, unlike the gently convex frontals of *Homotherium serum* illustrated by Rawn-Schatzinger (1992) and *Smilodon fatalis* by Merriam and Stock (1932). There is a depression at the nasal-frontal

Figure 6.1. *A*, Locality where the Birch Creek specimen, *Homotherium ischyrus* IMNH 900-11862, was discovered. *B*, Close-up photograph of the skull and some of the limb bones of *H. ischyrus*, IMNH 900-11862, in situ prior to being collected. (courtesy Jonena Hearst)

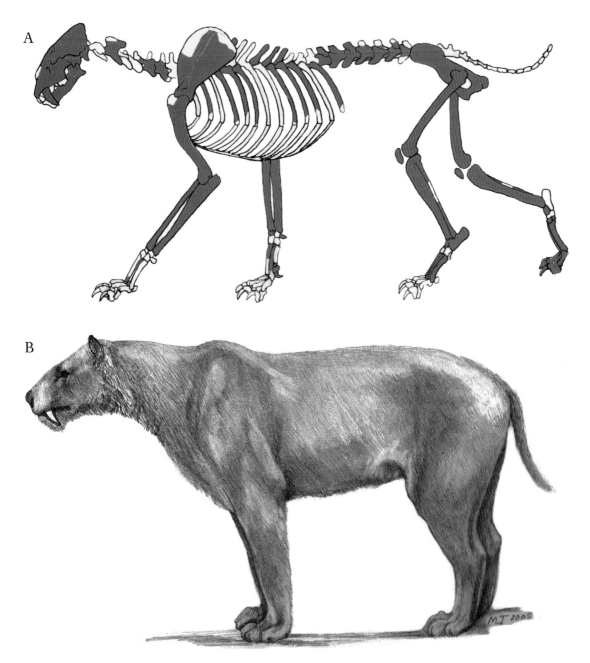

Figure 6.2. *A*, Skeleton of *Homotherium ischyrus*, IMNH 900-11862, assembled in a life posture. The elements that were collected are indicated in red. *B*, Reconstruction of the Idaho homothere in a life position. (drawing Mary Tanner)

suture that is more pronounced than in *H. serum* and slightly more than that illustrated for *H. crenatidens* (Ballesio, 1963, fig. 10A). Anterior to the frontal-parietal suture, the margins of the frontals are convex in cross section, giving rise to a midline furrow (deepest portion 11–12 mm) running at the junction of the nasal-frontal and nasal sutures. In the Idaho homothere, the premaxillary-maxillary suture, just dorsal to the canine root, is delineated by a groove slightly more pronounced than that in *Smilodon*. The frontal bones are

more convex posteriorly, where they bear strong frontal-temporal crests. Anterior to the point of merger of these crests into the sagittal crest is a deep fossa that continues into the previously mentioned furrow. The sagittal crest is lower than in *Smilodon* and deflected downward as it joins the occipital crest. Overall, the sutures are poorly displayed, but the nasal-frontal suture is just slightly posterior to the maxillary-frontal suture and is nearly straight, resembling *H. crenatidens* in this respect. The nasals are tilted anteriorly in

Discovery and Recovery

The discovery and recovery of the first nearly complete skeleton of *Homotherium ischyrus* was the result of a dissertation project by Jonena Hearst under the advisement of Dr. Larry Martin of the University of Kansas in 1998. The field area was in the Glenns Ferry Formation between Birch and Poison Creeks in southwest Idaho. The fauna of the area was previously brought to her attention by Dr. Bill Akersten of the Idaho Museum of Natural History while working on a master's degree.

The Glenns Ferry formation of Southwestern Idaho was deposited in a subsiding basin formed by extensional faulting initiated in the early Miocene (Zoback and Thompson, 1978). From Late Miocene (nine M.Y.A.) through the Early Pleistocene (1.4 M.Y.A.), thick sequences of lacustrine sediments of the Idaho Group accumulated in this basin. Interbedded with the lake sediments are fluvial and floodplain deposits and minor volcaniclastic sediments. This complex of sedimentary beds attests to active subsidence for seven million years and persistence of large lakes within the basin. The Glens Ferry Formation is the product of sedimentation associated with the third large lake system in the Snake River graben. Most of the vertebrate fossils of the Birch Creek fauna are associated with the regressive phase of Glenns Ferry Lake within a diatomite and mud facies, occurring as couplets less than 70 cm thick with siliclastic units (diatomite, sand, and clay) (Hearst, 1998).

The somewhat scattered fossil remains of *Homotherium* were discovered while scouting areas to measure stratigraphic columns of the formation in the upper reaches of Birch Creek. Initially fragments of long bones encased in concretions were found, followed by a claw and canine fragments projecting from the ground. After checking the lateral extent of the outcrop, it was verified that the bone concentration was an isolated occurrence with associated elements and others that were found partially articulated. Most of the lower limb elements were exposed; some were partially encased in sandstone concretions.

A shallow excavation in loose and unconsolidated sediments revealed the main bone layer and the bulk of the cat's skeleton lying on its left side on an underlying clay layer. The pelvis, vertebral column, ribs, and upper limb elements all lay in roughly the same horizontal plane and were overlain by fine sand. The long axis of the cranium was at an oblique angle to the postcranial skeleton, with the skull and jaws partially embedded in the underlying clay. Isolated elements of other taxa including those of a frog (acetabulum), a muskrat *Pliopotamys meadensis* (maxilla), and a large fresh water fish *Mylocheilus,* were found associated with the *Homotherium* skeletal remains. Several unionid mollusk casts were also found clustered in the abdominal area at the sand and clay interface.

---

an exaggerated way and may have suffered injury as indicated by a possible round puncture hole through the right nasal bone. The frontals are very broad, with massive, blunt postorbital processes that nearly overlap the postorbital process on the jugal bone. The muzzle is narrow anteriorly so that the broad zygoma and frontals reorient the eyes toward the anterior to give better binocular vision than in *Xenosmilus hodsonae* or *Smilodon.* In dorsal view, the relative width of the infraorbital and postorbital regions are broader than those of *Smilodon* and closely approximate the dimensions in *H. serum.* A large infraorbital foramen continues anteriorly into a moderately deep fossa. The foramen is smaller than the corresponding foramen in *Smilodon.* Dorsal to the infraorbital foramen and anterior to the orbit is a roughly circular depression. This feature (the location for the origin of the lip elevator muscle M. levator nasolabialis) is more circular and deeper than is ordinary for a scar of origin of that muscle. The weathering of the face on the left side has obliterated this feature, and its true nature cannot be confirmed. The zygomatic bone is more gracile than that of *Smilodon,* with a prominent zygomatic apex closely resembling *H. crenatidens* (Ballesio, 1963, fig. 10, plate 1) rather than

S. fatalis (Merriam and Stock, 1932, fig. 7) or *H. serum* (Rawn-Schatzinger, 1992, fig. 1). The external narial opening is narrow and square, being broader than in *Xenosmilus,* and the internal narial opening is also larger. The premaxillary bone and the incisor arcade are badly damaged, but the arcade is rounded. There does not appear to have been any appreciable diastema between the incisor arcade and the canines. The incisors and the incisor arcade are narrower, and the premaxilla is less massive than in *Xenosmilus.*

In posterior view, the occipital crest in the Idaho homothere flares laterally, although not as much as in *Xenosmilus,* in which the occipital crest essentially forms a rectangle that is broadly rounded dorsally (fig. 6.3C; tables 6.A.1, 6.A.2). In *H. serum,* the occipital crests converge dorsally, giving the occiput a triangular shape. In *Smilodon,* the occipital crests show a more rounded profile and also converge dorsally but not as much as in *H. serum.* In posterior view, the Idaho homothere has lateral edges of the occipital crests that are constricted at about the midpoint and flare laterally as they continue ventrally, approaching the mastoid process. In *Smilodon,* the lateral edges of the occipital crest diverge ventrally.

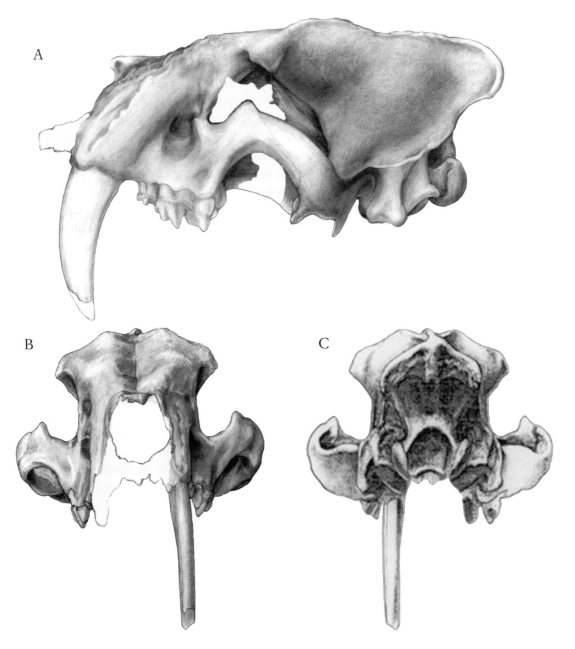

Figure 6.3. The skull of *Homotherium ischyrus*, IMNH 900-11862: *A*, lateral view; *B*, anterior view; *C*, posterior view. (drawing Mary Tanner)

In lateral view, the overlap of the occipital crests above the occipital condyles is less in the Idaho specimen than in *H. serum, Xenosmilus,* and *Smilodon,* which all show a greater degree of overlap. This feature is the most pronounced in *Xenosmilus.*

In *Smilodon* and *H. serum,* just below the occipital ridge and between the dorsal and ventral nuchal lines there are muscle scars for the M. splenius capitus, which assists in rotation of the cervical spine. In the Idaho homothere, the scar for the insertion of this muscle only extends across the horizontal portion of the occipital crest. In *Xenosmilus,* this

muscle scar is larger, extending ventrally about midway along the occipital crest. Ventral to the insertion of the M. splenius capitus is the scar of insertion of the M. obliquus capitus superioris. This muscle insertion is more ventral and medial in the Idaho homothere and *H. serum* than in *Xenosmilus.* In *Smilodon,* this insertion is more lateral than in the other species, arising from a round bulge in the occipital crest. The muscle originates from the tip of the transverse process of the axis vertebra, which is large and extends farther laterally in these animals than is usual. Such a line of action gives this muscle an especially advantageous orienta-

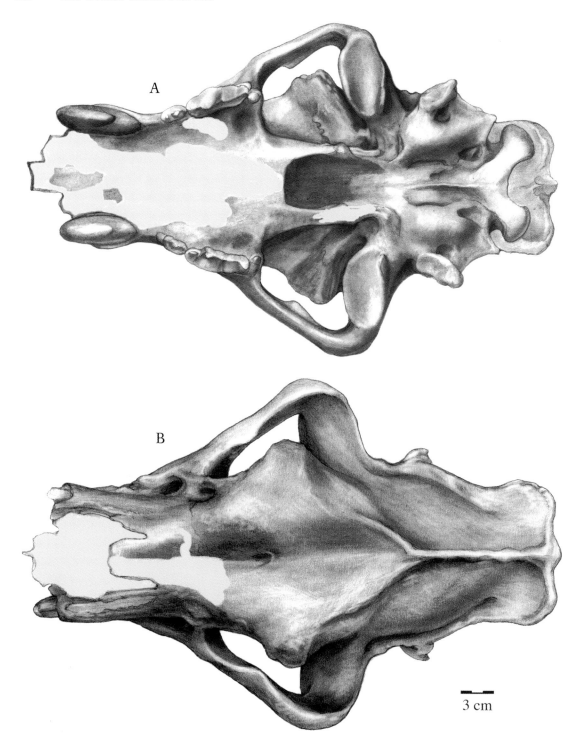

Figure 6.4. The skull of *Homotherium ischyrus*, IMNH
900-11862: *A*, ventral view; *B*, dorsal view. (drawing
Mary Tanner)

tion for effecting cranial elevation and rotation, movements
essential to using the elongated canine teeth with precision.
The insertion of the M. trapezius in the Idaho homothere
and *H. serum* extends into a pair of dorsal and medial elongated oval depressions ventral to the external occipital pro-

tuberance. The left and right sides are divided by a relatively
broad, short nuchal ridge. The insertion of the M. trapezius
in *Xenosmilus* differs significantly from that of the other homotheres in that the occipital crest is continuous dorsal to
the insertion, with no apparent external occipital protuberance. In *Xenosmilus,* the scar of insertion for the M. trapezius
is broad and teardrop-shaped, with the upper end broader,
and the median nuchal crest that divides the right and left

muscles is reduced and represented dorsally only by a small raised knob. In *Smilodon,* M. trapezius differs from all the compared species by being larger, inserting into a pair of deeply incised elongated oval pits that reach the dorsalmost aspect of the occipital crest, dividing the insertions of the M. splenius capitus. These scars are separated at the dorsal midline by a prominent external occipital protuberance, continuing ventrally as a robust, raised median nuchal crest. In the Idaho homothere, the insertion of this muscle is bordered ventrolaterally by a pair of large oval depressions representing the origin for the M. semispinalis capitus. The insertion of this muscle is deeper in the Idaho homothere than in all of the compared cats, although the arrangement in *H. serum* cannot be discerned from available figures and descriptions. In *Smilodon,* this insertion is also oval but relatively smaller as well as more dorsal, and in *Xenosmilus* it is oval but more ventral. Ventral and lateral to the insertion of the M. semispinalis capitus is a smaller oval depression for the M. rectus capitus posterior major and minor. The scars for these muscle insertions in the Idaho homothere are deep pits dorsal to the foramen magnum. In *Xenosmilus,* these scars are oval, obliquely oriented, and deep, surrounded by raised ridges. The insertions of these muscles in *Smilodon* are also oval but more shallow. The insertion of the M. rectus capitus lateralis in the Idaho homothere is an oval depression, with the longer axis oriented vertically and lateral to the insertion of the M. rectus capitus posterior major and minor and dorsolateral to the foramen magnum. In *Smilodon,* the insertion for the M. rectus capitus lateralis is more lateral and extends ventral to the lateral aspect of the occipital condyles. It is larger overall than in the other genera. In *Xenosmilus,* the scar of origin for this muscle is a smooth oval depression, mostly lateral to the occipital condyle.

In lateral view, the edges of the occipital crest in the Idaho homothere, and probably in *H. serum,* are thicker than in *Xenosmilus* and *Smilodon,* and the depression for the M. temporalis origin is more depressed (fig. 6.3A). The ventral portion of the temporalis muscle origin is separated as a distinct depression from the upper portion that is immediately dorsal to the mastoid process. There is some indication of this feature in the other cats, but in none of them is it so clearly indicated. There seems to be a small, triangular paroccipital process extending posteroventrally from the posterior aspect of the auditory bulla. This process is reduced in *Xenosmilus* and *Smilodon.* A very similar triangular paroccipital process occurs in the clouded leopard (*Neofelis nebulosa*), but in *Panthera* it is blunt and oriented downward. In the Idaho homothere, the paroccipital process continues to the mastoid process as a ridge, which encloses a shallow rectangular lateral depression. In *Xenosmilus,* a somewhat similar depression occurs on the posterolateral side of the mastoid process, extending vertically as far as the occipital crest. In the Idaho homothere, the occipital condyles are relatively smaller than in *Xenosmilus,* and the foramen magnum is more rounded.

The reduced size of the occipital condyles reflects the smaller body mass of the Idaho homothere when compared to *Xenosmilus.*

The auditory bullae in the Idaho specimen are depressed and flattened posteriorly as opposed to *Xenosmilus,* in which they are more inflated, and the stylomastoid pit is larger and shallower in the Idaho specimen than in *Xenosmilus.* The auditory bullae are more rounded ventrally and tend to be flattened on the anterior surface. The mastoid process is large, tilted anterolaterally, and divided into a lower teardrop-shaped posterior region and a higher oval surface (fig. 6.4A). The combined surface is 27.7 mm by 22.4 mm. The mastoid process is nearly a centimeter posterior to the glenoid process. The external auditory meatus is bounded posteriorly by the mastoid process and dorsally by a broad ridge coming off the back of the zygomatic arch. The back of the zygomatic arch is broad and extends ventrally into a long rounded postglenoid process so that the glenoid faces partially forward. The preglenoid process is reduced in comparison to *Panthera.* The glenoid articulation of the Idaho specimen is larger and wider anteroposteriorly than in *Xenosmilus* (transverse dimension 45.9 mm, anterior-posterior dimension 23.2 mm). *Xenosmilus* has an enlarged mastoid process that is on a level with the postglenoid process. The Idaho homothere has the glenoid extending much farther ventrally than the mastoid process. In *Xenosmilus,* the mastoid process and the tooth row are nearly at the same level, while in the Idaho homothere the mastoid process terminates considerably above the tooth row. The foramen magnum is more rounded than in *Smilodon,* and the shape and placement agree with that illustrated for *H. crenatidens* (Ballesio, 1963, fig. 10E) and *H. serum* (Rawn-Schatzinger, 1992, fig. 2).

The condyloid and posterior lacerate foramina in the Idaho homothere share a single opening. On the basiocipital, there is a high ridge separating two elongate depressions for the origins of the M. longus capitus. There is a deep groove in the posterior orbital region, similar to that in *Xenosmilus,* containing the orbital fissure and the foramen for the optic nerve.

The Palate

The palatal region terminates immediately posterior to the tooth row. The pterygoid wings are rounded laterally and terminate in prominent hamulae. There are shallow embrasure pits medial to the carnassials. There is an indication of distinct palatal ridges, but the palatal surface is too weathered for a detailed description (fig. 6.4A).

The Dentition

The superior dentition is badly damaged anteriorly, with a single first incisor in situ, indicating a posteriorly recurved conical tooth. The single in situ upper left incisor (I^1) is conical and recurved backward. There is an isolated right second incisor (I²) that is large and triangular, with a central cusp flanked by two accessory tubercles. In dorsal view, the medial tubercle rises slightly anterior to the lateral tubercle, and a faint cemento-enamel junction (fig. 6.5A) also runs anteriorly and ventrad from the medial to the lateral edges. Both tubercles bear faint serrations, suggesting that the incisors were serrated as in other homotheres. There is an isolated left I³ that is large and oval in cross-section and has a small accessory cuspule on the medial face (fig. 6.5B). On the posterior labial surface, a triangular wear facet runs from the crown to nearly the base of the enamel. There is a similar but smaller wear facet along the lateral edge, beginning at the crown and running halfway to the base of the enamel. These wear facets show that the upper and lower incisors interlocked when the mouth was fully closed. The canines are in extremely poor condition. The badly fragmented left canine is missing most of the enamel on the margins; therefore, it is impossible to tell much about serrations, although a small section of the posterior enamel at least indicates their presence. The left canine is well enough preserved to show that the canines were larger and more elongate than in *Xenosmilus* or *H. serum*. This agrees with the canines of other North American Blancan homotheres that have commonly been assigned to the genus *Ischyrosmilus*. The upper premolars are much reduced (figs. 6.3–6.4, 6.5C) but retain a tooth pattern consisting of a large central cusp, bordered anteriorly and posteriorly by two smaller cusps and surrounded laterally by a thick cingulum that presents posteriorly as a small additional cuspule. The upper P³ is double rooted. The P⁴ is enlarged and elongated. It lacks cingula, and the protocone is reduced to a slight bump supported by a root. There is a well-developed parastyle with a tiny cuspule on its anterior surface in the position where preparastyles sometimes develop in saber-tooth cats. There is a distinct narrow notch between the parastyle and the paracone, and a shallow carnassial notch between the paracone and the metacone blade. The M¹ is greatly reduced, and only the single small root of the left is preserved (anteroposterior diameter 6.35 mm). Unlike the condition in *H. serum* or *Smilodon,* the root is more directly posterior to the carnassial and slopes medially. Ballesio (1963) did not describe or illustrate M¹ for *H. crenatidens,* and it is not developed in *Xenosmilus*.

THE MANDIBLE

The anterior portions of both mandibular rami are missing, including the mental foramina (fig. 6.5D–G; tables 6.A.3, 6.A.4). The presence of a mandibular flange is suggested by a ventral flare of the ramal border below the anterior root of P₃. The left mandible retains the alveolus for P₃ and the surface of the diastema. The dorsal edge of the diastema is thin, and the lateral edge of the mandible in this region is concave to accommodate the upper canine. The diastema slopes upward toward the lower canine, in which the alveolus would have been at about the same level as the occlusal edge of the carnassial. The masseteric fossa is deeply impressed and shallows toward its anterior margin, slightly posterior to the carnassial notch on M₁. The ventral surface of the masseteric fossa projects as a shelf lateral to the coronoid and slightly undercuts the anterior lateral edge of the mandibular condyle. It continues onto the lateral surface of the angular process, which is distinctly curved medially and terminates as a blunt projection underneath the condyle. In *H. serum,* the masseteric fossa does not become shallower anteriorly, but instead becomes enclosed as a distinct pocket. This is an unusual morphology but is seen in some living animals, for instance, kangaroos. Anteromedial to the angular process is a flattened area bordered by a ridge for the attachment of the medial pterygoid muscle. There is a short gap between the posterior edge of the carnassial and the coronoid process, but it is longer than that in *Xenosmilus*. In extant cats, this gap is much longer, and the coronoids tend to slant backward, as opposed to being nearly vertical as in the Idaho homothere and most other saber-tooth cats. The dorsal edge of the ramus, between the carnassial and the coronoid process, slants laterally, making it easier for the jaw to slide by the posterior edge of the maxillary. The coronoid process is rounded and bends slightly laterally at the apex. The horizontal ramus is more gracile than that of *Smilodon* and lacks the labial bulge below the tooth row. The mandibular condyle is conical, widening medially and narrowing to a point at the lateral tip. Its anterior edge is separated from the rest of the ramus by a groove. The P₃ consists of three cusps arranged linearly along the midline of the tooth. The middle cusp is the largest and bears faint serrations on the anterior and posterior margins. The tooth widens posteriorly and has an internal cingulum. It has a broad, single root. The P₄ also widens posteriorly, has large anterior and posterior cusps that are equal in size, and a triangular middle cusp. The tooth crown is relatively high, lacks cingula, and there is a double root. The P₄ slightly overlaps the anterior edge of the carnassial. The cusps bear faint serrations. The long axes of the cusps are canted slightly posteriorly, although they are more nearly vertical than the cusps of *H. serum* or *Smilodon*. Wear facets occur on the labial face of all cusps: on the posterior face of the middle cusp and on the anterior and posterior faces of the posterior cusps. Facets on the anterior portion of the tooth extend one-half to two-thirds of crown height, while the posterior facet extends to the base of the tooth. The carnassial is large and elongate, broader across the paraconid than across the protoconid. The two cusps are separated by a shallow carnassial notch. The posterior margin of the tooth is straight with no trace of a metaconid. The tooth is only slightly tilted posteriorly. The posterior edge of the carnassial is slightly serrated. The wear facets bear nearly vertical striations.

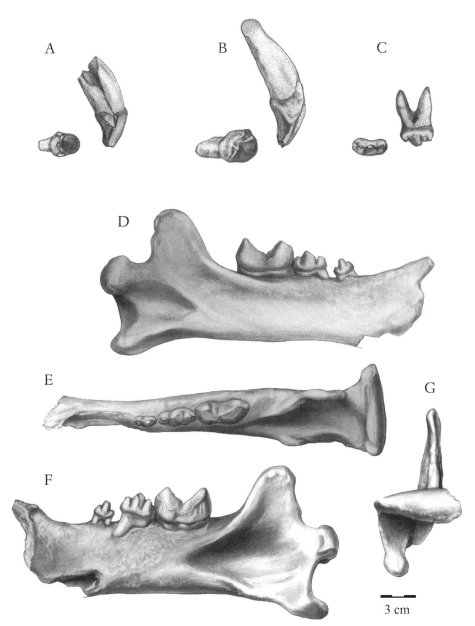

Figure 6.5. Teeth from the upper dentition of *Homotherium ischyrus*, IMNH 900-11862: *A*, Right upper second incisor, I^2, in lateral and occlusal views. *B*, Left upper third incisor, I^3, in medial and occlusal views. *C*, Left upper third premolar, P^3, in lateral and occlusal views. The mandible of the same individual (left hemimandible): *D*, medial view; *E*, dorsal view; *F*, lateral view; *G*, posterior view. (drawing Mary Tanner)

3 cm

THE POSTCRANIAL SKELETON
The Axial Skeleton
The Cervical Vertebrae

THE ATLAS

The atlas differs slightly from that of *H. serum* illustrated by Rawn-Schatzinger (1992, fig. 5; fig. 6.6A–D; table 6.1). The external borders of the transverse process are straighter, with a more rounded anterior tip. They rise more abruptly, resulting in a greater angle of the alar notch than is seen in *H. serum*. The ventral tubercle is larger than that of *H. serum*. The shape of the transverse process is similar to that of *Smilodon* (Merriam and Stock, 1932, plate 16). The transverse processes are more triangular, straighter, and longer posteriorly than in *H. crenatidens* as illustrated by Ballesio (1963, fig. 16).

THE AXIS

The axis is incomplete, missing the entire right transverse process and the spinous process (fig. 6.6E–F, table 6.2). The odontoid process is more robust, with a more pointed tip than in *Smilodon,* and in this respect resembles *H. crenatidens* and *H. serum*. The intervertebral canal resembles *H. serum* and differs from *Smilodon*, occupying a more dorsal position. Small hyperapophyses arise directly from the body of the posterior neural arch, anteroventral to the postzygapophyses. This condition is not described for other homotheres and is possibly pathological.

CERVICAL VERTEBRA THREE

Cervical vertebra three has a posteriorly situated low spinous process, and the neural canal is deeply embayed into the

Table 6.1. Atlas measurements (mm): *Homotherium ischyrus* IMNH 900-11862. Compared specimens (observed range), *H. serum* (Meade, 1961; Rawn-Schatzinger, 1992), TMM 933-2276, 2280, 3007, *3231, 3382; *H. crenatidens* (Ballesio, 1963), FSL 210991. (*e* = estimate, *Meade)

	H. ischyrus	*H. serum*	*H. crenatidens*
Greatest width across transverse processes	139.0	140.6–*150.0*e*	140.0
Greatest width of anterior end across articulation for condyles of skull	78.4	60.6–*70.0	—
Greatest width across outer borders of articulation for axis	66.8	65.1–*75.0	—
Length from anterior end of articulation for condyles to posterior end of articulation for axis	68.9	61.6–72.0	—
Length of neural arch along median line	34.7	30.8–36.1	33.0
Length of inferior arch along median line	26.8	25.7–29.4	—
Greatest length of transverse process, taken oblique to fore and aft axis of vertebra	67.7	76.0	—
Greatest height from ventral surface of inferior arch to dorsal surface of neural arch	46.5	43.0–*50.3	52.0

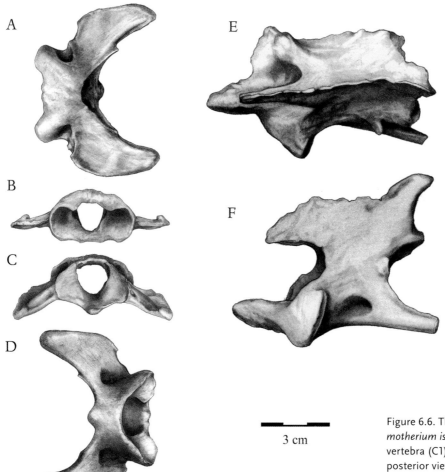

3 cm

Figure 6.6. The first two cervical vertebrae of *Homotherium ischyrus*, IMNH 900-11862. The atlas vertebra (C1): *A*, dorsal view; *B*, anterior view; *C*, posterior view; *D*, ventral view. The axis vertebra (C2): *E*, dorsal view; *F*, lateral left view. (drawing Mary Tanner)

Table 6.2. Axis measurements (mm): *Homotherium ischyrus* IMNH 900-11862. Compared specimens (observed range), *H. serum* (Meade, 1961; Rawn-Schatzinger, 1992), TMM 933-2276, 2280, 3007, *3231, 3382; *H. crenatidens* (Ballesio, 1963), FSL 210991. (*a* = approximate, *Meade)

	H. ischyrus	*H. serum*	*H. crenatidens*
Greatest length of neural spine	—	92.2	—
Greatest width of posterior end of neural spine	—	92.2	—
Length of centrum along median line measured parallel to lower surface from posterior end to tip of odontoid process	87.2	79.9–95.0	90.0
Depth of centrum measured normal to floor of neural canal and across posterior epiphysis	20.0	24.0–*27.0	—
Greatest transverse width across posterior epiphysis of centrum	—	*39.4	
Width across articulating surfaces for atlas	66.4	67.5–72.7	63.0
Width across outer ends of transverse processes	—	—	—
Width across posterior zygapophyses	44.6*a*	43.0	
Distance measured in median line from lower border of posterior epiphysis of centrum to top of posterior end of neural spine	—	73.4–87.1	—
Length of odontoid process	21.1	—	—
Width of odontoid process at base	9.4	—	19.0
Greatest depth of odontoid process at base	15.0	—	26.0

dorsal surface (fig. 6.7A–E, table 6.3). The prezygapophyses are large, flat, and tilted medially, turning down toward the anteromedial corner until they almost contact the centrum. In anterior view, the neural canal is roughly square and significantly smaller than in *Xenosmilus*. It is rectangular in posterior view. As in *Xenosmilus,* the dorsal edge of the centrum projects anterior to the ventral edge. The dorsal margin of the ventral edge is depressed more than in *Xenosmilus* to accommodate the neural canal. This results in a broader strut that is concave in the Idaho homothere and convex in *Xenosmilus*. The transverse process sweeps backward, mirroring the condition in the axis vertebra. Its ventral edge is extended ventrally and laterally delineates a deep pocket on the ventral surface of the centrum (fig. 6.7E). This pocket is paired by another, on the opposite side, and they are separated by a triangular ridge on the ventral midline of the centrum. The neural arch is flat dorsally and expands posteriorly over the postzygapophyses. The articular surfaces of the postzygapophyses are roughly square, with their lateral edge continuous with the dorsal surface of the neural arch. There is a deep pocket in the ventral surface of the neural arch, just anterior to the postzygapophysis. On the right side, the lateral edge of the neural arch turns down over this excavation. The posterior dorsal surface of the neural arch is extended into distinct short, spinous processes just above the posteromedial corners of the postzygapophyses. These are called hyperapophyses by Merriam and Stock (1932, fig. 20). The posterior face of the centrum is tilted forward and roughly heart-shaped. The foramen for the vertebral artery opens almost even with the articular surface of the centrum. The centrum terminates where the postzygapophyses begin, and the neural arch overhangs it.

CERVICAL VERTEBRA FOUR
The fourth cervical vertebra is missing.

CERVICAL VERTEBRA FIVE
The fifth cervical vertebra has the neural arch and most of the right side missing (fig. 6.7F–I, table 6.4). The centrum is narrower than in *Xenosmilus,* and its articular facets are more rounded. The neural canal is smaller, and the strut over the transverse foramen carrying the vertebral artery is wider. The inferior lamina of the transverse process, as shown in Merriam and Stock (1932, fig. 20), is shallower and more elongate than in *Xenosmilus,* and the lateral facet of the transverse process is more elongate. The transverse process extends posteriorly past the centrum, while in *Xenosmilus* it is about even with the centrum. It is shorter than in *Xenosmilus* and more arched. The ventral edge of the inferior lamella is about level with the ventral edge of the centrum, but this feature extends ventrally beyond the centrum in *Xenosmilus*.

CERVICAL VERTEBRA SIX
In the sixth cervical vertebra in the Idaho homothere, the centrum is heart-shaped in anterior view (fig. 6.8A–D, table 6.5). The dorsal edge of the anterior surface of the centrum extends beyond the ventral edge. The prezygapophyses are rectangular and tilted inward. The centrum extends beyond the anterior edge of the neural arch. There is a small, triangular, centrally located neural spine. The neural canal is oval and smaller than in *Xenosmilus*. The transverse foramen for the vertebral artery is smaller than in *Xenosmilus*. The inferior laminae of the transverse process are swept backward and are triangular, resembling those of *Smilodon*. The

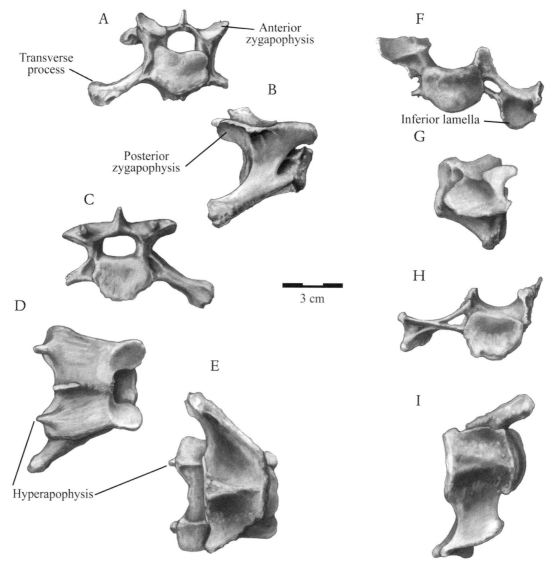

Figure 6.7. Cervical vertebrae C3 and C5 of *Homotherium ischyrus*, IMNH 900-11862. Cervical vertebra three: *A*, anterior view; *B*, lateral right view; *C*, posterior view; *D*, dorsal view; *E*, ventral view. Cervical vertebra five: *F*, anterior view; *G*, lateral left view; *H*, posterior view; *I*, ventral view. (drawing Mary Tanner)

base of the transverse process is shorter and broader than in *Xenosmilus*. The postzygapophyses are rectangular. In *Xenosmilus*, the postzygapophyses are larger and oval. The medially tilted prezygapophyses are higher than the postzygapophyses, and this feature is even more emphasized in *Xenosmilus*. The posterior centrum is rounded as opposed to oval in *Xenosmilus*. The ventral surface has an anteroposteriorly oriented central ridge that is not as sharp as that in *Xenosmilus*.

CERVICAL VERTEBRA SEVEN

In the Idaho homothere cervical vertebra seven is missing.

The Thoracic Vertebrae

THORACIC VERTEBRA ONE

Thoracic vertebra one is missing from the Idaho homothere specimen.

THORACIC VERTEBRA TWO

In anterior view, the second thoracic vertebra in the Idaho homothere has a heart-shaped neural canal and a tall triangular dorsal process with a sharp anterior edge (fig. 6.8E–H, table 6.6). The anterior edge of the neural canal is almost level with the centrum, while it is distinctly posterior to the centrum in *Xenosmilus*. The postzygapophyses are al-

Table 6.3. Third cervical vertebra measurements (mm): *Homotherium ischyrus* IMNH 900-11862. Compared specimens (observed range), *H. serum* (Meade, 1961; Rawn-Schatzinger, 1992), TMM 933-2821, 2943, *3231; *H. crenatidens* (Ballesio, 1963), FSL 210991. (*e* = estimate, *Meade)

	H. ischyrus	*H. serum*	*H. crenatidens*
Length from end of anterior zygapophysis to end of posterior zygapophysis	62.4	*50.6	62.0
Greatest length from ends of anterior zygapophyses to ends of hyperapophyses	—	—	—
Length of centrum measured normal to posterior face and along median line	42.4	30.5–*38.5	42.0
Width across anterior zygapophyses	52.0	45.6–*47.0	—
Width across posterior zygapophyses	67.7	46.0–*48.9	—
Greatest width of neural canal at anterior end	19.6	18.6–23.6	—
Greatest transverse width of posterior epiphysis of centrum	37.3	29.1–*35.0	40.0
Greatest width across outer ends of transverse processes	112.4	*116.0	114.0
Greatest length of transverse process from outer end to end of anterointernal projection of inferior lamella	61.0*e*	—	—
Height from median posterior border of centrum to top of neural crest	—	*55.0	—
Depth of centrum measured normal to floor of neural canal and across posterior epiphysis	26.4	*27.4	29.0

Table 6.4. Fifth cervical vertebra measurements (mm): *Homotherium ischyrus* IMNH 900-11862. Compared specimens (observed range), *H. serum* (Meade, 1961; Rawn-Schatzinger, 1992), TMM 933-2300, *3231, 3384, 3507, 3509; *H. crenatidens* (Ballesio, 1963), FSL 210991. (*e* = estimate, *Meade)

	H. ischyrus	*H. serum*	*H. crenatidens*
Length from end of anterior zygapophysis to end of posterior zygapophysis	—	*55.0	56.0
Length of centrum measured normal to posterior face and along median line	37.0	32.2–*36.0	36.0
Greatest width across anterior zygapophyses	61.0*e*	50.1–*59.5	—
Width across posterior zygapophyses	—	*49.4	—
Greatest width of neural canal at anterior end	25.5	21.5–23.2	—
Greatest transverse width of posterior epiphysis of centrum	37.8	31.8–34.2	39.0
Depth of centrum measured normal to floor of neural canal and along median line of posterior epiphysis	26.7	*36.0	30.0
Greatest width across outer ends of transverse processes	103.0*e*	*102.0	—
Greatest distance from end of transverse process to lower posterior end of lamella	42.0*e*	—	—
Height from middle of ventral border of posterior epiphysis of centrum to top of neural spine	—	—	—

most even with the centrum in *Homotherium* and are distinctly posterior to the centrum in *Xenosmilus*. The posterior surface of the neural spine has a midline ridge, while the surface is concave in *Xenosmilus*. The posterior margin of the neural arch is expanded laterally in *Xenosmilus* but not in the Idaho homothere. The ventral surface of the centrum is flat in *Xenosmilus* and rounded in the Idaho homothere.

THORACIC VERTEBRAE THREE AND FOUR
Thoracic vertebrae three and four are missing from the Idaho *Homotherium* specimen.

THORACIC VERTEBRA FIVE
The fifth thoracic vertebra in the Idaho *Homotherium* has the neural arch inclined backward (fig. 6.9A–D, table 6.7). The prezygapophyses are more widely separated than in *Smilodon*.

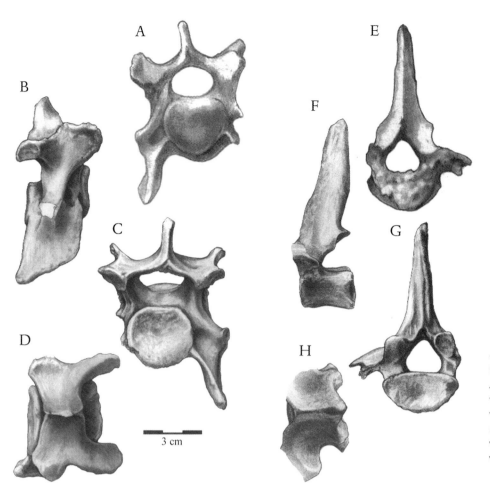

Figure 6.8. Vertebrae C6 and T2 of *Homotherium ischyrus*, IMNH 900-11862. Cervical vertebra six: *A*, anterior view; *B*, lateral right view; *C*, posterior view; *D*, dorsal view. Thoracic vertebra two: *E*, anterior view; *F*, lateral left view; *G*, posterior view; *H*, ventral view. (drawing Mary Tanner)

Table 6.5. Sixth cervical vertebra measurements (mm): *Homotherium ischyrus* IMNH 900-11862. Compared specimens (observed range), *H. serum* (Meade, 1961; Rawn-Schatzinger, 1992), TMM 933-307, 2218, 3282; *H. crenatidens* (Ballesio,1963), FSL 210991. (*e* = estimate)

	H. ischyrus	H. serum	H. crenatidens
Length from end of anterior zygapophysis to end of posterior zygapophysis	47.3	—	55.0
Length of centrum measured normal to posterior face and along median line	34.4	32.2–35.5	35.0
Greatest width across anterior zygapophyses	31.3	51.2–51.6	—
Width across posterior zygapophyses	65.0*e*	45.0	—
Greatest width of neural canal at anterior end	29.6	22.4–24.1	—
Greatest height of neural canal at anterior end	17.7	19.8–20.9	—
Greatest transverse width of posterior epiphysis of centrum	35.3	31.8–35.1	37.0
Depth of centrum measured normal to floor of neural canal and along median line of posterior epiphysis	30.4	—	28.0
Greatest width across outer ends of transverse processes	—	—	—
Greatest width across inferior lamella	70.0*e*	—	—
Greatest length of inferior lamella measured along lower border	42.0	—	—

Table 6.6. Second thoracic vertebra measurements (mm): *Homotherium ischyrus* IMNH 900-11862. Compared specimen (observed range), *H. serum* (Meade, 1961; Rawn-Schatzinger, 1992), TMM 933-2816, 3385, *3231. (*Meade)

	H. ischyrus	*H. serum*
Greatest length from end of anterior zygapophyses to end of posterior zygapophyses	—	*44.0
Length of centrum measured normal to posterior face along median line	30.5	29.5–31.4
Greatest width across anterior zygapophyses	—	53.3–54.0
Width across posterior zygapophyses	39.8	—
Greatest height of neural canal at anterior end	20.0	15.6–18.4
Greatest width of neural canal at anterior end	22.0	15.6–18.4
Greatest transverse width of posterior face of centrum across capitular facets	47.0	*55.6
Depth of centrum measured normal to floor of neural canal and along median line of posterior epiphysis	23.8	26.2–28.5
Greatest width across outer ends of transverse processes	—	*94.8
Greatest anteroposterior diameter of outer end of transverse process	—	*22.6
Height from middle of ventral border of posterior epiphysis of centrum to top of neural spine	—	*128.0

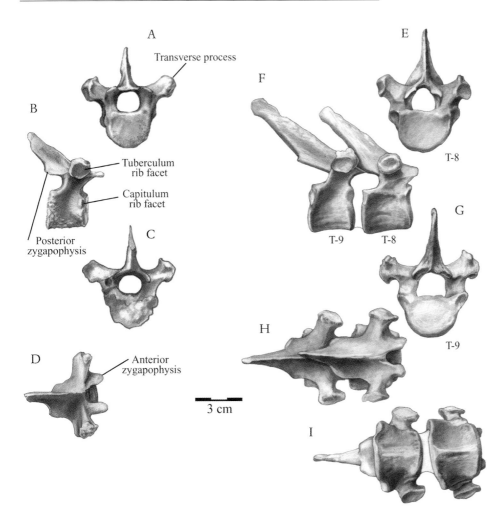

Figure 6.9. Three thoracic vertebrae, T5, T8, and T9, of *Homotherium ischyrus*, IMNH 900-11862. Thoracic vertebra five: *A*, anterior view; *B*, lateral right view; *C*, posterior view; *D*, dorsal view. Thoracic vertebrae T8 and T9 (right to left) were found and left in articulation. The eighth and ninth thoracic vertebrae: *E*, anterior view of T8; *F*, right lateral view of T8 and T9; *G*, posterior view of T9; *H*, dorsal view of T8 and T9; *I*, ventral view of T8 and T9. (drawing Mary Tanner)

Table 6.7. Fifth thoracic vertebra measurements (mm): *Homotherium ischyrus* IMNH 900-11862. Compared specimens (observed range), *H. serum* (Meade, 1961; Rawn-Schatzinger, 1992), TMM 933-372, *3231, 3357, 3459; *H. crenatidens* (Ballesio, 1963), FSL 210991. (*e* = estimate, *Meade)

	H. ischyrus	*H. serum*	*H. crenatidens*
Greatest length from end of anterior zygapophyses to end of posterior zygapophyses	43.2	*39.0	—
Length of centrum measured normal to posterior face along median line	33.5	24.0–30.8	31.0
Greatest width across anterior zygapophyses	34.0	25.3–30.2	—
Width across posterior zygapophyses	26.3	*26.0	—
Greatest height of neural canal at anterior end	16.7	17.0–21.2	17.0
Greatest width of neural canal at anterior end	17.6	19.8	21.0
Greatest transverse width of posterior face of centrum across capitular facets	44.0e	28.9–*51.6	46.0
Depth of centrum measured normal to floor of neural canal and along median line of posterior epiphysis	26.3	24.3–30.5	29.0
Greatest width across outer ends of transverse processes	71.8	75.4–83.0	64.0
Greatest anteroposterior diameter of outer end of transverse process	17.9	19.7–24.0	—

The transverse processes are short, with a roughly square facet for the rib articulation, and are splayed laterally. The neural canal is round in both anterior and posterior view, and the centrum is short when compared to *Smilodon* and *Panthera*.

THORACIC VERTEBRAE SIX AND SEVEN

The sixth and seventh thoracic vertebrae are missing from the Idaho homothere specimen.

THORACIC VERTEBRAE EIGHT AND NINE

The eighth and ninth thoracic vertebrae are similar in shape. They were found in articulation and will be described together (fig. 6.9E–I, tables 6.8–6.9). The eighth and ninth thoracic vertebrae have short posteriorly inclined neural spines. The prezygapophyses extend past the centrum and are closer together than on the fifth thoracic vertebra. The transverse processes are shorter and more dorsally inclined than on the fifth thoracic vertebra. The centra of the eighth and ninth thoracic vertebrae are broad; the postzygapophyses arise posterior to the centra, and the articulation for the ribs on the transverse processes are mostly lateral, extending slightly above the level of the prezygapophyseal facets, as in the fifth thoracic vertebra.

THORACIC VERTEBRAE TEN AND ELEVEN

The tenth and eleventh thoracic vertebrae were also found in articulation and were left in that state. The tenth thoracic vertebra in *Homotherium* closely resembles the ninth, with a short, highly posteriorly inclined neural spine that is more gracile and shorter than in *Smilodon* or *Panthera* (fig. 6.10A–E, table 6.10). In anterior view, the neural canal is heart-shaped.

THORACIC VERTEBRA ELEVEN

The eleventh thoracic vertebra begins the transition to the lumbar series. The distal half of the neural spine was broken off and not recovered, but the transverse processes and the prezygapophyses and postzygapophyses are intact (fig. 6.10A–E, table 6.11). The transverse processes are closely appressed to the neural arch, and the tubercular rib facets are laterally positioned and angled downward and slightly medially. These facets are lower than the facets of the preceding vertebra, with the dorsal edge at about the same level as the ventral edge of the postzygopohyseal facet. The tubercular rib facet of the transverse process is oblong and points posteroventrally. The capitular facets are also somewhat oblong or kidney bean shaped but are oriented vertically along the lateral edge of the centrum and across from the neural canal. When the vertebrae are articulated (T10 and T11), these facets form a heart-shaped pocket. There is a distinct posterior spine-like anapophysis. The prezygapophyses terminate just slightly anterior to the centrum and are flattened, as in the preceding thoracic vertebrae. The postzygapophyses wrap around and face laterally, as they do in the lumbar series. The gap between the tubercular and capitular rib facets of both the centrum and transverse process for the interarticular ligament is narrow, separated by only four mm, while this feature on the preceding vertebra (T10) is much wider, measuring nine mm. The centrum body is broader posteriorly than in the other more anterior vertebrae. There is a distinct keel on the midline of the ventral surface.

Table 6.8. Eighth thoracic vertebra measurements (mm): *Homotherium ischyrus* IMNH 900-11862. Compared specimens (observed range), *H. serum* (Meade, 1961; Rawn-Schatzinger, 1992), TMM 933-966, 3231; *H. crenatidens* (Ballesio, 1963), FSL 210991. (*e* = estimate)

	H. ischyrus	H. serum 933-966	H. serum 933-3231	H. crenatidens
Greatest length from end of anterior zygapophyses to end of posterior zygapophyses	52.0	—	—	—
Length of centrum measured normal to posterior face along median line	32.7	29.3	28.3	31.4
Greatest width across anterior zygapophyses	28.9	30.5	—	—
Width across posterior zygapophyses	28.3*e*	—	—	—
Greatest height of neural canal at anterior end	15.0*e*	19.4	17.5	16.0
Greatest width of neural canal at anterior end	17.0*e*	—		18.0
Greatest transverse width of posterior face of centrum across capitular facets	50.3	—	47.8	51.0
Depth of centrum measured normal to floor of neural canal and along median line of posterior epiphysis	28.9*e*	30.3	31.0	30.0
Greatest width across outer ends of transverse processes	64.8	77.5	75.0	78.0
Greatest anteroposterior diameter of outer end of transverse process	20.1	17.8	26.0	—

Table 6.9. Ninth thoracic vertebra measurement (mm): *Homotherium ischyrus* IMNH 900-11862. Compared specimen (observed range), *H. serum* (Meade, 1961; Rawn-Schatzinger, 1992), TMM 933-*3231, 3391. (*Meade)

	H. ischyrus	H. serum
Greatest length from end of anterior zygapophyses to end of posterior zygapophyses	52.2	*45.0
Length of centrum measured normal to posterior face along median line	32.7	28.3
Greatest width across anterior zygapophyses	25.0	34.3–30.5
Width across posterior zygapophyses	24.4	24.3–30.5
Greatest height of neural canal at anterior end	—	24.3–30.5
Greatest width of neural canal at anterior end	—	17.4–18.3
Greatest transverse width of posterior face of centrum across capitular facets	48.9	16.1–18.3
Depth of centrum measured normal to floor of neural canal and along median line of posterior epiphysis	28.1	34.6–30.3
Greatest width across outer ends of transverse processes	68.2	32.5–27.1
Greatest anteroposterior diameter of outer end of transverse process	20.6	76.0
Height from middle of ventral border of posterior epiphysis of centrum to top of neural spine	97.7	—
Length of spine from middle of notch between anterior zygapophyses to top measured parallel to anterior end	76.8	—

THORACIC VERTEBRA TWELVE

In anterior view, the neural canal of the twelfth thoracic vertebra is round, while posteriorly, the opening is more oval (fig. 6.11A–E, table 6.12). The prezygapophyses curve upward and are slightly anterior to the centrum. The capitular rib facet is hourglass in shape and restricted to the centrum. The centrum has a distinct ventral keel. There is a short triangular neural spine and an anapophysis laterally on each side. The postzygapophyses above the anapophyses are laterally positioned and strongly curved upward.

THORACIC VERTEBRA THIRTEEN

In anterior view, the neural canal of the thirteenth thoracic vertebra is similar to T12, rounded anteriorly, and oval in posterior view (fig. 6.11F–J, table 6.13). The thirteenth thoracic vertebra has the prezygapophyses strongly curved

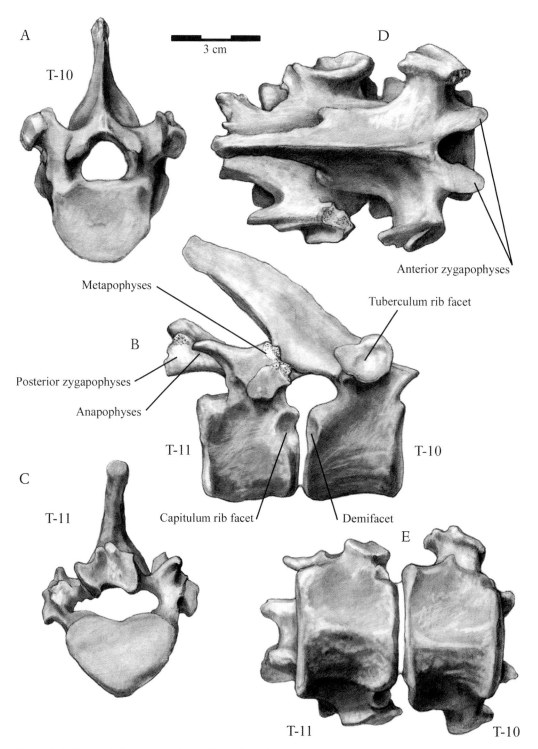

A T-10

3 cm

D

Anterior zygapophyses

Metapophyses

Tuberculum rib facet

B

Posterior zygapophyses

Anapophyses

T-11   T-10

C

T-11

Capitulum rib facet   Demifacet

E

T-11   T-10

Figure 6.10. The tenth thoracic vertebra and the eleventh (transitional) thoracic vertebra, T10 and T11, of *Homotherium ischyrus*, IMNH 900-11862, in articulation (*right to left*): *A*, anterior view of T10; *B*, right lateral view of T10 and T11; *C*, posterior view of T11; *D*, dorsal view of T10 and T11; *E*, ventral view of T10 and T11. (drawing Mary Tanner)

Table 6.10. Tenth thoracic vertebra measurements (mm): *Homotherium ischyrus* IMNH 900-11862. Compared specimen (observed range), *H. serum* (Meade, 1961; Rawn-Schatzinger, 1992), TMM 933-542, 1383, *3231, 3247. (*e* = estimate, *Meade)

	H. ischyrus	H. serum
Greatest length from end of anterior zygapophyses to end of posterior zygapophyses	51.5	*43.0
Length of centrum measured normal to posterior face along median line	33.6	26.9–31.3
Greatest width across anterior zygapophyses	29.7	27.0–31.0
Width across posterior zygapophyses	28.4	*31.3
Greatest height of neural canal at anterior end	16.2	16.0–19.6
Greatest width of neural canal at anterior end	14.6	16.0–19.6
Greatest transverse width of posterior face of centrum across capitular facets	47.7	32.0–34.0
Depth of centrum measured normal to floor of neural canal and along median line of posterior epiphysis	—	30.0–32.5
Greatest width across outer ends of transverse processes	67.4	—
Greatest anteroposterior diameter of outer end of transverse process	—	—
Height from middle of ventral border of posterior epiphysis of centrum to top of neural spine	—	*97.0
Length of spine from middle of notch between anterior zygapophyses to top measured parallel to anterior end	71.5e	—

Table 6.11. Eleventh thoracic vertebra measurements (mm): *Homotherium ischyrus* IMNH 900-11862. Compared specimen (observed range), *H. serum* (Meade, 1961; Rawn-Schatzinger, 1992), TMM 933-724, 2219, 2231, 2279, 2290, *3231. (*a* = approximate, *e* = estimate, + = damaged element, *Meade)

	H. ischyrus	H. serum
Greatest length from end of anterior zygapophyses to end of posterior zygapophyses	53.3e	*47.3
Length of centrum measured normal to posterior face along median line	33.6	28.1–32.6
Greatest width across anterior zygapophyses	27.4	29.4–33.9
Width across posterior zygapophyses	28.4e	*26.5
Greatest width across metapophyses	54.0e +	—
Greatest width across anapophyses	51.5e +	—
Greatest length from end of metapophyses to end of anapophyses	47.5e +	—
Greatest height of neural canal at anterior end	—	17.4–19.5
Greatest width of neural canal at anterior end	—	17.4–18.3
Greatest transverse width of posterior face of centrum across capitular facets	43.5	*45.4
Depth of centrum measured normal to floor of neural canal and along median line of posterior epiphysis	27.4	25.3–29.2
Greatest width measured from outer ends of facets for tubercle of rib	62.4	70.8–*76.8
Height from middle of ventral border of posterior epiphysis of centrum to top of neural spine	—	*97.0a
Length of spine from middle of notch between anterior zygapophyses to top measured parallel to anterior end	—	*66.0a
Greatest anteroposterior diameter of plate above facet for tubercle of rib	29.8	22.6–*37.7
Greatest height of neural canal at posterior end	15.0e +	—
Greatest width of neural canal at posterior end	19.0e	—

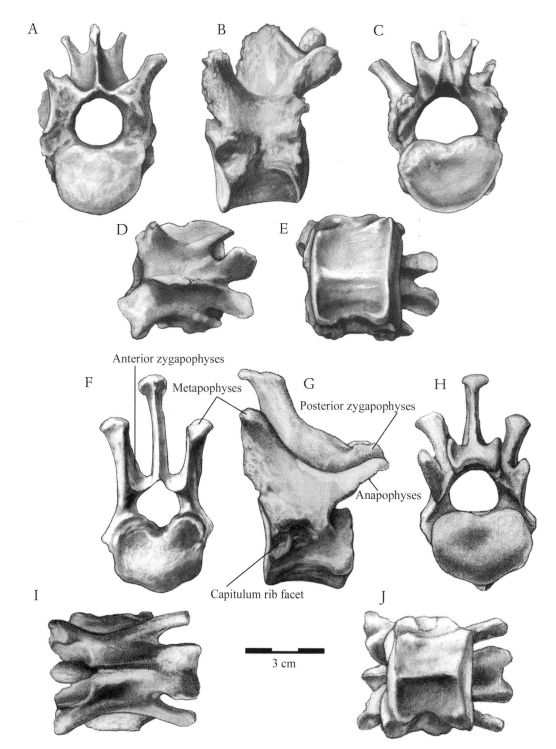

Figure 6.11. The most posterior two thoracic vertebrae, T12 and T13, of *Homotherium ischyrus*, IMNH 900-11862. Thoracic vertebra twelve: *A*, anterior view; *B*, left lateral view; *C*, posterior view; *D*, dorsal view; *E*, ventral view. The thirteenth thoracic vertebra: *F*, anterior view; *G*, left lateral view; *H*, posterior view; *I*, dorsal view; *J*, ventral view. (drawing Mary Tanner)

Table 6.12. Twelfth thoracic vertebra measurements (mm): *Homotherium ischyrus* IMNH 900-11862. Compared specimens (observed range), *H. serum* (Meade, 1961; Rawn-Schatzinger, 1992), TMM 933-**1870, *3231; *H. crenatidens* (Ballesio, 1963), FSL 210991. (*a* = approximate, + = damaged element, *Meade)

	H. ischyrus	**H. serum	*H. serum	H. crenatidens
Greatest length from end of anterior zygapophyses to end of posterior zygapophyses	56.3	—	57.0	—
Length of centrum measured normal to posterior face along median line	37.1	34.7	35.0	33.5
Greatest width across anterior zygapophyses	56.0a	30.0	—	—
Width across posterior zygapophyses	26.3	—	27.9	—
Greatest width across metapophyses	—	—	62.0a	—
Greatest length from end of metapophyses to end of anapophyses	50.0a +	—	*50.6	—
Greatest width across outer ends of anapophyses	54.6a	—	—	—
Greatest height of neural canal at anterior end	19.8	20.0	—	17.0
Greatest width of neural canal at anterior end	21.0	20.2	—	20.0
Greatest transverse width of posterior face of centrum across capitular facets	45.1	—	47.3	46.0
Depth of centrum measured normal to floor of neural canal and along median line of posterior epiphysis	27.6	30.7	29.3	33.0
Greatest width across outer ends of transverse processes at tubercle rib facet	48.7	64.1	62.0	60.0
Greatest anteroposterior diameter of facets for tuberculum of rib	12.3	*67.4	67.4	—
Height from middle of ventral border of posterior epiphysis of centrum to top of neural spine	—	—	95.0a	—

Table 6.13. Thirteenth thoracic vertebra measurements (mm): *Homotherium ischyrus* IMNH 900-11862. Compared specimens (observed range), *H. serum* (Meade, 1961; Rawn-Schatzinger, 1992), TMM 933-**2152, *3231; *H. crenatidens* (Ballesio, 1963), FSL 210991. (*a* = approximate, *e* = estimate, *Meade, **Rawn-Schatzinger)

	H. ischyrus	**H. serum	*H. serum	H. crenatidens
Greatest length from end of anterior zygapophyses to end of posterior zygapophyses	62.6	—	60.3	—
Length of centrum measured normal to posterior face along median line	33.4	34.9	37.8	36.5
Greatest width across anterior zygapophyses	23.3	23.5	—	—
Width across posterior zygapophyses	23.4	—	26.0	—
Greatest width across metapophyses	50.1	—	62.0a	—
Greatest length from end of metapophyses to end of anapophyses	60.1e	—	—	—
Greatest width across outer ends of anapophyses	48.0e	—	—	—
Greatest height of neural canal at anterior end	20.0	19.6	—	18.0
Greatest width of neural canal at anterior end	19.5	20.2	—	18.0
Greatest transverse width of posterior face of centrum across capitular facets	43.4	46.1	48.1	45.0
Depth of centrum measured normal to floor of neural canal and along median line of posterior epiphysis	29.0	25.8	30.0	—
Greatest anteroposterior diameter of facets for capitulum of rib	R 11.3 L 11.6	—	—	—
Height from middle of ventral border of posterior epiphysis of centrum to top of neural spine	102.0	—	90.2	85.0
Length of spine from middle of notch between anterior zygapophyses to top measured parallel to anterior end	45.0	41.9	47.0	36.0
Maximum anteroposterior diameter of distal protuberance of neural spine	20.0	—	27.5	—
Maximum width of distal protuberance of neural spine	12.0	—	—	—

upward, with short metapophyses. The posterior edge of the neural spine is anteriorly inclined, coming off the neural arch at the anterior end, and is somewhat taller than in *Smilodon*. The anterior edge of the neural spine is almost vertical. The front of the neural arch is only slightly anterior to the centrum. The anterior capitular rib facet is positioned at about the midline of the centrum and is underlain by a blunt ridge. There is no posterior rib facet. There is a strong midline keel on the ventral border of the centrum. The anapophysis is strongly developed and extends along the ventral border of the postzygapophyses, and the postzygapophyseal facets are essentially vertical. The postzygapophyses originate posterior to the centrum and occlude tightly with L1, forming a strong bond.

The Lumbar Vertebrae

LUMBAR VERTEBRAE ONE, TWO, THREE, AND FOUR

The first, second, and third lumbar vertebrae were found and kept in articulation (fig. 6.12A–E; tables 6.14–6.18). All of the lumbars (L1–L6) except the last (L7) have strong, sharp ventral keels that become progressively larger posteriorly. The fourth lumbar vertebra (L4) is shown in fig. 6.13A–E. The first four lumbar vertebrae have long, slender, spine-like anapophyses that become progressively reduced posteriorly in the series. The neural spines are inclined forward and distally and have broad, heart-shaped terminal facets. The facets of the prezygapophyses and postzygapophyses curve upward and become nearly vertical. The centra are both wider and longer than in the thoracic vertebral series. There are distinct metapophyses on the dorsolateral sides of the prezygapophyses and short, flat anterior and ventrally inclined transverse processes.

LUMBAR VERTEBRAE FIVE AND SIX

The fifth and sixth lumbar vertebrae were also found and left articulated (fig. 6.13F–J, table 6.19). The transverse processes are longer and wider than in the previous lumbars, although their tips do not curve medially as in *Smilodon*. The zygapophyses become more widely separated, and the anaphophyses are reduced to short projections, with the neural spine forming a short triangular process, not much higher than the zygapophyses, as opposed to the relatively high neural spines in *Smilodon*.

LUMBAR VERTEBRA SEVEN

The seventh and last lumbar vertebra makes the transition to the sacral series (fig. 6.14A–E, table 6.20). The transverse processes project horizontally, bending slightly downward at their tips, but the transverse process is not inclined downward as much as it is in *Smilodon*. The tip of the transverse process is triangular and inclined forward. The posterior margin is rounded and terminates in an elongate, expanded knob. Just behind the anterior edge of the transverse process

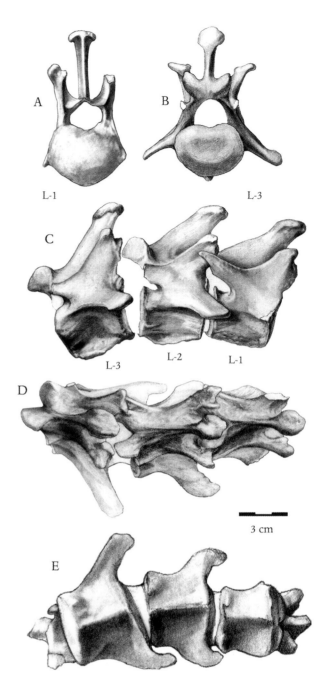

Figure 6.12. Lumbar vertebrae, L1, L2, and L3 (*right to left*) of *Homotherium ischyrus*, IMNH 900-11862, were found articulated: *A,* anterior view of L1; *B,* posterior view of L3; *C,* right lateral view of L1, L2, and L3; *D,* dorsal view of L1, L2, and L3; *E,* ventral view of L1, L2, and L3. (drawing Mary Tanner)

is a low rounded ridge that has an elongate depression behind it. There is a low rounded ventral keel, bordered on either side by an elongate shallow depression and a second small oval depression on either side of these ridges. The prezygapophyses turn upward as in the other lumbars.

Table 6.14. First lumbar vertebra measurements (mm): *Homotherium ischyrus* IMNH 900-11862. Compared specimens (observed range), *H. serum* (Meade, 1961; Rawn-Schatzinger, 1992), TMM 933-965, 2581, 2582, *3231; *H. crenatidens* (Ballesio, 1963), FSL 210991. (*a* = approximate, *e* = estimate, *Meade)

	H. ischyrus	H. serum	H. crenatidens
Length of centrum measured normal to face of posterior epiphysis and along median line	38.0	*37.0–40.3	39.5
Greatest length from anterior end of metapophyses to end of posterior zygapophyses	66.3	*66.5	—
Greatest width across metapophyses	45.2	*56.0*a*	—
Greatest width across anterior zygapophyses	24.0	27.4–28.1	—
Width across posterior zygapophyses	30.5	25.9–29.8	—
Maximum width of neural canal at anterior end	20.0	20.7–21.7	19.0
Maximum height of neural canal at anterior end	16.2	16.0–19.5	19.0
Depth of anterior epiphysis of centrum	30.7	—	—
Width of anterior epiphysis of centrum	46.0	—	—
Length of spine from middle of notch between posterior zygapophyses to anterior end of protuberance	68.8	—	—
Width of posterior epiphysis of centrum immediately above lateral projection of centrum	45.5	*49.6	45.0
Depth of centrum measured normal to floor of neural canal and across posterior epiphysis	35.2*e*	*30.4	31.0
Height from middle ventral border of anterior epiphysis of centrum to top of neural spine	85.0	*93.4	86.0
Maximum anteroposterior diameter of distal protuberance	15.6	—	—
Maximum width of distal protuberance	21.6	—	—
Greatest width across outer ends of transverse processes	49.2	—	—
Greatest anteroposterior diameter of transverse process parallel along lateral edge of centrum body	10.0*e*	—	—
Maximum width of neural canal at posterior end	—	—	22.0

Table 6.15. Second lumbar vertebra measurements (mm): *Homotherium ischyrus* IMNH 900-11862. Compared specimens (observed range), *H. serum* (Meade, 1961; Rawn-Schatzinger, 1992), TMM 933-*3231; *H. crenatidens* (Ballesio, 1963), FSL 210991. (*e* = estimate, + = damaged element, *Meade)

	H. ischyrus	*H. serum	H. crenatidens
Length of centrum measured normal to face of posterior epiphysis and along median line	41.6	41.0	41.5
Greatest length from anterior end of metapophyses to end of posterior zygapophyses	65.2	—	—
Greatest width across metapophyses	52.0	—	—
Greatest width across anterior zygapophyses	29.2	—	—
Width across posterior zygapophyses	28.5	30.0	—
Maximum width of neural canal at anterior end	—	—	19.0
Maximum height of neural canal at anterior end	—	—	19.0
Depth of anterior epiphysis of centrum	—	—	—
Width of anterior epiphysis of centrum	46.8*e*	—	—
Length of spine from middle of notch between posterior zygapophyses to anterior end of protuberance	69.0	—	—
Width of posterior epiphysis of centrum taken immediately above lateral projection of centrum	45.2*e*	50.8	46.0
Depth of centrum measured normal to floor of neural canal and across posterior epiphysis	37.0*e*	31.4	31.0
Height from middle ventral border of anterior epiphysis of centrum to top of neural spine	91.6	95.0	90.0
Maximum anteroposterior diameter of distal protuberance	21.3	—	—
Maximum width of distal protuberance	16.2	—	—
Greatest width across outer ends of transverse processes	84.0*e* +	—	—
Greatest anteroposterior diameter of transverse process parallel along lateral edge of centrum body	20.3	—	—
Maximum width of neural canal at posterior end	—	—	23.0

Table 6.16. Third lumbar vertebra measurements (mm): *Homotherium ischyrus* IMNH 900-11862. Compared specimens (observed range), *H. serum* (Meade, 1961; Rawn-Schatzinger, 1992), TMM 933-3231; *H. crenatidens* (Ballesio, 1963), FSL 210991. (*a* = approximate, *e* = estimate, + = damaged element, *Meade)

	H. ischyrus	*H. serum	H. crenatidens
Length of centrum measured normal to face of posterior epiphysis and along median line	42.8	44.7	43.0
Greatest length from anterior end of metapophyses to end of posterior zygapophyses	75.0	67.6	—
Greatest width across metapophyses	—	54.0*a*	—
Greatest width across anterior zygapophyses	26.0	—	—
Width across posterior zygapophyses	30.8	35.4	—
Maximum width of neural canal at anterior end	—	—	21.0
Maximum height of neural canal at anterior end	—	—	18.0
Depth of anterior epiphysis of centrum	42.8*e*	—	—
Width of anterior epiphysis of centrum	48.4	—	—
Length of spine from middle of notch between posterior zygapophyses to anterior end of protuberance	72.2	—	—
Width of posterior epiphysis of centrum taken immediately above lateral projection of centrum	47.0	52.5	46.0
Depth of centrum measured normal to floor of neural canal and across posterior epiphysis	33.7	33.0	32.0
Height from middle ventral border of anterior epiphysis of centrum to top of neural spine	91.3	96.6	92.0
Maximum anteroposterior diameter of distal protuberance	23.7	—	—
Maximum width of distal protuberance	18.1	—	—
Greatest width across outer ends of transverse processes	92.0*e* +	—	—
Greatest anteroposterior diameter of transverse process parallel along lateral edge of centrum body	22.0	—	—
Maximum width of neural canal at posterior end	26.7*e*	—	26.0
Maximum height of neural canal at posterior end	14.3	—	—

The Sacrum

A partial sacrum is preserved, containing the anterior two sacral vertebrae (fig. 6.14F–H, table 6.21). It is relatively broader than in *Smilodon*, with the anterior central facet essentially rectangular, rather than being curved ventrally. The prezygapophyses are widely separated from the anterolateral edge of the sacrum. The neural arch is crushed and badly damaged. The auricular face of the sacrum is a rounded rectangle, with a broad and rounded anterior extension projecting slightly anterior to the prezygapophyses. There is an oval depression anteriorly, with the posterior end surrounded by a roughened crescentic articular surface that incorporates several rugose facets, which are stepped laterally. These are wider than the anterior ovate, rugose depression.

The Caudal Vertebrae

The caudal vertebrae were not recovered, but we expect that the tail was short, as in other homotheres.

The Ribs

Several incomplete ribs belonging to the IMNH 900-11862 homothere were collected and compared to the rib cage of *Smilodon* and *Panthera atrox* in Merriam and Stock (1932, plate 20). The ribs of *Smilodon* are shorter and more curved and robust than in *P. atrox*. In the Idaho specimen the rib shafts in general are rather small and rounded as compared to the other two cats. They are more elongate, straighter, and probably longer than in *Smilodon*. The ribs closely resemble those of *P. atrox* rather than *Smilodon* and might even be considered more like the rib cage of cheetahs.

The Appendicular Skeleton
The Forelimb and Pectoral Girdle
THE SCAPULA

A proximal fragment of the right scapula, including the glenoid and the neck, is preserved (fig. 6.15C–E, table 6.22). The left scapula is remarkably well-preserved, and the characters described here are primarily based on that specimen. The scapular spine and proximal head are complete, although the posterior portion of the scapula has been eroded away. The scapular neck is fairly narrow, with the dorsal margin rising abruptly from the neck, forming a broad angle similar to that figured by Ballesio (1963, fig. 24). The proximal scapular spine does not curve ventrally as in modern cats. The humeral end of the scapula in the Idaho homothere is not as elongate as in *Xenosmilus*, although it is as wide transversely, and the glenoid is more cupped. The lateral margin

Table 6.17. Fourth lumbar vertebra measurements (mm): *Homotherium ischyrus* IMNH 900-11862. Compared specimen (observed range), *H. crenatidens* (Ballesio, 1963), FSL 210991. (*e* = estimate, + = damaged element)

	H. ischyrus	H. crenatidens
Length of centrum measured normal to face of posterior epiphysis and along median line	48.0	45.0
Greatest length from anterior end of metapophyses to end of posterior zygapophyses	65.8e +	—
Greatest width across metapophyses	—	—
Greatest width across anterior zygapophyses	33.5	—
Width across posterior zygapophyses	33.4e	—
Maximum width of neural canal at anterior end	23.8	25.0
Maximum height of neural canal at anterior end	21.3	17.0
Depth of anterior epiphysis of centrum	34.4	—
Width of anterior epiphysis of centrum	46.2	—
Length of spine from middle of notch between posterior zygapophyses to anterior end of protuberance	—	—
Width of posterior epiphysis of centrum taken immediately above lateral projection of centrum	47.1	47.0
Depth of centrum measured normal to floor of neural canal and across posterior epiphysis	33.6	31.0
Height from middle ventral border of anterior epiphysis of centrum to top of neural spine	—	94.0
Maximum anteroposterior diameter of distal protuberance	—	—
Maximum width of distal protuberance	—	—
Greatest width across outer ends of transverse processes	108.0e +	—
Greatest anteroposterior diameter of transverse process parallel along lateral edge of centrum body	25.6	—
Maximum width of neural canal at posterior end	35.1	30.0

Table 6.18. Fifth lumbar vertebra measurements (mm): *Homotherium ischyrus* IMNH 900-11862. Compared specimen (observed range), *H. crenatidens* (Ballesio,1963), FSL 210991. (*e* = estimate, + = damaged element)

	H. ischyrus	H. crenatidens
Length of centrum measured normal to face of posterior epiphysis and along median line	52.1	45.5
Greatest length from anterior end of metapophyses to end of posterior zygapophyses	79.0	—
Greatest width across metapophyses	58.0e +	—
Greatest width across anterior zygapophyses	41.0	—
Width across posterior zygapophyses	34.0	—
Maximum width of neural canal at anterior end	27.3	28.0
Maximum height of neural canal at anterior end	21.5	16.0
Depth of anterior epiphysis of centrum	37.8	—
Width of anterior epiphysis of centrum	45.8	—
Length of spine from middle of notch between posterior zygapophyses to anterior end of protuberance	—	—
Width of posterior epiphysis of centrum taken immediately above lateral projection of centrum	47.6	50.0
Depth of centrum measured normal to floor of neural canal and across posterior epiphysis	—	32.0
Greatest width across outer ends of transverse processes	122.0e +	—
Greatest anteroposterior diameter of transverse process parallel along lateral edge of centrum body	35.0	—
Maximum width of neural canal at posterior end	—	34.0

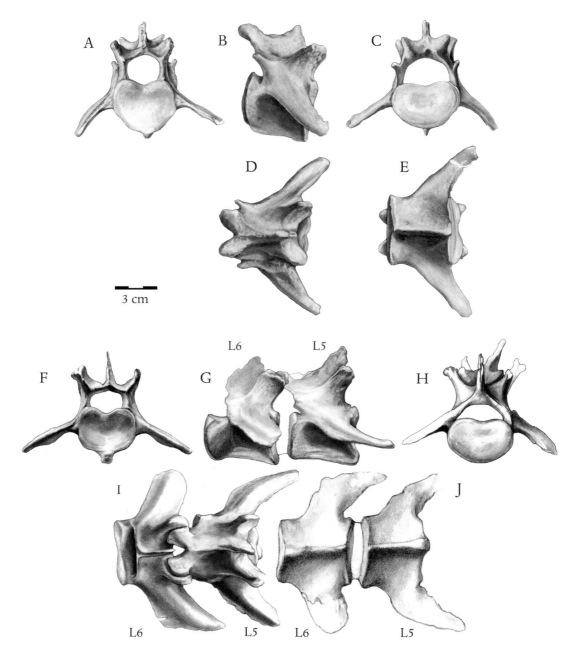

3 cm

Figure 6.13. Three lumbar vertebrae, L4, L5, and L6 of *Homothe-rium ischyrus*, IMNH 900-11862. Lumbar vertebra four: *A*, anterior view; *B*, right lateral view; *C*, posterior view; *D*, dorsal view; *E*, ventral view. Lumbar vertebrae five and six are shown in articulation (*right to left*): *F*, anterior view of L5; *G*, right lateral view of L5 and L6; *H*, posterior view of L6; *I*, dorsal view of L5 and L6; *J*, ventral view of L5 and L6. (drawing Mary Tanner)

of the ovoid glenoid fossa forms a sigmoid curve with a convex ventral and concave dorsal border. The medial margin is convex. The lateral face of the coronoid process is flatter than the medial face. The supraglenoid tuberosity is larger than that figured by Rawn-Schatzinger (1992, fig. 15) for *H. serum*. The scapular tuberosity is neither as rounded nor as rugose as that of *H. serum*. The humeral end in *Panthera*

is equally long and cupped, but the glenoid is wider transversely than in *Homotherium*. In *Smilodon*, the glenoid is much wider, being almost circular. The scapular spine is high in both *Panthera* and the Idaho homothere (the height is approximately 52 mm) but starts a little more dorsally in the latter, and the acromion process in the Idaho specimen extends farther ventrally. The acromion process is nearly

Table 6.19. Sixth lumbar vertebra measurements (mm): *Homotherium ischyrus* IMNH 900-11862. Compared specimens (observed range), *H. serum* (Meade, 1961; Rawn-Schatzinger, 1992), TMM 933-2023, 2108; *H. crenatidens* (Ballesio, 1963), FSL 210991. (*e* = estimate, *Meade, **Rawn-Schatzinger)

	H. ischyrus	**H. serum*	***H. serum*	*H. crenatidens*
Length of centrum measured normal to face of posterior epiphysis and along median line	54.4	43.5	42.8	45.0
Greatest length from anterior end of metapophyses to end of posterior zygapophyses	78.1	—	—	—
Greatest width across metapophyses	51.4	—	—	—
Greatest width across anterior zygapophyses	35.5	—	25.6	—
Width across posterior zygapophyses	34.2	36.6	32.6	—
Maximum width of neural canal at anterior end	—	—	—	25.0
Maximum height of neural canal at anterior end	—	—	—	16.0
Width of anterior epiphysis of centrum	45.6	—	—	—
Length of spine from middle of notch between posterior zygapophyses to anterior end of protuberance	—	—	—	—
Width of posterior epiphysis of centrum taken immediately above lateral projection of centrum	48.3	38.6	51.0	51.0
Depth of centrum measured normal to floor of neural canal and across posterior epiphysis	30.7	25.4	29.9	32.0
Height from middle ventral border of anterior epiphysis of centrum to top of neural spine	—	—	—	92.0
Greatest width across outer ends of transverse processes	128.0*e*	—	—	—
Greatest anteroposterior diameter of transverse process parallel along lateral edge of centrum body	32.8	21.1	30.1	—
Maximum width of neural canal at posterior end	38.0	—	—	35.0
Maximum height of neural canal at posterior end	13.3*e*	—	—	—

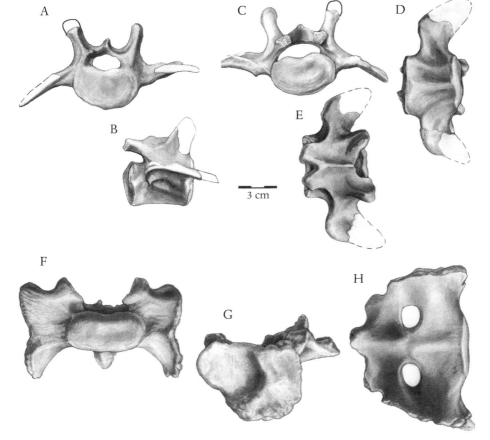

3 cm

Figure 6.14. The last lumbar vertebra, L7, and the sacrum of *Homotherium ischyrus*, IMNH 900-11862. Lumbar vertebra seven: *A*, anterior view; *B*, right lateral view; *C*, posterior view; *D*, ventral view; *E*, dorsal view. The sacrum: *F*, anterior view; *G*, left lateral view; *H*, ventral view. (drawing Mary Tanner)

Table 6.20. Seventh lumbar vertebra measurements (mm): *Homotherium ischyrus*
IMNH 900-11862. Compared specimens (observed range), *H. serum* (Meade, 1961; Rawn-Schatzinger, 1992), TMM 933-376, 540, 2109, 2817, 2818; *H. crenatidens* (Ballesio, 1963), FSL 210991. (*e* = estimate, + = damaged element)

	H. ischyrus	*H. serum*	*H.crenatidens*
Length of centrum measured normal to face of posterior epiphysis and along median line	48.5	37.7–43.5	39.0
Greatest length from anterior end of metapophyses to end of posterior zygapophyses	—	—	—
Greatest width across metapophyses	5.3e +	—	—
Greatest width across anterior zygapophyses	32.8	28.4–39.4	—
Width across posterior zygapophyses	—	43.2–53.2	—
Maximum width of neural canal at anterior end	27.4	26.8–32.4	25.0
Maximum height of neural canal at anterior end	13.7	11.8–12.6	14.0
Depth of anterior epiphysis of centrum	31.5	—	—
Width of anterior epiphysis of centrum	46.6	—	—
Width of posterior epiphysis of centrum taken immediately above lateral projection of centrum	47.8	—	17.0
Depth of centrum measured normal to floor of neural canal and across posterior epiphysis	30.2	—	31.0
Greatest width across outer ends of transverse processes	130.0e +	—	—
Greatest anteroposterior diameter of transverse process parallel along lateral edge of centrum body	26.2	22.4–29.8	—
Maximum width of neural canal at posterior end	41.7	—	34.0
Maximum height of neural canal at posterior end	13.7e	—	—

Table 6.21. Sacrum (sacral vertebrae) measurements (mm): *Homotherium ischyrus* IMNH 900-11862. Compared specimens (observed range), *H. serum* (Meade, 1961; Rawn-Schatzinger, 1992), TMM 933-492, 1772, 2080, 2277, 2303, *3231; *H. crenatidens* (Ballesio, 1963), FSL 210991. (*a* = approximate, *e* = estimate, + = damaged element, * = Meade)

	H. ischyrus	*H. serum*	*H. crenatidens*
Greatest length measured parallel to median line	—	87.2e–*125.0	92.0
Greatest width at anterior end and from outer sides of surfaces for ilia	118.7	*106.0a	94.0
Greatest width of third sacral vertebra across transverse processes	—	32.6–*67.3	—
Greatest width between dorsal borders of anterior zygapophyses	45.8	*54.0–54.7	46.0
Width across posterior zygapophyses of third sacral vertebra	—	—	—
Depth of centrum of first sacral vertebra measured normal to floor of neural canal and across anterior epiphysis	27.4	23.8–*28.4	26.0
Depth of centrum of third sacral vertebra measured normal to floor of neural canal and across posterior epiphysis	—	13.4	14.0
Greatest width of third sacral vertebra across centrum	—	—	26.0
Greatest width of anterior first centrum	50.5	49.7	42.0
Greatest distance from dorsal margin of anterior zygapophysis to ventral border of surface joining with ilium	73.7	80.3	—
Height from median ventral surface of first sacral vertebra to top of first neural spine	—	83.0a	—
Greatest height of anterior neural canal	—	11.2–13.5	15.0
Greatest width across anterior neural canal	31.6e +	22.3–27.9	24.0
Height of posterior neural canal	—	7.5–8.3	8.0
Greatest transverse diameter across posterior neural canal	—	17.6–18.5	16.0

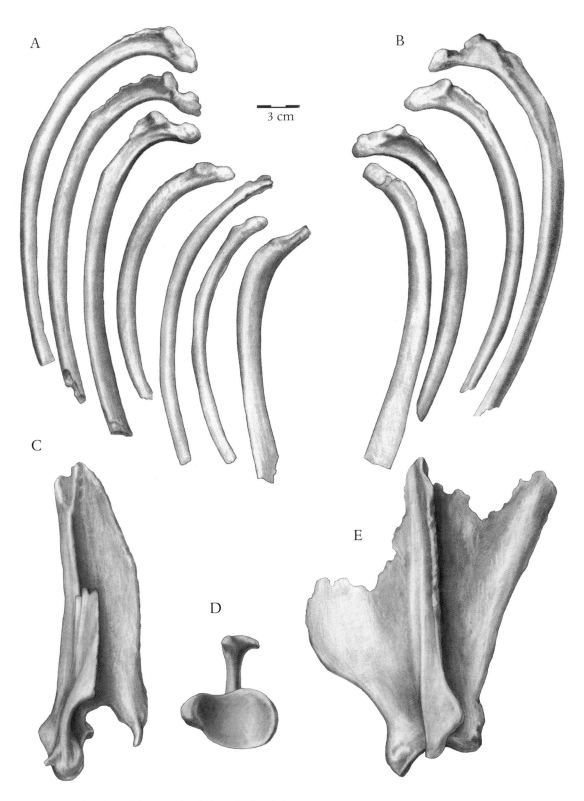

Figure 6.15. The ribs and the scapula of *Homotherium ischyrus*, IMNH 900-11862. *A*, ribs from the left side; *B*, ribs from the right side. The left scapula: *C*, dorsal view; *D*, distal end view, showing the glenoid cavity; *E*, lateral (superficial posterior) view. (drawing Mary Tanner)

Table 6.22. Right scapula measurements (mm): *Homotherium ischyrus* IMNH 900-11862. Compared specimens (observed range), *H. serum* (Meade, 1961; Rawn-Schatzinger, 1992), TMM 933-8, *3231, 3387, 3540; *H. crenatidens* (Ballesio, 1963), FSL 210991. (*e* = estimate, + = damaged element, *Meade)

	H. ischyrus R	H. ischyrus L	*H. serum R	H. serum	H. crenatidens R	H. crenatidens L
Greatest length coracoid process to top of scapula measured along axis of spine	—	240.0*e* +	322.0	*299.5–322.0	282.0	282.0
Width of scapular blade measured obliquely across spine	—	—	169.7	169.7–192.0	168.0	168.0
Greatest anteroposterior diameter of articulating end across glenoid cavity	56.7	58.9	80.5	62.8–*78.3	63.0	63.0
Greatest transverse diameter across glenoid cavity (fossa)	—	34.6	46.0	37.2–49.8	27.0	26.7
Distance from inner border of glenoid cavity to top of spine	—	75.8	89.0*e*	—	—	—
Least width of neck across articulating end	—	67.5	66.0	—	—	—
Width of supraspinous fossa	—	75*e* +	72.8	*71.7–*72.8	74.0	—
Width of infraspinous fossa		101.1*e* +	124.5	124.5–108.8	—	—

straight in the Idaho homothere and tilted forward in *Panthera.* It is smaller in *Smilodon* and inclined farther forward, as it is in *Panthera.* The lateral surface of the acromion process is constricted near its tip in *Panthera,* as is true in the Idaho homothere, but the metacromion process in the latter is enormously broadened. This feature could allow for an enlarged origin for the M. atlantoscapularis, a muscle that can contribute to strengthening the movements depressing the head. The scapular blade is broader than in either *Smilodon* or *Panthera. Smilodon* has a relatively narrow scapular blade so that the scapula in the Idaho homothere appears more constricted near the glenoid. The anterior part of the scapular blade is especially expanded in the Idaho homothere. Both the supraspinous and infraspinous fossae are deeper in the Idaho homothere than in *Smilodon* and *Panthera.* These enlarged features allow greater room for fleshy origins of the supraspinatus and infraspinatus muscles.

THE HUMERUS

The posterior proximal surfaces of the humeri are damaged, and the proximal shaft of the left humerus has been crushed (fig. 6.16, table 6.23). This distortion lengthened the left humerus by 17 mm when compared to the right humerus. There has also been erosion of the medial and lateral surfaces of the lesser and greater tuberosities in both humeri. The lesser tuberosity is more posterior than in *H. serum,* and the greater tuberosity of the Idaho specimen is more transversely expanded than in *H. serum.* The capitulum is highly curved, with the curvature extending farther ventrally than in *H. serum.* The deltoid tuberosity is well-delineated and straight as in other *Homotherium.* The posterior face of the humerus distally has a large deep triangular olecranon fossa. The distal humeral shaft separating the trochlear surface and the olecranon fossa is only a thin wall of bone. The triangular olecranon fossa is shallower and not as deeply pocketed dorsally as in *Xenosmilus.* The shape and position of the olecranon fossa indicates that the foreleg could be extended a few

degrees farther than was possible in *Xenosmilus.* Even such a small difference could have a significant effect on the amount of total extension of the leg in a long-legged, somewhat cursorial animal, such as a homothere (Gonyea, 1976). The entepicondylar foramen is large and oval, arising slightly more distally than that figured for *Homotherium crenatidens* by Ballesio (1963, fig. 25).

In the Idaho homothere, the greater tuberosity of the humerus slopes downward anteriorly more than in *Xenosmilus.* In *Xenosmilus,* this feature is thicker mediolaterally, more squared, and projects farther dorsally. The smaller size and lesser robustness of this feature in *Homotherium* indicate that the insertions of the M. supraspinatus, M. infraspinatus, and M. teres minor on the greater tuberosity are smaller than in *Xenosmilus.* The insertion of the M. subscapularis on the lateral aspect of the lesser tubercle in the Idaho homothere is also smaller than in *Xenosmilus, Panthera,* and *Smilodon.* These muscles form the rotator cuff that controls the position of the humeral head, and their relative size and attachments indicate that the humerus had potentially a greater range of motion in the Idaho homothere than in *Xenosmilus, Panthera,* or *Smilodon.* The greater downward anterior slope of the greater tubercle in *Homotherium* also indicates that the humerus had a greater range of motion at shoulder extension than in *Xenosmilus.* The deltopectoral crest in the Idaho homothere is more gracile than in *Xenosmilus, Panthera,* or *Smilodon,* indicating that these muscles would generate less forceful limb extension and protraction movements, although the range of motion allowed at the glenohumeral joint may have been greater. The posterior site of origin of the parts of the M. triceps brachii on the posterior humeral shaft is relatively smaller in the Idaho homothere than in *Xenosmilus, Smilodon,* and *Panthera.* The origin of the common flexor tendons on the distal medial humeral condyle in the Idaho homothere is a distinct rounded knob. It is less robust than that of *Xenosmilus, Panthera,* and *Smilodon.* When coupled with the greater relative length of the radius and

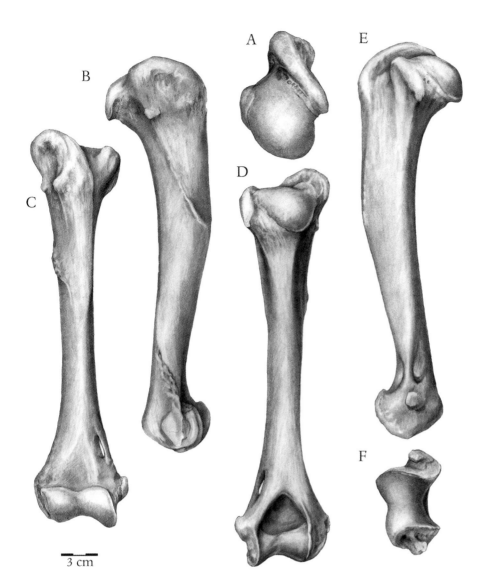

3 cm

Figure 6.16. The right humerus of *Homotherium ischyrus*, IMNH 900-11862: *A*, proximal end view; *B*, lateral view; *C*, anterior view; *D*, posterior view; *E*, medial view; *F*, posterior distal view. (drawing Mary Tanner)

Table 6.23. Right and left humerus measurements (mm): *Homotherium ischyrus* IMNH 900-11862. Compared specimens (observed range), *H. serum* (Meade, 1961; Rawn-Schatzinger, 1992), TMM 933-73, 333, *2206, 2292, 2661, *2506, 2805, *3231, 3272, 5704; *H. crenatidens* (Ballesio, 1963), FSL 210991; *Ischyrosmilus johnstoni* (Mawby, 1965), W.T. 625, UCMP 66487. (*e* = estimate, + = damaged element, *Meade, **W.T. 625, ◆UCMP 66487)

| | H. ischyrus | | *H. serum | | H. serum | H. crenatidens | | I. johnstoni |
	R	L	R	L		R	L	
Greatest length measured parallel to longitudinal axis	328.0	331.0e +	358.0	356.0	*287.4–*358.0	354.0	354.0	**319.0
Greatest transverse diameter (mediolateral) of proximal extremity	71.0	70.0	74.5	76.0	*67.5–76.0	65.0	64.0	—
Greatest anteroposterior diameter of proximal head	95.7	99.7e	104.0	104.0	*91.8–*104	97.0	97.0	—
Transverse diameter at mid shaft	24.9	24.6e +	33.0	32.0	27.5–*33.0	29.0	28.0	**23.3
Anteroposterior diameter at mid shaft	45.4	46.5e +	49.0	49.0	*39.4–*49.0	44.0	43.0	**35.9
Greatest width of distal extremity	80.5e	83.7	82.5	84.4	76.4–*84.4	86.0	85.0	**70.4–◆71.7
Least anteroposterior diameter of articulating surface for (trochlea) ulna	26.8	27.1	30.7	30.7	25.3–*30.7	—	—	**27.0–◆25.0
Maximum width of entepicondylar foramen	—	7.0	—	—	6.6–8.3	—	—	—
Maximum length of entepicondylar foramen	13.5	16.5	—	—	13.2–20.9	—	—	—
Greatest breadth across distal articular surface	56.3	55.7	—	—	—	56.0	57.0	—

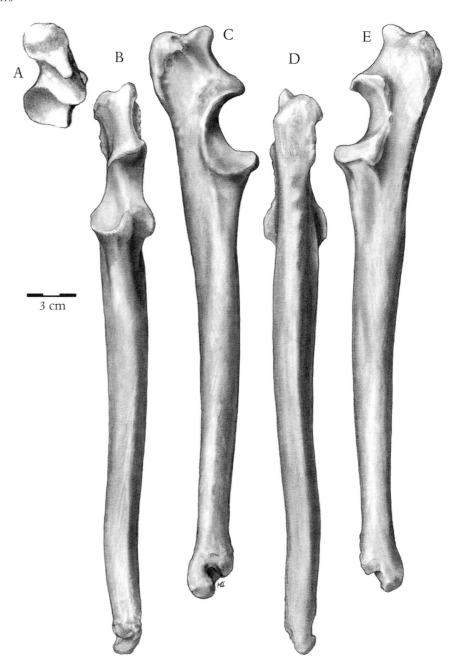

Figure 6.17. The left ulna and right radius of *Homotherium ischyrus*, IMNH 900-11862. The left ulna: *A*, proximal end view; *B*, anterior view; *C*, medial view; *D*, posterior view; *E*, lateral view. The radius: *F*, proximal end view; *G*, medial view; *H*, distal end view; *I*, posterior view; *J*, lateral view; *K*, anterior view. (drawing Mary Tanner)

3 cm

ulna in the Idaho homothere, these muscles had a reduced mechanical advantage but were more capable of generating an increased speed of movement than in the relatively shorter-limbed *Xenosmilus*, *Panthera*, and *Smilodon*.

THE ULNA

The right ulna is nearly complete, missing only the distal end, while the left ulna retains a complete distal end but lacks the olecranon process (fig. 6.17A–E, table 6.24). This limb element is elongated compared to that of *Smilodon* and *Xenosmilus*. In the Idaho homothere, the posterior surface of the olecranon process has a smaller anterior-posterior diameter than in either *Smilodon* or *Xenosmilus*. This narrow, rectangular surface, which serves as the insertion point of the ex-

tensor muscles M. anconeus, M. flexor carpi ulnaris, and the M. triceps brachii, is even more reduced than that in *Homotherium crenatidens*. This reduction in diameter results in less mechanical advantage for the insertion of the M. triceps brachii and indicates a higher gear muscle than in both *Xenosmilus* and *Smilodon*, again emphasizing increased movement at the expense of force generated. *Homotherium serum* has a relatively short olecranon, more in line with its coronoid articulation, and this suggests a greater emphasis on lateral motion. *H. crenatidens* (Ballesio, 1963, fig. 26) has a long olecranon process, resembling the Idaho homothere. The semilunar notch is slightly shallower than that of *H. serum*. The articular facet on the dorsomedial surface of the semilunar notch is posteriorly directed, and not sigmoid, as

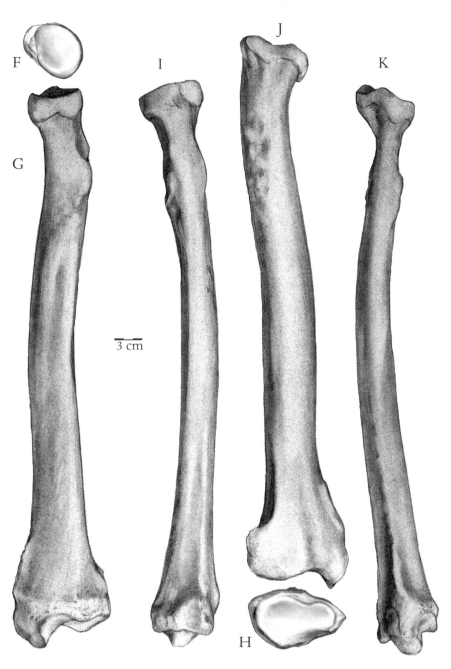

illustrated for *H. serum.* Rawn-Schatzinger (1992) describes the coronoid articular surface of the ulna of *H. serum* as sigmoid-shaped and broadly flared. The broad flaring is a measure of the lateral motion possible at the elbow. We would expect more cursorial forms to have this flaring less developed, and the notch less sigmoid, as is the case in the Idaho homothere. In *Smilodon,* and especially in *Xenosmilus,* this flaring is even more pronounced than in *H. serum.* In the Idaho homothere, just distal and lateral to the semilunar notch, there is a narrow radial notch. This facet scribes an arc, allowing rotation of the radioulnar part of the elbow joint. The radial notch in *Xenosmilus* and *Smilodon* is much broader, and the arc greater, allowing more lateral rotation than in the Idaho homothere.

The ulnar shaft in the Idaho homothere is more bowed than in *Homotherium serum, Xenosmilus,* or *Smilodon,* more closely resembling *Panthera.* The anterolateral tubercle is more prominent and pointed in the Idaho homothere than in *Xenosmilus* or *H. serum,* in which it is low and rounded. In this regard, it resembles *Homotherium crenatidens.* The styloid process on the distal end is short and rounded, as in *Xenosmilus,* but is even shorter and more flattened in *H. serum.* In *Smilodon,* it is longer and pointed. The radial articular facet in the homotheres is not widely separated from the styloid process as it is in *Smilodon,* although the homotheres have a deep groove separating it from the base of the styloid process. This groove undercuts the styloid process in the Idaho homothere and, to a lesser degree, in *Xenosmilus* and *H. serum.*

Table 6.24. Right and left ulna measurements (mm): *Homotherium ischyrus* IMNH 900-11862. Compared specimens (observed range), *H. serum* (Meade, 1961; Rawn-Schatzinger, 1992), TMM 933-811, 989, 2740, *3231, 3534, 5745; *H. crenatidens* (Ballesio,1963), FSL 210991; *Ischyrosmilus johnstoni* (Mawby, 1965), W.T. 628. (*e* = estimate, + = damaged element, *Meade, **Rawn-Schatzinger, ◆W.T. 628)

	H. ischyrus R	L	*H. serum R	L	H. serum	H. crenatidens R	L	◆I. johnstoni R
Greatest length measured parallel to longitudinal axis of ulna	355.0e +	—	376.0	380.0	376.0–*380.0	370.0	372.0	—
Greatest width of posterior surface of olecranon process	—	31.7	—	59.5	30.3–*59.5	59.0	59.0	30.3
Greatest transverse width of greater sigmoid cavity	32.5	30.5	45.2	45.0	34.3–*45.1	37.0	37.0	31.7
Length of radial notch	35.7	36.2	—	—	—	—	—	32.0
Height of trochlear notch (inside)	35.6	37.0	**45.2	**45.0	34.3–45.2	—	—	33.8
Anteroposterior diameter of shaft at proximal end of tendon scar	33.5	35.0	37.0	37.0	24.0–*34.4	—	—	—
Transverse diameter (mediolateral) of shaft at proximal end of tendon scar	22.4	20.6	28.0	24.0	16.8–24.0	—	—	—
Greatest anteroposterior diameter of distal extremity	26.6	27.0	34.6	34.0	*34.0–*34.6	35.0	35.0	—
Greatest width of distal extremity	18.2	20.0	23.7	24.0	23.7–24.0	23.0	24.0	—
Anteroposterior diameter of shaft at mid shaft	29.7e +	30.7	—	—	—	34.0	33.0	—
Transverse diameter at mid shaft	23.7e +	22.6	—	—	—	21.0	19.0	—

Table 6.25. Right and left radius measurements (mm): *Homotherium ischyrus* IMNH 900-11862. Compared specimens (observed range), *H. serum* (Meade, 1961; Rawn-Schatzinger, 1992), TMM 933-304, **485, 988, 2299, **2421, **2565, 2935, *3231; *H. crenatidens* (Ballesio, 1963), FSL 210991. (*e* = estimate, *Meade, **Rawn-Schatzinger)

	H. ischyrus R	L	*H. serum R	L	*H. serum		H. crenatidens R	L
Length measured along internal border	296.0e	303.0	323.0	328.0	295.7–323.0		317.0	317.0
Long diameter of proximal end (mediolateral diameter)	35.5	35.8	37.0e	38.5	*35.3–57.4		34.0	34.0
Greatest diameter at right angles to long diameter of proximal end	28.4	29.0	32.0	32.0	*30.0–38.1		29.0	29.0
Width of shaft at middle (mediolateral diameter)	27.0	27.0	30.0	29.0	24.3–*30.0		28.0	28.0
Thickness of shaft at middle (anteroposterior diameter)	21.1	21.3	22.0	21.5	15.6–*22.0		23.0	24.0
Greatest width at distal end taken to internal face	42.0	42.5	61.4	57.0	33.2–*61.4		55.0	55.0
Greatest thickness of distal end	57.2	56.0	36.7	36.0	28.9–*36.7		38.0	39.0
Greatest breadth of distal articular surface (facet)	40.7	41.0	—	—	—		—	—

The shaft of the ulna tapers more distally in the Idaho homothere than in *H. crenatidens* or in *H. serum,* suggesting that the Idaho specimen might have been even more adapted for running. Although neither the right nor left ulna is complete, the length of the complete element may be estimated at 360 mm. This is slightly less than the 370–372 mm reported for the European *H. crenatidens* (Ballesio, 1963), contrasting with the yet shorter ulnae found in both *Xenosmilus* (320 mm) and *Smilodon* (345 mm; Merriam and Stock, 1932).

THE RADIUS

The radius is more elongated than that of *H. serum* and similar to that of *H. crenatidens* as described by Ballesio (1963), although the proximal end (capitulum) of the radius is more ovoid than that figured by Ballesio (1963, fig. 28; fig. 6.17F–K; table 6.25). Dorsally, the ovate capitular depression is slightly concave and articulates with the lateral epicondyle of the humerus. This feature is much larger anterolaterally and

posteromedially in both *Smilodon* and *Xenosmilus.* The shaft in the Idaho homothere is more bowed mediolaterally than in the other homotheres and more closely resembled *Panthera.* The Idaho homothere radius is strongly bowed anteriorly, more so than in *Panthera* or *H. crenatidens.* In contrast, that in *H. serum* is exceptionally straight, while those of *Xenosmilus* and *Smilodon* are slightly more bowed. The interosseus crest is well developed and runs the entire length of the shaft. The radial neck is narrow, with an extremely expanded and rugose radial tuberosity. The nutrient foramen is in the base of the upper one-third of the shaft. The anterior margin of the distal radial articulation differs from that of *H. serum* in that the articular facet of the styloid process is convex. In *H. serum,* the edge of the distal end extends past the distal process and is about even with it in the Idaho homothere. The Idaho homothere more closely resembles *H. crenatidens* in that respect. The portion of the distal articular facet for the scapholunar wraps onto the posteromedial surface.

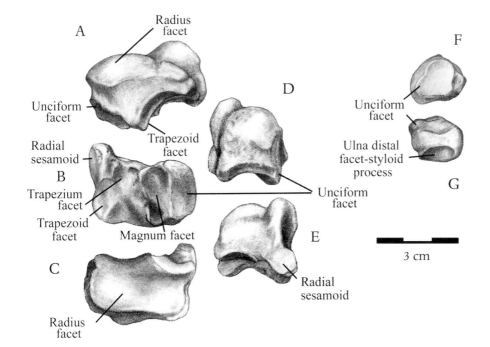

Figure 6.18. The right scapholunar of *Homotherium ischyrus*, IMNH 900-11862: *A*, dorsal view; *B*, distal view; *C*, proximal view; *D*, lateral view; *E*, medial view. The right cuneiform: *F*, medial view; *G*, dorsal view. (drawing Mary Tanner)

Table 6.26. Right scapholunar measurements (mm): *Homotherium ischyrus* IMNH 900-11862. Compared specimens (observed range), *H. serum* (Meade, 1961; Rawn-Schatzinger, 1992), TMM 933-968, 1085, 2088, 2833, *3231, 3311, 3620; *H. crenatidens* (Ballesio, 1963), FSL 210991. (*Meade)

| | H. ischyrus | | *H. serum | H. crenatidens |
	R	L		
Greatest transverse diameter (mediolateral) measured normal to external border of proximal surface	46.0	—	49.2–42.1	50.0
Greatest dorsopalmar length	39.0	—	25.2–*37.7	39.0
Greatest proximal distal diameter	33.0	—	32.3–*42.0	31.0

THE WRIST (CARPAL BONES)

THE SCAPHOLUNAR

The right scapholunar was recovered (fig. 6.18A–E, table 6.26). The posterior radial ridge of the scapholunar in the Idaho homothere is higher than in either *Smilodon* or *Panthera* and resembles that in *Homotherium crenatidens*. The main radial articular surface is narrower than in *Smilodon* and *Panthera* and more closely resembles that in *H. crenatidens* and *H. serum*. The facets for the magnum and unciform are dropped farther distally than in *Smilodon* and *Panthera*, as is also true in *H. crenatidens*. The trapezoid and magnum facets of the scapholunar are broadly joined anteroposteriorly and separated by a large ridge. The joined length of the two facets is greater than in the European *H. crenatidens*. The trapezium and trapezoid facets are larger in the Idaho specimen than those figured by Ballesio (1963), extending nearly to the plantar surface. The magnum facet is deeply concave, with a shallow ligamental pit at the anteromedial margin. As in

H. serum, the unciform facet is broad and rounded dorsally and distinct from the magnum facet. Seen in dorsal view, the body of the scapholunar is semirectangular, with a very pronounced dorsal process, posterodorsal to a large plantar process. In the European *H. crenatidens*, the termination of this latter process is rounded and is more angular in the Idaho specimen. The orientation of the sesamoid articular facet on the plantar process is nearly vertical.

THE UNCIFORM

The unciform was not recovered for the Idaho homothere specimen.

THE CUNEIFORM

Both left and right cuneiform bones (table 6.27) were recovered in the Idaho *Homotherium*. The cuneiform has a concave medial surface with a well-developed articular facet for the unciform. The ventrolateral surface has two facets: the posterior facet articulates with the pisiform and the anterior

Table 6.27. Right cuneiform measurements (mm): *Homotherium ischyrus* IMNH 900-11862. Compared specimen, *H. crenatidens* (Ballesio, 1963), FSL 210991.

	H. ischyrus R L	*H. crenatidens*
Greatest length measured parallel to ridge separating pisiform and ulnar facets	22.4 —	29.0
Diameter normal to line of greatest length	17.6 —	18.0
Greatest thickness	16.7 —	20.0

facet with the styloid process of the ulna. In dorsal view, the external face of the anterior facet is strongly convex.

THE PISIFORM

The right pisiform (fig. 6.19A–F, table 6.28) is columnar, with two proximal articular facets that meet at an acute angle. The lateral proximal facet articulates with the styloid process of the ulna. The medial facet articulates with the cuneiform. The distal end terminates in a bony knob. The pisiform in *H. crenatidens* is shorter and more squared. It is slightly more elongate in *H. serum* but not as much as in the Idaho homothere.

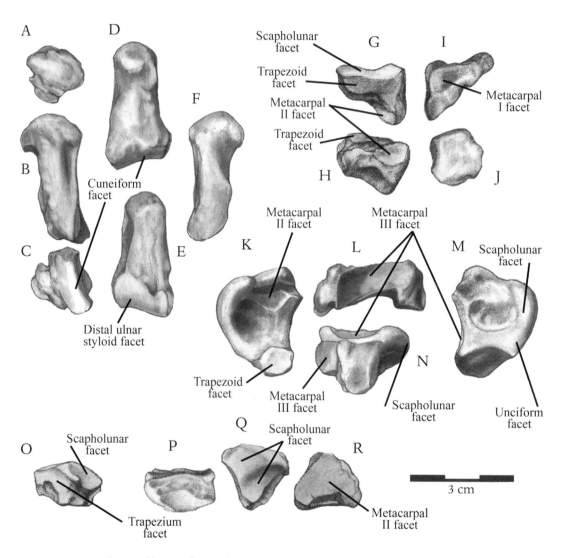

Figure 6.19. Four right carpal bones of *Homotherium ischyrus*, IMNH 900-11862. *A–F*, the pisiform; *G–J*, the trapezium; *K–N*, the magnum; *O–R*, the trapezoid. The pisiform bone: *A*, head in distal view; *B*, distal view of the shaft; *C*, proximal head view; *D*, radial view; *E*, ulnar view; *F*, anterior view. The trapezium: *G*, lateral view; *H*, distolateral view; *I*, mediodistal view; *J*, dorsal view. The magnum: *K*, medial view; *L*, distal view; *M*, lateral view; *N*, proximal view. The trapezoid: *O*, ventral view; *P*, dorsal view; *Q*, proximal view; *R*, distal view. (drawing Mary Tanner)

Table 6.28. Right pisiform measurements (mm): *Homotherium ischyrus* IMNH 900-11862. Compared specimens (observed range), *H. serum* (Rawn-Schatzinger, 1992), TMM 933-180, 1311, 1600, 2398, 3252, 3555; *H. crenatidens* (Ballesio, 1963), FSL 210991.

	H. ischyrus R	L	*H. serum*	*H. crenatidens*
Greatest length measured from and normal to the contact edge of ulnar and unciform facets to the head	35.8	—	37.2–39.8	34.0
Greatest (long) diameter of articulating end	20.8	—	21.1–23.4	23.0
Greatest (long) diameter of head	18.1	—	20.4–23.5	19.0
Greatest proximodistal diameter of head normal to long diameter	12.5	—	9.1–12.4	—
Height of head	12.0	—	14.4–18.1	—

Table 6.29. Right trapezium measurements (mm): *Homotherium ischyrus* IMNH 900-11862. Compared specimen, *H. crenatidens* (Ballesio, 1963), FSL 210991.

	H. ischyrus R	L	*H. crenatidens* R	L
Greatest dorsopalmar depth (anteroposterior)	21.8	—	—	21.0
Greatest dorsal width, transverse diameter	19.2	—	—	20.0
Greatest proximodistal diameter	18.0	—	—	24.0

Table 6.30. Right trapezoid measurements (mm): *Homotherium ischyrus* IMNH 900-11862. Compared specimen, *H. crenatidens* (Ballesio, 1963), FSL 210991.

	H. ischyrus R	L	*H. crenatidens*
Greatest dorsopalmar depth	20.1	—	23.0
Greatest dorsal width, transverse diameter	21.4	—	21.0
Greatest proximodistal diameter	15.3	—	18.0

Table 6.31. Right magnum measurements (mm): *Homotherium ischyrus* IMNH 900-11862. Compared specimens (observed range), *H. serum* (Rawn-Schatzinger, 1992), TMM 933-76, 3317; *H. crenatidens* (Ballesio, 1963), FSL 210991.

	H. ischyrus R	L	*H. serum*	*H. crenatidens*
Greatest dorsopalmar length	31.8	—	36.0–39.6	33.0
Greatest dorsal width	13.1	—	—	21.0
Greatest proximodistal diameter	25.5	—	24.9–26.0	26.0

Table 6.32. Right radial sesamoids measurements (mm): *Homotherium ischyrus* IMNH 900-11862.

	H. ischyrus R	L
Greatest proximodistal length measured along a line parallel to facet for scapholunar	12.2	11.9
Greatest dorsoventral length of scapholunar facet	11.9	11.8
Greatest transverse diameter (mediolateral)	7.6	8.0

THE TRAPEZIUM

The lateral surface of the right trapezium (fig. 6.19G–J, table 6.29) bears an L-shaped facet for articulation with the trapezoid. The long axis of the "L" runs anteroposteriorly. There is a small elliptical articular facet for articulation with the anterolateral portion of metacarpal II. The medial face is occupied by a broadly triangular facet for metacarpal I. A poorly defined dorsal facet articulates with the ventromedial portion of the scapholunar.

THE TRAPEZOID

The trapezoid element (fig. 6.19O–R, table 6.30) is broadly triangular. On its medial face it has a subtriangular trapezium facet and on the lateral face an L-shaped facet that is confluent with the ventral metacarpal II facet. The lateral face supports two scapholunar facets, the posterior of which extends well down the lateral side of the trapezoid.

THE MAGNUM

The right magnum (fig. 6.19K–N, table 6.31) was recovered in the Idaho homothere. The dorsal surface of the magnum is strongly convex. On the medial face, a portion of the articular surface is missing, but we can recognize two articular facets. The anterior facet is smaller, flat, and subtriangular. The posterior facet is larger and concave and would have articulated with the plantar process of metacarpal II. Ventrally, the articular surface for metacarpal II is concave and rectangular. The lateral face bears a vertically oriented anterior facet confluent with the dorsal scapholunar facet. In distal view, there is a large concave vertically oriented facet for metacarpal III.

THE WRIST SESAMOIDS

Both the right and left radial sesamoid (fig. 6.30G–M, table 6.32) that articulate with the scapholunar were recovered.

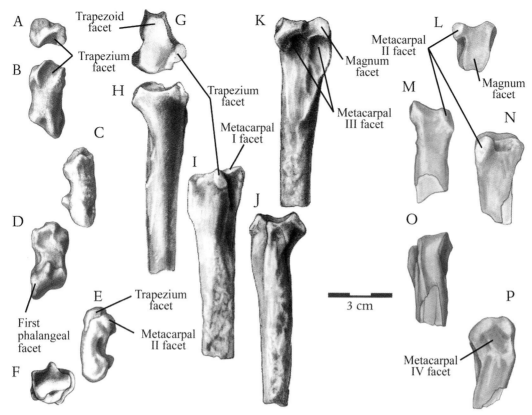

Figure 6.20. The left first metacarpal (Mc I), the right second metacarpal (Mc II), and the right third metacarpal (Mc III) of *Homotherium ischyrus*, IMNH 900-11862. The first metacarpal, (Mc I): *A*, proximal view; *B*, dorsolateral view; *C*, dorsal view; *D*, ventral (palmer) view; *E*, lateral view; *F*, distal view. The second metacarpal (Mc II): *G*, proximal end view; *H*, dorsal view; *I*, medial view; *J*, ventral view; *K*, lateral view. The third metacarpal (Mc III): *L*, proximal end view; *M*, dorsal view; *N*, medial view; *O*, ventral view; *P*, lateral view. (drawing Mary Tanner)

These are circular and have a circular convex medial facet for the scapholunar. In addition, six distal metacarpal/tarsal sesamoids were recovered. Although associated, they were not articulated, therefore their identity is uncertain.

THE MANUS (FOREFOOT)
METACARPAL I

Both the right and left metacarpal are present in the Idaho homothere (fig. 6.20A–F, table 6.33). They are more elongated than in *Smilodon* but not as elongated as in *H. serum* or *H. crenatidens*. The outer proximal articular facet is broader than in *H. crenatidens*. The convex medial facet articulates with the trapezium. The phalangeal articulation is canted, and the carina is reduced.

METACARPAL II

The proximal three-fourths of the right metacarpal II is present (fig. 6.20G–K, table 6.34). The second metacarpal is broader anteriorly than posteriorly, giving the shaft a subtriangular cross section. Proximally there are two articular facets: a small medial elliptical facet that articulates with metacarpal I and a small ovoid lateral facet that articulates with the trapezium. Posterior to the ovoid facet on the plantar process is a slightly convex magnum facet. Below the anterior medial process is a strong concavity that articulates with the convex anteromedial process of metacarpal III.

METACARPAL III

The proximal one-half of the right metacarpal III was recovered (fig. 6.20L–P, table 6.35). The primary articular facet with the magnum is rectangular and inclined medially. The lateral face of the metacarpal has two facets for the fourth metacarpal. The more dorsal of the two is gently crescentic; ventrad to this is a concavity with which the medial convex process of the fourth metacarpal articulates. The anteromedial facet for metacarpal II is more extended, and the anterior proximal faces of metacarpal II and metacarpal III

Table 6.33. Left metacarpal I measurements (mm): *Homotherium ischyrus* IMNH 900-11862. Compared specimens (observed range), *H. serum* (Rawn-Schatzinger, 1992), TMM 933-1880, *3231; *H. crenatidens* (Ballesio, 1963), FSL 210991. (*Meade)

	H. ischyrus R	L	*H. serum*	*H. crenatidens*
Greatest length	35.0	35.5	41.2	42.0
Greatest transverse diameter (mediolateral) of proximal end	20.5	19.3	18.9–*19.3	20.0
Greatest dorsoventral (dorsopalmar) diameter of proximal end	16.6	16.0	17.1–*17.6	16.0
Transverse diameter at mid shaft	13.7	13.2	—	15.0
Dorsoventral diameter at mid shaft	11.9	11.7	—	14.0
Greatest transverse diameter at distal end of shaft	17.0	16.2	19.1–*19.4	17.0
Greatest dorsoventral diameter at distal end	16.3	16.8	17.3	17.0

Table 6.34. Right metacarpal II measurements (mm): *Homotherium ischyrus* IMNH 900-11862. Compared specimens (observed range), *H. serum* (Rawn-Schatzinger, 1992), TMM 933-600, 1278, 1643, 2134, 2567, 2569, 2598, 2823, 3941; *H. crenatidens* (Ballesio, 1963), FSL 210991. (*e* = estimate, + = damaged element)

	H. ischyrus R	L	*H. serum*	*H. crenatidens*
Greatest length	—	—	108.8–116.1	110.0
Greatest transverse diameter (mediolateral) of proximal end	25.6	—	24.6–29.8	23.0
Greatest dorsoventral (dorsopalmar) diameter of proximal end	26.0	—	27.4–33.7	27.0
Transverse diameter at mid shaft	12.0*e* +	—	15.5–17.6	15.0
Dorsoventral diameter at mid shaft	14.4*e* +	—	15.5–18.8	18.0
Greatest transverse diameter at distal end of shaft	—	—	20.3–23.2	23.0
Greatest dorsoventral diameter at distal end	—	—	20.5–24.3	21.0

Table 6.35. Right metacarpal III measurements (mm): *Homotherium ischyrus* IMNH 900-11862. Compared specimens (observed range), *H. serum* (Rawn-Schatzinger, 1992), TMM 933-391, 598, 599, 2133, 2568, 2573, 2710; *H. crenatidens* (Ballesio, 1963), FSL 210991. (+ = damaged element)

	H. ischyrus R	L	*H. serum*	*H. crenatidens*
Greatest length	22.3	—	112.0–127.5	122.0
Greatest transverse diameter (mediolateral) of proximal end	23.2	—	184.0+–27.4	22.0
Greatest dorsoventral (dorsopalmar) diameter of proximal end	—	—	26.1–28.5	24.0
Transverse diameter at mid shaft	—	—	17.7–21.1	16.0
Dorsoventral diameter at mid shaft	—	—	16.0–20.8	15.0
Greatest transverse diameter at distal end of shaft	—	—	23.1–25.6	24.0
Greatest dorsoventral diameter at distal end	—	—	23.0–27.7	22.0

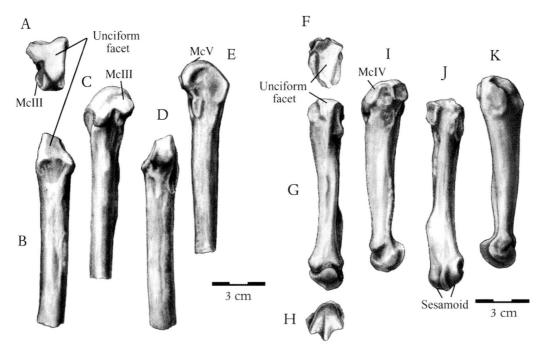

Figure 6.21. The fourth right metacarpal (Mc IV) and the fifth right metacarpal (Mc V) of *Homotherium ischyrus*, IMNH 900-11862. The fourth metacarpal (Mc IV): *A*, proximal view; *B*, dorsal view; *C*, medial view; *D*, ventral view; *E*, lateral view. The fifth metacarpal (Mc V): *F*, proximal end view; *G*, dorsal view; *H*, distal end view; *I*, medial view; *J*, ventral view; *K*, lateral view. (drawing Mary Tanner)

Table 6.36. Right metacarpal IV measurements (mm): *Homotherium ischyrus* IMNH 900-11862. Compared specimens (observed range), *H. serum* (Rawn-Schatzinger, 1992), TMM 933-596, 1773, 2566, 2572, 3344, 3345, 3451; *H. crenatidens* (Ballesio, 1963), FSL 210991. (*e* = estimate, + = damaged element)

| | *H. ischyrus* | | *H. serum* | *H. crenatidens* |
	R	L		
Greatest length	—	—	118.0–129.5	120.0
Greatest transverse diameter (mediolateral) of proximal end	25.4	20.9	16.9+–22.7	20.0
Greatest dorsoventral (dorsopalmar) diameter of proximal end	20.9	24.4	26.1–31.8	25.0
Transverse diameter at mid shaft	14.2*e* +	13.7	14.5–17.7	15.0
Dorsoventral diameter at mid shaft	13.7	12.7	16.2–19.6	14.0
Greatest transverse diameter at distal end of shaft	—	—	20.8–23.8	22.0
Greatest dorsoventral diameter at distal end	—	—	23.4–26.7	22.0

are broader than in *H. crenatidens*. A distal end, presumably belonging to the left metacarpal III but lacking a contact to the proximal part, shows a strongly developed posterior carina and a raised anterior surface separated from the shaft by a prominent groove; the articular surface is canted laterally and bordered by two strong ligamental attachments.

METACARPAL IV

The proximal three-fourths of the fourth right metacarpal is present (fig. 6.21A–E, table 6.36). The fourth metacarpal is large. Proximally, it has three facets separated by small ridges, and all the facets are strongly convex. The plantar

Table 6.37. Right metacarpal V measurements (mm): *Homotherium ischyrus* IMNH 900-11862. Compared specimens (observed range), *H. serum* (Rawn-Schatzinger, 1992), TMM 933-272, 601, 1097, 1649, 2570, 2711; *H. crenatidens* (Ballesio, 1963), FSL 210991.

| | *H. ischyrus* | | *H. serum* | *H. crenatidens* |
	R	L		
Greatest length	96.3	—	102.2–110.3	102.0
Greatest transverse diameter (mediolateral) of proximal end	18.3	—	23.7–29.9	24.0
Greatest dorsoventral (dorsopalmar) diameter of proximal end	23.7	—	23.0–26.5	26.0
Transverse diameter at mid shaft	11.3	—	12.5–16.4	15.0
Dorsoventral diameter at mid shaft	16.0	—	14.5–17.7	13.0
Greatest transverse diameter at distal end of shaft	19.0	—	19.5–21.5	22.0
Greatest dorsoventral diameter at distal end	20.3	—	20.3–23.4	21.0

process of this element is somewhat reduced relative to those of the other metacarpals. The anteromedial facet articulates with the third metacarpal and is separated from the lateral unciform facet by a strong ridge. The unciform facet is broadly convex and is separated from the posteromedial magnum facet by a subdued ridge. The distal one-fifth of the right metacarpal IV is preserved but lacks contact with the shaft. The anterior articular surface is strongly raised and separated from the shaft by a groove, and the carina is strongly developed posteriorly. The articular surface is relatively narrow.

METACARPAL V

The right fifth metacarpal (fig. 6.21F–K, table 6.37) is missing the plantar process but is otherwise complete. It retains two proximal facets, one medially for articulation with the fourth metacarpal and a strongly convex, triangular unciform facet. The anterior articular surface of the distal end is raised and bulbous, separated from the shaft by a groove, and bordered by two strong ligamental attachments. The medial palmar side of metacarpal V has a crest that extends along the distal one-third of the shaft. This ridge corresponds to the insertion of the opponens digiti minimi muscle, which would act to pull this metacarpal bone forward and, with a slight rotational movement, would assist the flexor digiti minimi muscle to deepen the cup of the palm and flex the carpometacarpal joint. This would help the cat resist forces that would spread the metacarpals, such as those caused by struggling prey. The proximal articular surfaces between metacarpal IV and V in *Homotherium* also indicate little lateral rotational capability, although movement toward the palmar surface would still be possible. This feature would help the cat resist forces that would spread the metacarpals, although the opposition movement is not as well developed as in *Xenosmilus*. The crest on the medial palmar side of metacarpal V is not present in either *Smilodon* or *Panthera*.

THE PHALANGES

THE PROXIMAL FIRST PHALANX The right proximal first phalanx (fig. 6.22A–J, table 6.38) of the first digit is asymmetrical and closely resembles those figured by Ballesio (1963, fig. 40) and Rawn-Schatzinger (1992, fig. 30) for *H. crenatidens* and *H. serum* respectively, although it is slightly more robust than in *H. serum*. The phalanx has a pronounced groove on the ventral surface as in the European *H. crenatidens* but not in *H. serum*. The *H. serum* phalanx is more elongated than in either the Idaho homothere or *H. crenatidens*.

THE DISTAL FIRST PHALANX A large thumb claw is present in the Idaho homothere (fig. 6.22A–B). As in *Smilodon,* it is the largest of the series.

THE SECOND DIGIT FIRST PHALANX The left first phalanx (fig. 6.22) of the second digit is present but is badly damaged, missing the distal end and a portion of the proximal articulation. It is relatively curved dorsoventrally and slightly bowed medially.

THE SECOND DIGIT SECOND PHALANX The second phalanx of the second digit was not recovered.

THE SECOND DIGIT THIRD PHALANX The third phalanx of the second digit was not recovered.

THE THIRD AND FOURTH DIGITS FIRST PHALANGES The first phalanges for digits III and IV are not preserved.

THE THIRD DIGIT SECOND PHALANGES AND FOURTH DIGIT SECOND PHALANX There is a second phalanx and claw from the right digit III and a second phalanx from the left digit III. These phalanges are relatively large, short, and deeply bowed medially, much more so than those shown by Rawn-Schatzinger (1992) for *H. serum;* the claw is relatively larger than in *H. serum*. The second phalanx for the right digit IV is also preserved and is slightly more slender than that for digit III. Digit IV seems to be only represented by a claw that is relatively smaller than any others in the hand.

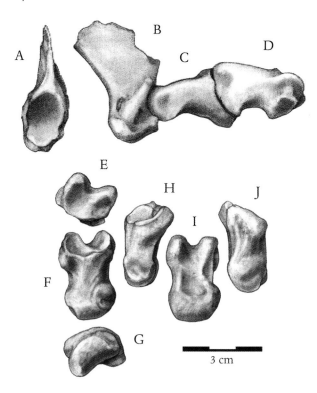

Figure 6.22. The elements of the right first digit of *Homotherium ischyrus*, IMNH 900-11862, individually and in articulation: *A*, proximal end view; *B*, medial view of the distal (terminal) phalanx; *C*, medial view of the first phalanx; *D*, medial view of the first metacarpal (Mc I). First phalanx: *E*, proximal end view; *F*, dorsal view; *G*, distal end view; *H*, medial view; *I*, ventral view; *J*, lateral view. (drawing Mary Tanner)

Table 6.38. Right manus first phalanx digit I measurements (mm): *Homotherium ischyrus* IMNH 900-11862. Compared specimen, *H. crenatidens* (Ballesio, 1963), FSL 210991.

	H. ischyrus	*H. crenatidens*
Greatest length	32.6	34.0
Greatest dorsoventral (dorsopalmar) diameter of proximal end	17.3	—
Greatest transverse diameter (mediolateral) of proximal end	20.3	22.0
Dorsoventral diameter at mid shaft	11.3	12.0
Transverse diameter at mid shaft	15.6	16.0
Greatest transverse diameter at distal end of shaft	20.0	2.0
Greatest dorsoventral diameter of distal end	12.1	13.0

The Hindlimb Skeleton and Pelvic Girdle
THE INOMINATE

Both left and right inominates were mostly complete (fig. 6.23, table 6.39). The left inominate lacks a portion of the pubis, ischium, and the iliac crest. The right inominate is heavily corroded. The iliac blade is shallowly concave, similar to that of *H. crenatidens* and *H. serum* rather than that of *Smilodon*. Like the former two, the ischiadic spine of the Idaho specimen is reduced compared to that of *Smilodon*, with a round obturator foramen. Anterodorsal to the acetabulum is a rugose ridge for the attachment of the rectus femoris muscle. Medial to this is a shallow sulcus that separates the attachment ridge from the iliopectineal eminence. The latter is greatly reduced. The pubis is a more gracile element than that of *Smilodon* (Merriam and Stock, 1932), *H. serum* (Meade, 1961), or *H. crenatidens* (Ballesio, 1963). The pubis of homotherine cats lacks a spine on the posterodorsal inner border of the obturator foramen. The ischium is more massive than the pubis, with a well-developed tuberosity that is more strongly expanded dorsolaterally to ventromedially than it is in *Smilodon*.

THE FEMUR

The femur (fig. 6.24, table 6.40) is more elongate than in *H. serum* and more closely resembles *H. crenatidens* in both this respect and in having the femoral head at a more nearly right angle from the shaft. *H. serum* has a more elongated femoral neck and more elevated head, making it more bear-like than in other *Homotherium*. The femoral shaft is slender and straight in all the *Homotherium*, and the femoral head is large and rounded. The fovea capitus pit for the round ligament in the femoral head is more posterodorsal than in *H. crenatidens*, in which it is more centered. The lesser trochanter is prominent and more laterally situated than in *H. crenatidens* or *H. serum*. A distinct ridge for the obturator externus insertion anteromedial to the lesser trochanter is more developed than in *H. serum* and more closely resembles *H. crenatidens*. The nutrient foramen is located near the middle of the shaft, as it is in *H. crenatidens*. The linea aspera is broad proximally but not strongly marked on the posterior shaft of the bone. The distal end is broader than in *H. crenatidens*, with the inner condyle more developed posteriorly (extending past the outer). The patellar groove extends farther proximally. The intercondylar notch extends farther anteriorly than in *H. serum* or *H. crenatidens*. The fabella attachment is more strongly developed than in *H. crenatidens* or *H. serum*.

THE PATELLA

The patella (fig. 6.25A–C, table 6.41) is distinctly smaller than that of *Xenosmilus*. The proximal portion shows roughening for attachment of the quadriceps tendon. The distal apex of the patella is broader and more squared than in *H. crenatidens* and is triangular and pointed in *Xenosmilus* and *H. serum*. The articular surface with the femur is more rounded than in *H. crenatidens*.

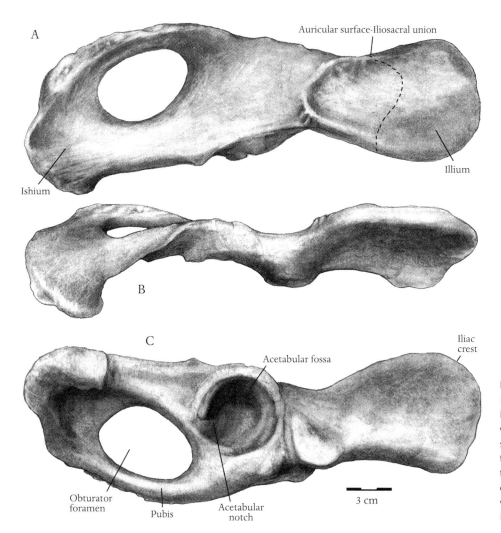

A

Auricular surface-Iliosacral union

Illium

Ishium

B

C

Iliac crest

Acetabular fossa

Obturator foramen

Pubis

Acetabular notch

3 cm

Figure 6.23. The right inominate (pelvis) of *Homotherium ischyrus*, IMNH 900-11862: *A*, medial view, demonstrating the auricular surface (the area of articulation to the sacrum) surrounded by the ridge and continued anteriorly by the dotted line; *B*, dorsal view; *C*, lateral view. (drawing Mary Tanner)

Table 6.39. Right inominate (pelvis) measurements (mm): *Homotherium ischyrus* IMNH 900-11862. Compared specimens (observed range), *H. serum* (Meade, 1961), TMM 933-*3231; *H. crenatidens* (Ballesio, 1963), FSL 210991. (*a* = approximate, *e* =estimate, + = damaged element, *Meade)

	H. ischyrus R	*H. ischyrus* L	**H. serum*	*H. serum*	*H.crenatidens*
Length from anterior end of ilium to posterior border of ischium	284.0e +	292.0	326.0	322.2–*326.0	300.0a
Greatest depth of ilium	81.0a e	82.0e +	107.0a	—	95.0
Length of pubic symphysis	—	72.0e +	—	—	88.0
Width of ischium measured from ischial tuberosity to posterior end of ischial symphysis	105.0e +	100.0	—	—	108.0
Diameter of acetabulum fossa measured at right angles to long axis of internal notch	45.6	52.3	47.0	—	50.0
Long diameter of obturator foramen	64.1	63.2	80.0	65.0–75.5	63.0
Greatest diameter of obturator foramen taken normal to long diameter	55.5	51.6	62.0	—	55.0
Greatest width at acetabulum measured between anteroposterior margins	73.9	74.2e +	—	90.1–86.6	—

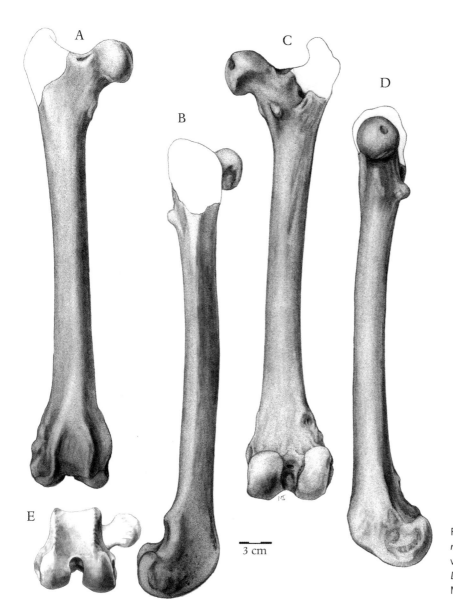

Figure 6.24. The right femur of *Homotherium ischyrus*, IMNH 900-11862: *A*, anterior view; *B*, lateral view; *C*, posterior view; *D*, medial view; *E*, distal end view. (drawing Mary Tanner)

3 cm

Table 6.40. Right femur measurements (mm): *Homotherium ischyrus* IMNH 900-11862. Compared specimens (observed range), *H. serum* (Meade, 1961; Rawn-Schatzinger, 1992), TMM 933-491, 582, 1570, 2140, 2284, *3231, 3498; *H. crenatidens* (Ballesio, 1963), FSL 210991; *Ischyrosmilus johnstoni* (Mawby, 1965), W.T. 1066, UCMP 66488. (*a* = approximate, *e* = estimate, *Meade, **W.T. 1066, ♦ = UCMP 66488)

	H. ischyrus R	*H. ischyrus* L	** H. serum* R	** H. serum* L	*H. serum*	*H. crenatidens* R	*H. crenatidens* L	**♦*I. johnstoni*
Greatest length from top of greater trochanter to distal condyles, measured parallel to long axis of femur	377.0	380.0e	353.0	351.0	323.3–*353.0	356.0	357.0	405.0a —
Transverse diameter of proximal end, outer face of greater trochanter to inner side of head, taken normal to median longitudinal plane	95.6	95.6e	92.0	93.0	81.7–94.4	89.0	90.0	95.0a —
Greatest anteroposterior diameter of head	40.8	41.0	42.8	42.4	40.3–42.9	41.0	41.0	42.9 43.2
Transverse diameter (mediolateral) of mid shaft	32.1	31.4	31.3	31.7	26.7–34.8	33.0	32.0	37.7 36.1a
Anteroposterior diameter at mid shaft	28.0	26.7	28.8	28.7	27.0–29.6	29.0	28.0	29.4 29.4
Greatest width of distal extremity	73.2	73.8	73.5	72.0	64.3–*73.5	70.0	70.0	73.9 —
Greatest anteroposterior diameter of the distal extremity at right angles to longitudinal axis of femur	70.0	70.0	68.7	71.0	63.9–*71.0	70.0	70.0	75.2 —
Greatest width of patellar groove (trochlea)	32.2	32.0	—	—	38.5–42.9	35.0	37.0	34.4 —
Greatest width of intercondylar notch	18.0	18.0	15.5	14.8	12.7–17.8	18.0	18.0	15.0 —
Greatest width of articular surface of medial (inner) condyle	28.7	28.7	32.5	32.0	*28.0–32.5	—	—	28.9 —

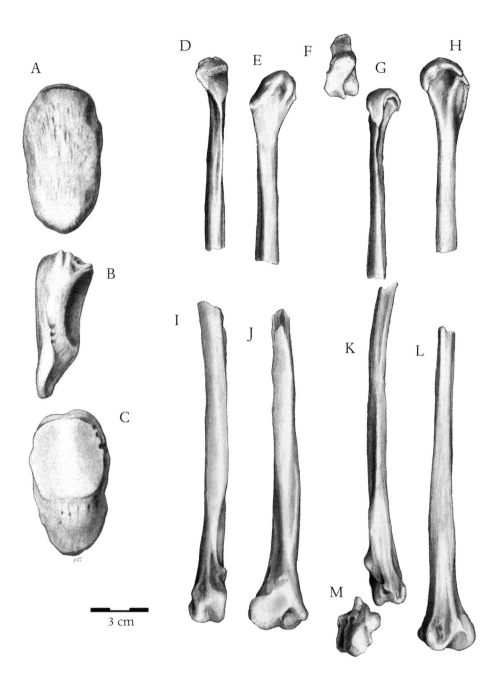

Figure 6.25. *Homotherium ischyrus*, IMNH 900-11862: *A–C*, left patella; *D–H*, the proximal unconnected portion of the right fibula; *I–L*, distal unconnected portion of the right fibula. Patellar views: *A*, anterior; *B*, lateral; *C*, posterior. Fibular views: *F*, proximal end; *D and I*, anterior; *E and J*, medial; *G and K*, posterior, *H and L*, lateral; *M*, distal end. (drawing Mary Tanner)

Table 6.41. Left patella measurements (mm): *Homotherium ischyrus* IMNH 900-11862. Compared specimens (observed range), *H. serum* (Meade, 1961; Rawn-Schatzinger, 1992), TMM 933-2519, 2942, *3231, 3458; *H. crenatidens* (Ballesio, 1963), FSL 210991. (*a* = approximate, + = damaged element, *Meade)

| | *H. ischyrus* | | **H. serum* | | *H. serum* | *H. crenatidens* | |
	R	L	R	L		R	L
Greatest proximodistal diameter (length)	62.3*a* +	67.0	67.0	66.8	59.3–*67.0	71.0	69.0
Greatest transverse (mediolateral) width	37.8	38.2	40.9	40.1	36.4–*40.9	37.0	39.0
Anteroposterior diameter through middle of articulating surface	22.4	22.3	—	—	22.9–24.9	21.0	21.0

Table 6.42. Right and left tibia measurements (mm): *Homotherium ischyrus* IMNH 900-11862. Compared specimens (observed range), *H. serum* (Meade, 1961; Rawn-Schatzinger, 1992), TMM 933-482, 580, 671, 723, 1641, 2741, *3231; *Ischyrosmilus johnstoni* (Mawby, 1965), UCMP 66489; *H. crenatidens* (Ballesio, 1963), FSL 210991. (*Meade, **UCMP 66489)

	H. ischyrus		*H. serum		H. serum	H. crenatidens		**I. johnstoni
	R	L	R	L		R	L	
Greatest length measured parallel to long axis	321.0	320.0	297.0	297.0	285.5–*297.0	308.0	315.0	345.0
Greatest transverse diameter (mediolateral) of proximal end	74.0	74.0	76.0	—	67.4–*76.0	69.0	—	78.7
Transverse diameter at mid shaft	26.5	26.1	27.5	28.0	26.3–29.3	29.0	30.0	30.5
Greatest transverse diameter of distal end	50.9	51.8	54.5	55.6	47.6–*55.6	50.0	51.0	52.0
Greatest anteroposterior diameter at distal end	49.0	40.0	39.0	38.0	36.6–*39.0	41.0	40.0	36.0
Greatest anteroposterior diameter at proximal end	80.0	80.0	—	—	62.8–64.4	80.0	—	—
Greatest anteroposterior diameter at mid shaft	28.2	31.0	—	—	25.9–29.5	34.0	34.0	—

THE TIBIA

The tibia (fig. 6.26, table 6.42) is more elongate than in *H. serum* or *H. crenatidens*. The shaft is slightly bowed, and the tibial crest is short. On the proximal end, there is a prominent tubercle at the posterior edge of the intertubercular sulcus that serves as the distal attachment of the posterior cruciate ligament. In *Homotherium,* the external condyle is larger than the internal while in *Smilodon,* the opposite is true. In the Idaho homothere, the tubercle for the distal attachment of the iliotibial tract and some of the fibers of the vastus lateralis muscle is prominently enlarged and causes the proximal aspect of the tibia to appear squared as opposed to the triangular shape in *H. crenatidens. H. serum* also shows a tendency to enlargement of this feature. The tibial tuberosity in the Idaho homothere for the attachment of the patellar ligament is narrower than in *Xenosmilus* and more proximal than in *Smilodon.* These features are adaptations for increased cursoriality. The depression for the insertion of the sartorius, gracilis, and semitendinosus muscles on the lateral anterior proximal face of the tibia of the Idaho homothere is shallower than that in *Xenosmilus* and more closely resembles the condition in *Panthera.* Distally, the medial trochlea is much deeper than the lateral. The knob-like projections proximal to the distal articulation that occur in *H. serum* (Rawn-Schatzinger, 1992, fig. 33D) appear to be unique to that species. The malleolar process in the Idaho homothere and *H. crenatidens* is about level with the rest of the distal end of the bone and extends farther distally in *H. serum.*

THE FIBULA

The right fibula (fig. 6.25D–M, table 6.43) lacks a clear contact between the proximal and distal segments, but when fitted to the tibia, only a small segment of the shaft appears to be missing. The proximal end of the fibula is rounded and bent toward the tibia. The shaft of the fibula fits tightly to the contour of the tibia.

THE ANKLE (TARSAL BONES)

Substantial portions of both hind feet in the Idaho homothere are preserved, with the right metatarsals (II–V) and ankle bones, ectocuneiform, and mesocuneiform in articulation. The metatarsals in the naturally articulated Birch Creek specimen are closely appressed, as is also the case in all living cats, demonstrating that the widely spread spatial arrangement illustrated by other authors (Meade, 1961; Rawn-Schatzinger, 1992) is probably incorrect.

THE CALCANEUM

The calcaneum (fig. 6.27, table 6.44) is more slender and elongate than in the European *Homotherium* or *H. serum.* There is a long posterior process on the calcaneum as in *H. serum.* The astragular facet is more elongated posteriorly than in *Smilodon* and resembles *H. crenatidens.* The internal astragular facet is also very similar to that of *H. crenatidens,* except that it is more extended medially and forms a continuous facet as in *Smilodon* and the other homotheres, but in *Panthera* it is divided into distinct posterior and anterior parts. The cuboid facet is large and shaped about as it is in *Smilodon.* In *Homotherium serum,* it is more compressed, and in *H. crenatidens* it is more rounded. The navicular facet is similar in position to that in *H. crenatidens.* The navicular facet is also found in *Smilodon* but according to Ballesio (1963) not in modern cats.

THE ASTRAGALUS

The tibial face of the astragalus (fig. 6.28A–D, table 6.45) is narrower and longer than in *Panthera atrox* and more so than in *Smilodon fatalis,* although it is a little more similar to the latter. It is more like those of the other homotheres although more compressed than in either *Homotherium crenatidens* or *H. serum.* The fibular edge is nearly straight as in *H. crenatidens,* and it is slightly curved inward in *H. serum.* The outer trochlea of the tibial facet extends slightly posterior to the

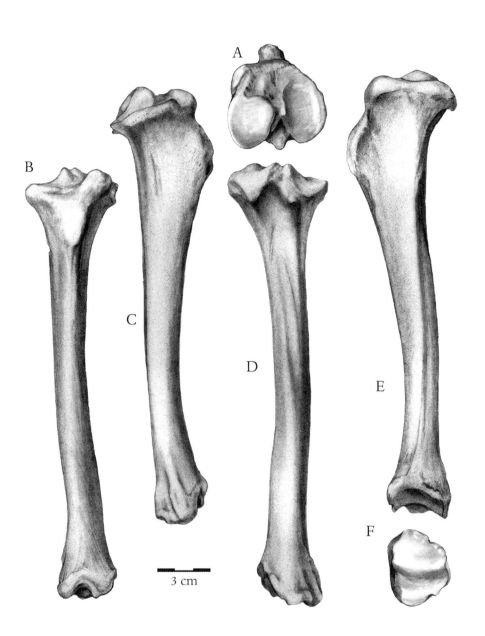

Figure 6.26. The left tibia of *Homotherium ischyrus*, IMNH 900-11862: *A*, proximal end view; *B*, anterior view; *C*, medial view; *D*, posterior view; *E*, lateral view; *F*, distal end view. (drawing Mary Tanner)

Table 6.43. Right and left fibula measurements (mm): *Homotherium ischyrus* IMNH 900-11862. Compared specimens (observed range), *H. serum* (Meade, 1961; Rawn-Schatzinger, 1992), TMM 933-761, 762, 2084, *3231, 3033, 4058; *H. crenatidens* (Ballesio, 1963), FSL 210991. (*a* = approximate, *e* = estimate, + = damaged element, *Meade)

| | *H. ischyrus* | | *H. serum* | | *H. serum* | *H. crenatidens* |
	R	L	R	L		R
Greatest length	294.0*a* +	291.0*a* +	287.0	287.2	*287.2–*287.3	280.0
Greatest anteroposterior diameter of proximal end	24.8	24.9*e*	26.0	22.2	*22.2–25.7	—
Greatest anteroposterior diameter of distal end	28.3	28.6	29.0	29.1	26.9–30.7	28.0
Greatest transverse diameter (mediolateral) of distal end	15.0	15.1	14.0	14.0	*14.0–*14.0	26.0
Anteroposterior diameter at mid shaft	7.7*e*	10.0*e* +	8.5	—	11.4–12.3	—
Transverse diameter of mid shaft	8.7*e* +	11.6*e* +	10.5	—	8.8–10.8	—

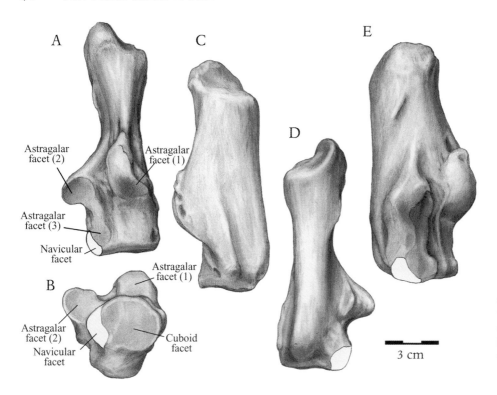

Figure 6.27. The left calcaneum of *Homotherium ischyrus*, IMNH 900-11862: *A*, proximal (astragalar) view; *B*, distal end view; *C*, lateral view; *D*, posterior (ventral) view; *E*, medial view. (drawing Mary Tanner)

Table 6.44. Left calcaneum measurements (mm): *Homotherium ischyrus* IMNH 900-11862. Compared specimens (observed range), *H. serum* (Meade, 1961; Rawn-Schatzinger, 1992; Mawby, 1965), TMM 933-106, 107, 108, 148, 1079, ◆2139, 2304, 2623, 2938, *3231; *H. crenatidens* (Ballesio, 1963), FSL 210991. (*Meade, **Rawn-Schatzinger, ◆Mawby)

	H. ischyrus R	*H. ischyrus* L	**H. serum*	*H. serum*	*H. crenatidens*
Greatest length	—	93.9	85.0	74.8–*85.0	85.0
Greatest width measured across astragalar facets As¹ and As²	40.4	40.1	37.5	33.0–*37.5	38.0
Greatest width across cuboid surface measured from astragalar facet As³ to lateral (outer) side	33.9	27.7	29.7	27.1–29.9	—
Greatest anteroposterior diameter of lateral face measured normal (perpendicular) to plantar border and astragalar facet As¹ (neck)	—	41.4	**◆45.0	39.9–45.0	45.0
Greatest width across cuboid facet measured from medial (inner) border of facet (As³) to lateral side	29.8	24.7	**29.7	26.9–29.9	—
Greatest mediolateral width of calcaneal tuberosity measured normal to plantar surface	—	28.0	—	26.8–30.7	—

inner but less so than in *Smilodon*. The tibial trochleae are rather high, separated by a distinct groove, and parallel to each other. When compared to *H. serum* and *H. crenatidens*, this increase in articular surface area with the tibia reflects a more cursorial adaptation. The neck is shorter and narrower than in *H. serum*, resembling more closely *H. crenatidens*, and the navciular attachment is broader than in *H. serum*. These differences are accentuated in *Xenosmilus*. In the Idaho homothere, the outer calcaneal attachment is larger than the inner, and they are about equal in size in *H. crenatidens* and *H. serum*. The inner tibial trochleae in *P. atrox* have a distinct medial notch that is missing from *Homotherium*, *Xenosmilus*, and *Smilodon*.

THE NAVICULAR

The navicular (fig. 6.28E–H, table 6.46) is small, as in *Homotherium*, but is large in *Xenosmilus*. The astragular facet is deeply cupped as it is in *Xenosmilus* and swings upward as in *Panthera*. *Smilodon* has a more flat dorsal outline, as does *H. serum*. The cuboid facet is large and round in the Idaho specimen, as in other *Homotherium*. In dorsal view, the navicular is roughly rectangular, as it is in *H. crenatidens*, and is more triangular in *H. serum*. It is more similar to *Panthera* than *Smilodon* in this respect. The ectocuneiform and mesocuneiform facets are subequal in size, with the ectocuneiform slightly larger. The entocuneiform and mesocuneiform facets are continu-

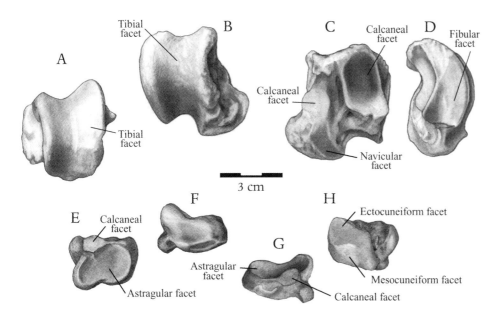

Figure 6.28. The right astragalus and left navicular of *Homotherium ischyrus*, IMNH 900-11862. The astragalus: *A*, anterior view; *B*, tibial (proximal) view; *C*, posterior (ventral) view; *D*, lateral view. The navicular: *E*, proximal view; *F*, medial view; *G*, lateral view; *H*, distal view. (drawing Mary Tanner)

Table 6.45. Right astragalus measurements (mm): *Homotherium ischyrus* IMNH 900-11862. Compared specimens (observed range), *H. serum* (Meade, 1961; Rawn-Schatzinger, 1992), TMM 933-208, 2150, 2296, 2419, 2977, *3231; *H. crenatidens* (Ballesio, 1963), FSL 210991. (*Meade)

| | *H. ischyrus* | | **H. serum* | *H. serum* | *H. crenatidens* |
	R	L			
Greatest length	48.6	49.6	52.5	49.3–56.1	49.0
Greatest width	46.9	46.2	41.0	33.9–*41.0	42.0
Least width of neck	—	18.7	22.0	17.5–*22.0	—
Greatest diameter of head	29.3	29.7	29.6	35.4–30.5	—
Length of neck	—	7.7	—	6.0–7.2	—
Width of groove of interosseous ligament	6.7	6.4	—	9.1–10.1	—

Table 6.46. Right and left navicular measurements (mm): *Homotherium ischyrus* IMNH 900-11862. Compared specimens (observed range), *H. serum* (Rawn-Schatzinger, 1992), TMM 933-835, 2618, 3617; *H. crenatidens* (Ballesio, 1963), FSL 210991.

| | *H. ischyrus* | | *H. serum* | *H. crenatidens* | |
	R	L		R	L
Greatest dorsoplantar length	36.7	36.6	34.7–36.1	39.0	38.0
Greatest transverse diameter (mediolateral)	26.5	28.3	29.6–31.2	30.0	31.0
Greatest proximodistal diameter	15.3	15.2	—	17.0	17.0

ous, but they are separated in *Panthera* and *Xenosmilus*. The medial face bears a single dorsal facet that articulates with the calcaneum, similar to that found in *Xenosmilus* and unlike the double articulation in *H. serum*.

THE ENTOCUNEIFORM

The entocuneiform was not recovered.

THE ECTOCUNEIFORM

The ectocuneiform and mesocuneiform (fig. 6.29A–D) are in articulation, with the second and third metatarsals obscuring their distal surfaces. The ectocuneiform (fig. 6.29A–D, table 6.47) is large and more rectangular than in *Smilodon* or *Panthera*, resembling the other homotheres in this respect. It has a large plantar process that bears a large lateral groove for the attachment of the longitudinal tendon of the fibularis

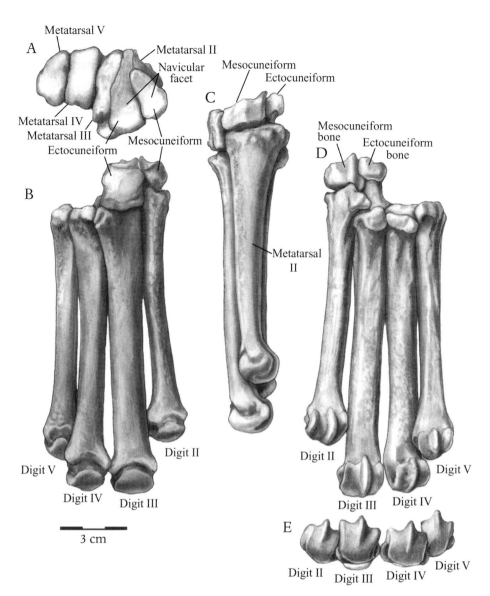

Figure 6.29. The articulated right metatarsals (Mt II through Mt V) of *Homotherium ischyrus*, IMNH 900-11862, with an articulated ectocuneiform and mesocuneiform. *A*, proximal end view; *B*, dorsal view; *C*, medial view; *D*, ventral view; *E*, distal end view. (drawing Mary Tanner)

Table 6.47. Right ectocuneiform measurements (mm): *Homotherium ischyrus* IMNH 900-11862. Found in articulation with mesocuneiform and MT II and III. Compared specimens (observed range), *H. serum* (Rawn-Schatzinger, 1992), TMM 933, 3256, 4308, 2115, 3001; *H. crenatidens* (Ballesio, 1963), FSL 210991. (*e* = estimate)

	H. ischyrus R	L	*H. serum*	*H. crenatidens*
Greatest dorsoplantar length	36.4	—	35.0–36.8	37.0
Greatest proximodistal diameter	18.7	—	15.1–17.6	18.0
Greatest width across metatarsal III articulation	20.5	—	17.0–19.6	20.0
Greatest dorsoplantar length of metatarsal III articular facet	30.1*e*	—	—	—
Greatest length of navicular facet	24.6	—	—	—

longus muscle. This groove is not as hook-shaped as in *Smilodon*, but more so than in *H. crenatidens* or *H. serum*. The lateral surface of the ectocuneiform bears a single poorly developed triangular facet on the anteroventral margin and a second more dorsal flattened area. In *H. crenatidens,* this region is occupied by a single facet. It is more compressed posteriorly than in *H. serum* and more like *H. crenatidens*.

THE MESOCUNEIFORM

The mesocuneiform (fig. 6.29A–D, table 6.48) is fairly large and is an elongate ovoid, with a mildly concave subtriangular proximal facet. It bears a small circular, slightly concave facet on its medial face that probably articulates with a small hemispherical sesamoid.

THE CUBOID

The right cuboid (fig. 6.30A–F, table 6.49) is tall and narrow as compared to *Smilodon* or *Panthera* and resembles that in

Table 6.48. Right mesocuneiform measurements (mm): *Homotherium ischyrus* IMNH 900-11862. Found in articulation with MT II and ectocuneiform.

	H. ischyrus	
	R	L
Greatest long diameter as measured along outer surface	25.3	—
Greatest proximodistal diameter	12.0	—
Greatest width of articulating surface for metatarsal II	11.8	—

H. crenatidens. That of *H. serum* is more like *Smilodon*. The surface of the proximal facet of the cuboid is gently convex and is broader transversely than anteroposteriorly. The dorsal surface of the cuboid terminates in a scalloped margin with the bone projecting farther ventrally and anteromedially than anterolaterally. Medially, there is no articular facet

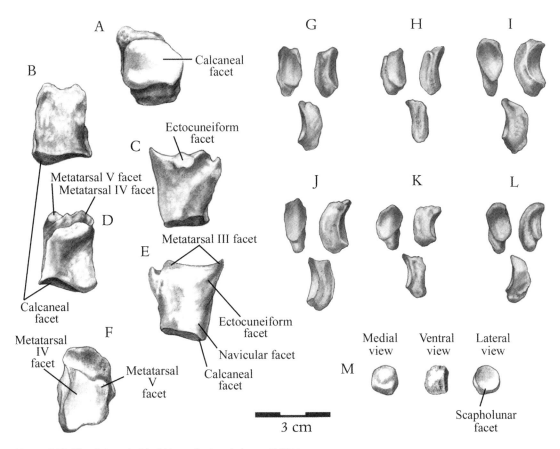

Figure 6.30. The right cuboid of *Homotherium ischyrus*, IMNH 900-11862: *A,* proximal end view; *B,* dorsal view; *C,* lateral view; *D,* ventral view; *E,* medial view; *F,* distal end. *G–M,* the sesamoids. None of the sesamoids found were articulated, but these elements are suggested to be from: *G,* the right side, right (Mt IV); *H,* the right side, right (Mt V); *I,* the right side, right (Mt III); *J,* the left side, left (Mt III); *K,* the left side, left (M II); *L,* the right side, right (Mt II). *M,* The radial sesamoid from the right scapholunar in medial view (*left*); ventral view (*center*); and lateral view (*right*). (drawing Mary Tanner)

Table 6.49. Right cuboid measurements (mm): *Homotherium ischyrus* IMNH 900-11862. Compared specimens (observed range), *H. serum* (Rawn-Schatzinger, 1992), TMM 933-629, 1304, 9251; *H. crenatidens* (Ballesio, 1963), FSL 210991.

	H. ischyrus R	*H. ischyrus* L	*H. serum*	*H. crenatidens*
Greatest length (proximodistal)	31.3	—	26.2–26.5	31.0
Greatest transverse diameter (mediolateral)	25.3	—	24.4–27.4	22.0
Greatest dorsoplantar diameter	30.5	—	29.8–33.7	28.0

for the ectocuneiform, although the medial surface of the cuboid is quite rough and the facet may be absent as a result of corrosion of the bone. On the ventral face, just below the facet for Metatarsals IV and V, there is a very large interosseous groove that runs diagonally from the dorsolateral surface of the bone to the medial side of the ventrodistal face. This groove would have served as the passageway for the tendon of the fibularis longus muscle, which would assist in stabilizing the foot and preventing over-inversion. The metatarsal facets are more compressed than in *H. serum,* and the facet for metatarsal IV is much more elongated than that for metatarsal V.

THE PES (HIND FOOT)
METATARSAL I

Metatarsal I was not found.

METATARSAL II

Metatarsal II (figs. 6.29, 6.31A–F; table 6.50) is more compressed than in *Panthera* but contrasts with *Smilodon* in being elongate. The lateral margin of the articular surface is straight and inset medially. The anterodorsal edge is rounded as in *Smilodon* rather than straight as in *Panthera*. The shaft diverges mediodistally. The proximal articular facet is narrower than in either *Smilodon* or *Panthera,* and the posterior process is larger. The posterior keel on the distal articular surface is pronounced and extends above the trochlear ridges as in *H. crenatidens.*

METATARSAL III

The proximal articular surface of metatarsal III (figs. 6.29, 6.31G–I; table 6.51) is more waisted than in *H. crenatidens.* The anterior facet has straight parallel sides and is not rounded or as wide as in *Panthera*. The posterior facet is triangular in *Smilodon* and *Panthera* and square in the homotheres. The bone is more slender than in *Panthera,* and the posterior keel on the distal articulation does not extend as far ventrally. The proximal articular surface is essentially horizontal in the Idaho *Homotherium* but slopes anteriorly in *Panthera* and posteriorly in *Smilodon.*

METATARSAL IV

The proximal articular facet of metatarsal IV (figs. 6.29, 6.32A–F; table 6.52) is broader posteriorly in the homotheres and anteriorly in *Smilodon* and *Panthera*. It is also more slender in the homotheres. The lateral facets are elongated front to back rather than up and down as in *Smilodon* and *Panthera*. The posterior keel on the distal articulation terminates on a level with the trochlea but extends more proximally in *Smilodon* and *Panthera*.

METATARSAL V

Metatarsal V (figs. 6.29, 6.32G–L; table 6.53) is straighter and more slender in the Idaho homothere than in either *Panthera* or *Smilodon*. Its proximal articular surface is narrower than in *Smilodon* or *H. crenatidens.* The distal process is elongated. The general shape is more like *Smilodon* than *Panthera*. There is a distinct distal medial ridge on the posterior edge of the shaft that is present on Ballesio's figure (Ballesio, 1963). The distal articulation seems broader and more rounded.

THE PHALANGES

Most of the phalanges present (figs. 6.30G–L, 6.33; tables 6.54, 6.55) appear to be elements of the rear feet. These phalanges and sesamoids were not found in articulation with any metapodials, so it is difficult to discern whether they belong to the manus or pes, but from their size and shape, most phalanges appear to be from the hind foot. They are robust and fairly long but not nearly as robust as those of *Smilodon* or *Panthera*. The shafts are fairly straight, with only a slight dorsoventral curvature. They have a small depression distad of the distal articular surface. The shafts appear waisted, and the proximal breadth is significantly broader than the transverse diameter of the shaft. The claws were hooded, but the thin margins of the claw hoods were not preserved with the distal phalanges.

THE SECOND DIGIT FIRST PHALANX The first phalanx for digit II (fig. 6.33A–F, table 6.55) is the smaller of the hind foot first phalanges. It is strongly bowed, and the proximal end is about as high as it is wide. The proximal articular surface is inclined laterally. There is a large tubercle on the radial side of the

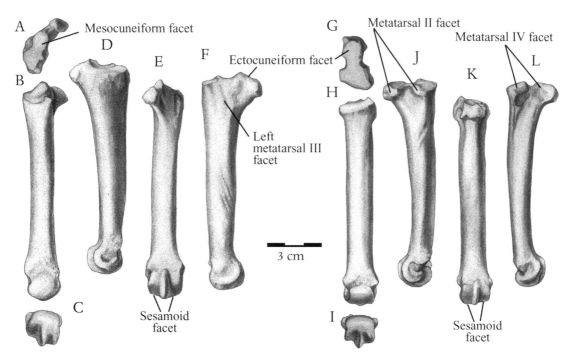

Figure 6.31. The left second metatarsal (Mt II) and left third metatarsal (Mt III) of *Homotherium ischyrus*, IMNH 900-11862. Second metatarsal (Mt II): *A,* proximal end view; *B,* dorsal view; *C,* distal end view; *D,* medial view; *E,* ventral view; *F,* lateral view. Third metatarsal (Mt III): *G,* proximal end view; *H,* dorsal view; *I,* distal end view; *J,* medial view; *K,* ventral view; *L,* lateral view. (drawing Mary Tanner)

Table 6.50. Left metatarsal II measurements (mm): *Homotherium ischyrus* IMNH 900-11862. Compared specimens (observed range), *H. serum* (Rawn-Schatzinger, 1992), TMM 933-602, 603, 721, 1234, 1448, 2627, 3449, 4503; *H. crenatidens* (Ballesio, 1963), FSL 210991. (+ = damaged element)

| | *H. ischyrus* | | *H. serum* | *H. crenatidens* |
	R	L		
Greatest length	109.4	108.1	102.6–118.0	105.0
Greatest transverse diameter (mediolateral) of proximal end	15.6	15.0	14.8–20.1	14.0
Greatest dorsoventral (dorsoplantar) diameter of proximal end	31.1	31.9	14.1+–31.6	—
Transverse diameter at mid shaft	12.6	10.7	12.7–17.6	14.0
Dorsoventral diameter at mid shaft	15.2	15.2	13.8–19.3	15.0
Greatest transverse diameter at distal end of shaft	18.2	19.0	18.0–21.1	21.0
Greatest dorsoventral dimension of distal end	19.7	19.0	19.9–22.3	20.0

proximal aspect of the phalanges that would serve as insertion for the lumbrical muscles that assist in flexion of the tarsometatarsal joints of the foot. The outer articular ridge of the distal end swings laterally, and the inner ridge is straight.

THE SECOND DIGIT SECOND PHALANX The second phalanx (fig. 6.33G–L) is strongly bowed laterally for the flexor tendon of the claw.

THE SECOND DIGIT THIRD PHALANX The third phalanx (fig. 6.33M–Q) that bears the claw is short, flat, and hooded, with a distinct ventral articular shelf.

THE THIRD DIGIT FIRST PHALANX The third digit first phalanx (fig. 6.33A–F, table 6.55) is the largest. It is strongly bowed and relatively slender with a single rounded articular surface.

THE THIRD DIGIT SECOND PHALANX The second phalanx (fig. 6.33G–L) is short and strongly curved for the flexor tendon.

THE THIRD DIGIT THIRD PHALANX The third phalanx bears the claw (fig. 6.33M–Q) and is similar to the third phalanx of the second digit.

Table 6.51. Left metatarsal III measurements (mm): *Homotherium ischyrus* IMNH 900-11862. Compared specimens (observed range), *H. serum* (Meade, 1961; Rawn-Schatzinger, 1992), TMM 933-1038, 2137, 3243, 3452, 3947, 3949, 3951, 4503; *H. crenatidens* (Ballesio, 1963), FSL 210991. (*e* = estimate)

	H. ischyrus R	*H. ischyrus* L	*H. serum*	*H. crenatidens*
Greatest length	122.8	121.2	105.6–123.0	120.0
Greatest transverse diameter (mediolateral) of proximal end	21.0*e*	21.1	16.2–22.6	23.0
Greatest dorsoventral (dorsoplantar) diameter of proximal end	36.2	34.5	27.0–31.0	—
Transverse diameter at mid shaft	15.0	15.0*e*	14.3–17.2	16.0
Dorsoventral diameter at mid shaft	14.3*e*	15.5	15.8–18.9	15.0
Greatest transverse diameter at distal end of shaft	22.0	21.7	18.2–20.9	22.0
Greatest dorsoventral dimension of distal end	21.7	20.6	21.6–23.0	20.0

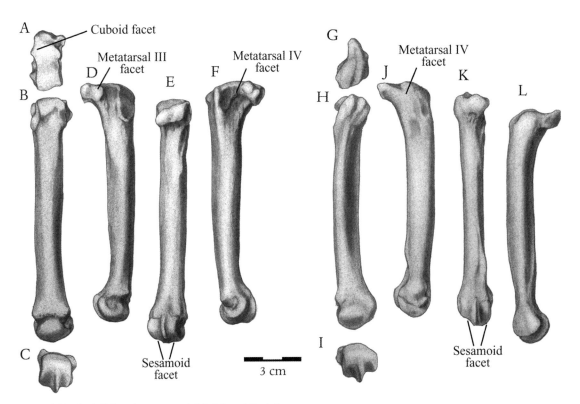

Figure 6.32. The left fourth metatarsal (Mt IV) and the left fifth metatarsal (Mt V) of *Homotherium ischyrus*, IMNH 900-11862. The fourth metatarsal (Mt IV): *A*, proximal end view; *B*, dorsal view; *C*, distal end view; *D*, medial view; *E*, ventral view; *F*, lateral view. The fifth metatarsal (Mt V): *G*, proximal end view; *H*, dorsal view; *I*, distal end view; *J*, medial view; *K*, ventral view; *L*, lateral view. (drawing Mary Tanner)

Table 6.52. Left metatarsal IV measurements (mm): *Homotherium ischyrus* IMNH 900-11862. Compared specimens (observed range), *H. serum* (Meade, 1961; Rawn-Schatzinger, 1992), TMM 933-318, 3351, 3450, 3950, 4503; *H. crenatidens* (Ballesio, 1963), FSL 210991.

	H. ischyrus R	*H. ischyrus* L	*H. serum*	*H. crenatidens*
Greatest length	122.3	120.5	104.2–113.8	121.0
Greatest transverse diameter (mediolateral) of proximal end	15.5	15.2	18.4–22.0	17.0
Greatest dorsoventral (dorsoplantar) diameter of proximal end	29.0	29.0	23.1–28.8	25.0
Transverse diameter at mid shaft	14.0	14.6	10.3–13.8	15.0
Dorsoventral diameter at mid shaft	13.2	13.9	14.1–16.7	15.0
Greatest transverse diameter at distal end of shaft	21.3	19.2	16.2–19.8	13.0

Table 6.53. Left metatarsal V measurements (mm): *Homotherium ischyrus* IMNH 900-11862. Compared specimens (observed range), *H. serum* (Meade, 1961; Rawn-Schatzinger, 1992), TMM 933-1544, 2135, 2625; *H. crenatidens* (Ballesio, 1963), FSL 210991.

	H. ischyrus R	*H. ischyrus* L	*H. serum*	*H. crenatidens*
Greatest length	108.8	106.7	91.1–107.0	108.0
Greatest transverse diameter (mediolateral) of proximal end	12.6	11.3	18.2–21.5	17.0
Greatest dorsoventral (dorsoplantar) diameter of proximal end	22.4	22.4	18.8–23.4	22.0
Transverse diameter at mid shaft	10.0	10.5	10.8–11.5	10.0
Dorsoventral diameter at mid shaft	11.4	10.9	15.2–15.5	14.0
Greatest transverse diameter at distal end of shaft	16.9	17.2	16.3–17.1	19.0
Greatest dorsoventral dimension of distal end	18.0	16.0	17.5–17.7	20.0

Table 6.54. Phalangeal sesamoids measurements (mm): *Homotherium ischyrus* IMNH 900-11862. Compared specimens *Smilodon fatalis* 2037-1, 2935-2 (Merriam and Stock, 1932).

	H. ischyrus	*S. fatalis*
Greatest length	15.2–19.8	27.3–32.0
Greatest width	9.0–9.8	11.1–12.4

THE FOURTH DIGIT FIRST PHALANX The first phalanx of the fourth digit (fig. 6.33A–F, table 6.55) is slightly smaller and more slender than that of the third. Its proximal articular surface is more divided and has a stronger carina than does the third.

THE FOURTH DIGIT SECOND PHALANX The fourth digit second phalanx (fig. 6.33G–L) is a little longer and more slender than that of the third, but it is distinctly bowed laterally for the flexor tendon.

THE FOURTH DIGIT THIRD PHALANX The third phalanx bears the claw (fig. 6.33M–Q) and is smaller but otherwise similar to the claw for the third digit.

THE FIFTH DIGIT FIRST PHALANX The first phalanx for the fifth digit (fig. 6.33A–F, table 6.55) is about as long as that for the fourth but distinctly more slender and curved inward.

THE FIFTH DIGIT SECOND PHALANX The second phalanx (fig. 6.33G–L) is shorter, straighter, and relatively broader than the other second phalanges.

THE FIFTH DIGIT THIRD PHALANX The third phalanx (fig. 6.33M–Q) bears the claw and is the smallest in the series. The third phalanges of *H. serum* are straighter and less modified for claw retraction than in the Idaho specimen.

CONCLUSIONS

The Idaho specimen can be assigned to the genus *Homotherium* on the basis of its broad forehead, enlarged narial openings, and elongated limbs. It resembles the other North American Pliocene *Homotherium*, which have often been grouped together as an endemic genus, *Ischyrosmilus*, in having larger and more elongated upper canines than in the

Digit II   Digit III   Digit IV   Digit V

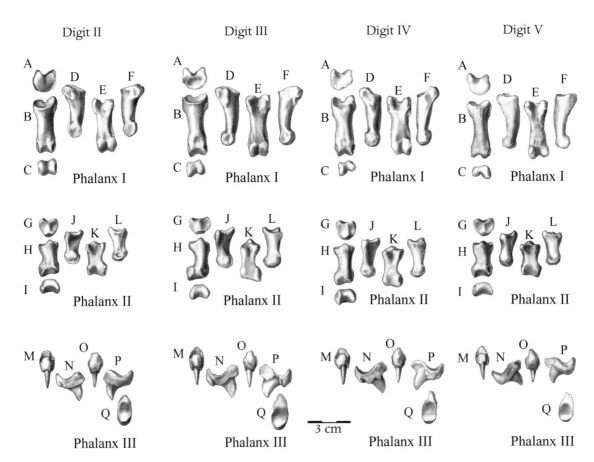

Figure 6.33. Digits II through V of the right pes, 1st, 2nd, and 3rd phalanges, of *Homotherium ischyrus*, IMNH 900-11862. *A–F*, the first phalanx of each digit: *A*, proximal end view; *B*, dorsal view; *C*, distal end view; *D*, lateral view; *E*, ventral view; *F*, medial view. *G–L*, the second phalanx of each digit: *G*, proximal view; *H*, dorsal view; *I*, distal view; *J*, lateral view; *K*, ventral view; *L*, medial view. *M–Q*, the third phalanx of each digit: *M*, dorsal view; *N*, lateral view; *O*, ventral view; *P*, medial view; *Q*, distal end view. (drawing Mary Tanner)

other species of *Homotherium*. In terms of overall size and configuration of the lower P$_4$–M$_1$, it resembles the holotype ramus of *I. ischyrus* enough that it can be reasonably assigned to that species, and as such, we have used it to ascertain the skeletal characters of the typical North American Blancan homothere. It resembles the European species *H. crenatidens* sufficiently to remove any doubt that it is congeneric with that species but differs enough that we regard it as a separate species. Beside the more elongate canines, it has a small M^1, missing from *H. crenatidens* described by Ballesio (1963), and differs from that species postcranially in a number of features that we have listed, including having longer transverse processes on the axis vertebra. Both of the Pliocene species *H. crenatidens* and *H. ischyrus* differ from the younger taxa, *H. latidens* in Europe and *H. serum* in North America, in the absence of a distinct anterior pocket in the anterior edge of

the masseteric fossa on the ramus. This is a derived feature that might indicate that the two younger species share a closer common ancestor to each other than to the Pliocene forms. The skeleton of the European *H. latidens* is not adequately known, but that of the American *H. serum* shows a tendency toward shortening of the legs. The legs and feet in the Idaho homothere are distinctly adapted toward a running lifestyle beyond that seen in lions, and the cat presumably would have been capable of pursuit of very cursorial prey such as horses. A nearly complete skeleton of a large Pliocene horse *Equus* ("*Plesippus stenonis anguinus*"), was found not too far from the Idaho homothere skeleton and represents a good example of what its preferred prey might have been. This species of horse also occurs in European localities that have also produced *H. crenatidens,* which suggests that those localities are roughly contemporary with the Idaho locality.

Table 6.55. Right pes, first phalanges digits II, III, IV, V measurements (mm): *Homotherium ischyrus* IMNH 900-11862. Compared specimen, *H. crenatidens* (Ballesio, 1963), FSL 210991.

	H. ischyrus				H. crenatidens			
	II	III	IV	V	II	III	IV	V
Greatest length	41.3	43.0	40.4	39.8	41.0	43.0	39.0	37.0
Greatest dorsoventral (dorsoplantar) diameter of proximal end	16.8	16.8	17.9	17.5	17.0	19.0	17.0	16.0
Greatest transverse diameter (mediolateral) of proximal end	17.0	17.5	16.3	16.0	18.0	18.0	17.0	16.0
Dorsoventral diameter at mid shaft	10.2	10.6	11.1	10.8	9.0	11.0	10.0	10.0
Transverse diameter at mid shaft	10.2	11.7	11.1	11.5	12.0	13.0	13.0	10.0
Greatest transverse diameter at distal end of shaft	13.8	12.9	13.8	13.4	15.0	17.0	15.0	14.0
Greatest dorsoventral diameter of distal end	10.6	10.9	10.6	10.5	11.0	13.0	12.0	10.0

SUMMARY

A nearly complete skeleton of *H. ischyrus* from the Late Pliocene of Idaho is compared to that of *H. crenatidens* from Europe and *H. serum* in the Late Pleistocene of North America. These comparisons show that it is clearly distinct from both these species, although it is more similar to the Pliocene form from Europe than the Pleistocene form, once again demonstrating the similarity in age between the Villafranchian of Europe and the Blancan of North America. *H. ischyrus* was extremely long-legged and probably more adapted for running than the Late Pleistocene *H. serum*.

Appendix: Measurement Comparison Tables

The tables in this appendix are comparison measurement tables of the skull and mandible of *Homotherium ischyrus*, *Ischyrosmilus crusafonti*, *Homotherium serum*, *Homotherium crenatidens*, and *Ischyrosmilus johnstoni* used in this study. The provenience identifications are listed under the respective table headings. All measurements were taken using dial or digital calipers from the actual specimens whenever possible; otherwise they were taken directly from the literature as noted in the caption heading. Measurements act as a reference to distinguish multiple variations among taxa, not only between genera, but even at the species level. These are needed whenever new genera or species are described.

Table 6.A.1. Skull measurements of *H. ischyrus*, IMNH 900-11862; *I. crusafonti*, AMNH 95297; *H. serum*, TMM 933-3582 and TMM 933-3231; *H. crenatidens*, FSL 210991; and *I. johnstoni*, W.T. 1860. (*e* = estimate, — = dimension not available)

No.	Measurement Points		*H. ischyrus* Martin, Naples, Babiarz, this vol. IMNH 900-11862	*I. crusafonti* Martin, Naples, Babiarz, this vol. AMNH 95297	*H. serum* Meade, 1961 TMM 933-3582	*H. serum* Meade, 1961 TMM 933-3231	*H. crenatidens* Ballesio, 1963 FSL 210991	*I. johnstoni* Mawby, 1965 W.T. 1860
1	1 & 2	Length from anterior end of alveolus of I^4 to posterior end of condyles	281.0	246.0	289.0	327.0	275.0	—
2		Basal length from anterior end of alveolus of I^4 to inferior notch between condyles	264.0	223.0	266.0	315.0	—	—
3	1 & 4	Length from anterior end of alveolus of I^1 to apex of occipital crest	297.0	265.0	324.0	262.0	—	—
4		Length from anterior end of alveolus of I^1 to anterior end of posterior nasal opening	145.0	113.0	142.0	—	—	—
5	1 & 6	Length of palate from anterior end of alveolus of I^1 to posterior border of palatine (anterior border of pterygoid fossa)	160.0*e*	146.0	—	—	—	—
6	7 & 2	Length from posterior end of glenoid fossa to posterior end of condyles	97.0	83.2	108.0	94.0	—	—
7	8 & 8.1	Maximum length of nasal bone (anterior posterior distance medial edge)	66.0*e*	56.7	—	—	—	—
8	8 & 9	Length of anterior nares (anterior nasal aperture)	52.0*e*	40.7	—	—	44.0	—
9	10 & 11	Width of anterior nares (anterior nasal aperture)	40.3	53.7	52.0	—	41.0	46.0*e*
10	12 & 13	Maximum width across the muzzle at canines	70.8	75.5	94.0	91.0	—	—
11	14 & 15	Minimum width between superior borders of orbits	80.8	71.7	93.0	—	—	—
12	16 & 17	Width across postorbital processes	124.8	87.6	—	—	—	—
13	18 & 19	Minimum width of postorbital constriction	67.0	60.5	—	—	—	—
14	20 & 21	Maximum width across zygomatic arches	181.0	153.0	177.0	—	185.0	—
15	22 & 23	Minimum anterior palatal width between superior canines	40.5	51.4	59.0	—	—	—
16		Width across palate between posterior ends of alveoli for superior carnassials	88.4	96.0	108.0	—	—	—
17		Length of auditory bulla from posterior lacerate foramen to external auditory meatus	35.5	30.2	—	—	—	—
18	28 & 29	Maximum width across mastoid processes	114.0	107.0	—	—	—	125.0
19	30 & 31	Maximum width across occipital condyles	64.5	53.0	61.0	62.0	—	56.9
20		Length from anterior end of alveolus of I^3 to posterior end of premaxillary process	86.0*e*	95.0*e*	—	—	—	—
21	6 & 3	Length from posterior border surfaces of palantine to inferior condylar notch	107.0*e*	80.0*e*	—	—	—	—
22	34 & 38	Perpendicular height from base of condyles to top of sagittal crest	117.0	86.2	111.0	—	—	—
23		Perpendicular height of skull from apex of occipital crest (inion) to dorsal border of foramen magnum	58.5	55.2	—	—	—	—
24	36 & 37	Maximum width of foramen magnum	32.6	33.2	—	—	—	—
25	35 & 38	Height of foramen magnum	26.7	—	—	—	—	—
26	39 & 40	Width of occiput just above level of condyles	74.5	72.3	—	—	—	—
27		Width between glenoid processes measured from medial edges of glenoid fossae	68.5	63.2	—	—	—	—

Table 6.A.2. Maxillary dentition measurements of *H. ischyrus*, IMNH 900-11862; *I. crusafonti*, AMNH 95297; *H. serum*, TMM 933-3582 and TMM 933-3231; *H. crenatidens*, FSL 210991; *I. johnstoni*, W.T. 1860. (*a* = approximate, *e* = estimate, + = damaged, — = dimension not available, *N/A* = not available/not developed)

No.	Measurement Points	*H. ischyrus* Martin, Naples, Babiarz, Hearst, this vol. AMNH 25297 900-11862 R	L	*I. crusafonti* Schultz, Martin, 1970 AMNH 95297 Holotype R	L	*H. serum* Meade, 1961 TMM 933-3582	*H. serum* Meade, 1961 TMM 933-3231	*H. crenatidens* Ballesio, 1963 FSL 210991	*I. johnstoni* Mawby, 1965 W.T. 1860
1	41 & 42	Length from anterior end of alveolus of C to posterior end of alveolus of P⁴	102.7	103.9	87.4		87.2	108.0	103.0	65.0 +
2	42 & 43	Length from anterior end of alveolus of P³ to posterior end of alveolus of P⁴	54.5	54.5	45.3		48.2	48.0	—	—
3	43 & 44	Length of diastema from posterior end of alveolus of C to anterior end of alveolus of P³	12.6	12.6	6.8		8.3	25.0	9.0	—
4	45 & 46	Width of incisor series measured between lateral borders of right and left I³ alveoli		45*e*		55.4	65.5	69.7	—	—
5	47.1 & 48.1	Maximum anterior posterior diameter of I¹ at alveolus (mesiodistal)		11.6*e*	11.9	11.4				10.7
6	47 & 48	Maximum width of I¹ at alveolus (buccolingual/ transverse diameter)		7.0*e*	6.1	6.1	10.0	7.0		6.5*a*
7	49.2 & 50.2	Maximum anterior posterior diameter of I² at alveolus	12.0	—	10.9	10.4	—			12.7
8	49 & 50	Maximum width of I² at alveolus	7.9	N/A	7.5	7.4	12.0	13.0		9.0*a*
9	46.3 & 51.3	Maximum anterior posterior diameter of I³ at alveolus		13.5	11.0	11.0				14.4
10	46 & 51	Maximum width of I³ at alveolus		10.0	12.0		14.5	13.0		11.0*a*
11	41 & 44	Anteroposterior length of C at alveolus (mesiodistal diameter)		39.3	30.5		32.8	26.5	33.0	36.5
12	23 & 52	Width of C at alveolus	13.3	13.3	12.5		14.0	15.0	14.0	16.5
13	43 & 53	Anteroposterior length of P³ at alveolus	13.3	13.4	8.3			9.0	9.0	—
14	54 & 55	Width of P³ alveolus	2.3	6.8	6.8			6.8	7.0	—
15	56 & 57	Anteroposterior length of P⁴ at base of crown	39.6	39.0	35.8		38.4	39.0	43.0	42.0
16	58 & 59	Maximum width of P⁴ across protocone	11.6	10.5	11.2		13.0	13.0		14.0
17	60 & 61	Anteroposterior length of P⁴ paracone at base of crown	14.5	14.2	11.0		12.3		14.8	—
18	42 & 53	Anteroposterior length of P⁴ at alveolus	38.0	37.5	33.7		—	—	41.0	—
19	60 & 56	Length of P⁴ parastyle from anterior base of paracone to anterior end of tooth	8.2	8.2		8.2	9.0	7.5	8.2	—
20	61 & 57	Length of P⁴ metacone blade from base of crown above carnassial notch to most posterior point on base of metacone	16.5	16.5	16.6	16.6	17.0	16.5	18.0	—
21	62 & 63	Anteroposterior length of M¹ at alveolus		7.5*a*	4.0		—	—	N/A	—
22	64 & 65	Width of M¹ at alveolus		7.0*a*	5.9		—	—	N/A	7.0

181

Table 6.A.3. Mandible measurements of *H. ischyrus*, IMNH 900-11962; *I. crusafonti*, UNSM 25493; *H. serum*, TMM 933-1 and TMM 933-3231; *H. crenatidens*, FSL 210991; *I. johnstoni*, W.T. 1239. (*a* = approximate, *e* = estimate, + = damaged element, — = dimension not available, *Churcher, CS 1984)

No.	Measurement Points	Measurement	*H. ischyrus* Martin, Naples, Babiarz, this vol. IMNH 900-11862 R	L	*I. crusafonti* Schultz, Martin, 1970 UNSM 25493 Holotype	*H. serum* Meade, 1961 TMM 933-1 R	L	*H. serum* Meade, 1961 TMM 933-3231	*H. crenatidens* Ballesio, 1963 FSL 210991	*I. johnstoni* Mawby, 1965 W.T. 1239 Holotype R	L
1	1 & 2	Length of ramus from anterior border of I$_1$ alveolus at symphysis to posterior end of condyle at center	—	—	—	197.0	205.0	227.0e	—	212.0+	—
2	2 & 3	Length of ramus from anterior end of outer flange to posterior end of condyle	—	—	—	186.0	191.0	208.0	—		
3	6 & 7	Distance from alveolus of C to ventral border of flange	—	—	—	—		—		—	—
4	1 & 26	Length of symphysis measured along anterior border	—	—	—	57.0	57.0	67.0e	—	76.0+	—
5	32 & 33	Minimum depth of ramus below diastema	—	35.0e	—	39.8	42.2	49.7a		49.0	
6	5 & 9	Depth of ramus below posterior border of M$_1$ alveolus	38.1	38.9	—	44.6	42.5	47.5	43.0		
7	10 & 11	Depth of ramus below anterior border of P$_4$ alveolus	36.6	35.4	*32.3	—		—		43.7	43.2
8	12 & 13	Thickness of ramus below M$_1$ (buccolingual diameter)	16.6	19.0	—	18.0	18.5	18.7	18.0	20.6	21.0
9	14 & 15	Height from inferior border of angular process to summit of condyle	—	44.6	—	41.8	40.3	45.0	—		
10	14 & 16	Height from inferior border of angular process to summit of coronoid process	—	78.6	—	76.5		83.0	—		82.0
11	17 & 18	Transverse width of condyle	—	48.1	—	42.0		50.5	—	—	44.0+
12	15 & 19	Maximum depth of condyle	—	19.3	—	18.0		19.0	—	—	—

Table 6.A.4. Mandibular dentition measurements of *H. ischyrus*, IMNH 900-11862; *I. crusafonti*, UNSM 25493; *H. serum*, TMM 933-1 and TMM 933-3231; *H. crenatidens*, FSL 210991; *I. johnstoni*, WT 1239. (*a* = approximate, *e* = estimate, + = damaged element, —— = dimension not available)

No.	Measurement	*H. ischyrus* Martin, Naples, Babiarz, this vol. IMNH 900-11862 R	L	*I. crusafonti* Schultz, Martin, 1970 UNSM 25493 Holotype	*H. serum* Meade, 1961 TMM 933-1 R	L	*H. serum* Meade, 1961 TMM 933-3231	*H. crenatidens* Ballesio, 1963 FSL 210991	*I. johnstoni* Mawby, 1965 W.T. 1239 Holotype R	L
1	21 & 5	Length from anterior border of C alveolus to posterior border of M_1 alveolus	118.0e+	—	117.5	—	130.0	120.0	130.0e	
2	4 & 8	Length of diastema from posterior border of C alveolus to anterior border of P_3 alveolus —	41.5e+	26.9[3]	43.6	—	—	37.0	42.0	40.0
3	4 & 10	Length of diastema from posterior border of C alveolus to anterior border of P_4 alveolus —	53.0e+	—	55.0	—	65.5	—	—	—
4	8.1 & 10	Length of diastema from posterior border of P_3 alveolus to anterior border of P_4 alveolus 4.2	4.2e	—	3.6	7.0	—	—	7.0	
5	8 & 5	Length from anterior border of P_3 alveolus to posterior border of M_1 alveolus 62.4	63.0e	59.5e[3]	58.0	65.0	—	67.0	72.5	69.8
6	10 & 5	Length from anterior border of P_4 alveolus to posterior border of M_1 alveolus 48.6	50.7	43.8[3]	47.2	47.6	48.3	—	48.2	48.9
7	1 & 20	I_1 maximum anteroposterior diameter at alveolus —	—	5.5e	—	—	—	9.0	—	—
8	1.1 & 27	I_1 maximum transverse diameter at alveolus —	—	6.0e	—	—	—	6.0	—	—
9	3.2 & 20.2	I_2 maximum anteroposterior diameter at alveolus —	—	6.0e	—	—	—	11.0	—	—
10	27.2 & 28	I_2 maximum transverse diameter at alveolus —	—	8.0e	8.3	—	6.5	8.0	—	—
11	3.3 & 20.3	I_3 maximum anteroposterior diameter at alveolus —	—	—	—	—	—	11.0	—	—
12	28.3 & 29	I_3 maximum transverse diameter at alveolus —	—	13.7[3]	11.2	11.3	9.8	11.0	—	—
13	30 & 31	C maximum anteroposterior diameter at base of enamel —	—	10.0[3]	16.1	16.7	15.8	—	—	—
14	30.1 & 31.1	C maximum transverse diameter at base of enamel —	—	—	—	—	—	—	—	—
15	34 & 35	C maximum transverse diameter at alveolus —	—	—	11.2	11.4	10.7	—	—	—
16	8 & 8.1	P_3 maximum anteroposterior diameter 10.5	10.0	10.0[3]	8.2	9.0	—	8.0	14 crown	13.0
17	36 & 37	P_3 maximum transverse diameter 17.6	18.0	6.0[3]	10.8	11.3	11.3	5.5		8 crown
18	10 & 23	P_4 maximum anteroposterior diameter 18.2	19.5	18.7[3]	18.5	19.2	20.4	22.0	22.3a	22.0a
19	38 & 39	P_4 maximum transverse diameter 8.0	7.8	9.6[3]	9.0	8.5	9.2	11.0	10.2a	10.3
20	25 & 23.1	M_1 maximum anteroposterior diameter 29.4	29.7	25.8[3]	28.0	28.0	30.6	32.0	28.0a	28.5a
21	40 & 41	M_1 maximum transverse diameter 14.0	14.3	12.5	10.5	10.8	12.8	13.0	13.7a	13.0a
22	23.1 & 24	M_1 length of paraconid blade at base of crown 14.7	14.6	—	—	—	—	—	—	—
23	24 & 25	M_1 length of protoconid blade at base of crown 14.3	17.0	—	13.8	14.0	15.0	—	—	—

A herd of mammoths attempting to drive off several individuals of *Homotherium serum* that had been stalking their young. Partial skeletons of young mammoths have been found with the remains of this scimitar-tooth cat, suggesting that they may have been a preferred prey species. (drawing Mark Hallett)

7

Revision of the New World Homotheriini

LARRY D. MARTIN

VIRGINIA L. NAPLES

JOHN P. BABIARZ

THE TRIBES HOMOTHERIINI and Smilodontini appear together in the Early Pliocene, with roots in the late Miocene (Zdansky, 1924). They have often been linked together in a subfamily, the Machairodontinae, along with *Machairodus*. Because *Machairodus* was one of the earliest saber-tooth genera to be described, many early species of saber-tooth carnivorans were assigned to it, often incorrectly. Subsequently, a myriad of new genera emerged from these species, some of which were closely related to the original type species and some of which were not. In North America, the original material described as an American *Machairodus* (*M. caticopis*) by Cope (1887) was proven by Martin and Schultz (1975) to belong to an endemic North American lineage of scimitar-tooth cat, *Nimravides*, derived from a Eurasian immigrant, *Pseudaelurus*. Like *Homotherium*, *Nimravides* had elongated limbs and shared North America with a short-legged dirk-tooth predator, *Barbourofelis*. These two genera became extinct at the end of the Kimballian (Lower Hemphillian) about seven M.Y.A. during one of the major extinctions of the Cenozoic. They were replaced in the Hemphillian (Upper Hemphillian) by a Eurasian immigrant, *Machairodus,* and later by *Amphimachairodus (A. tanneri),* as well as the ancestor of *Smilodon (Megantereon)* (Sardella, 1998; Christiansen and Adolfssen, 2007). At the beginning of the Blancan (Pliocene) in North America, a new scimitar-tooth cat appeared as a Eurasian immigrant. Originally it was described as an endemic North American genus, *Ischyrosmilus.*

The type species of *Ischyrosmilus, I. ischyrus* (UCMP 8140) from the Late Pliocene of Asphalto, California (fig. 7.1A), was found associated with a bone-crushing dog, *Borophagus.* Merriam (1917) included a second species, *I. osborni,* in his original paper on *Ischyrosmilus. I. osborni* was reassigned to the barbourofelid cat, *Barbourofelis,* by Schultz, Schultz, and Martin, (1970), who pointed out that its Late Miocene (Clarendonian) age made it some seven or eight million years older than any real *Ischyrosmilus.*

Merriam (1918) added a third species, *I. idahoensis* (fig. 7.1B), collected from a third locality, 3036C, near Froman Ferry, southwestern Idaho, on the Snake River, and seemingly equivalent to the Grand View local fauna (latest Blancan). More complete remains of *Ischyrosmilus* were described from Cita Canyon, Texas (figs. 7.1C, 7.2), and Pliocene Benson Local Fauna, by Mawby (1965), who named an additional species, *I. johnstoni.* Mawby commented on the similarity between *Ischyrosmilus* and *Homotherium* and speculated that they might be congeneric. Schultz and Martin

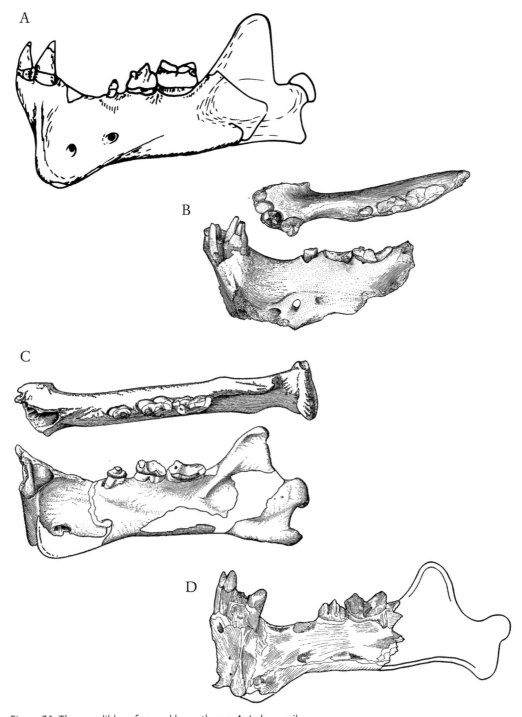

A

B

C

D

Figure 7.1. The mandibles of several homotheres: *A, Ischyrosmilus ischyrus*, UCMP 8140, from Asphalto, California (after Churcher, 1982). *B, I. idahoensis*, UC 22343 (Type), from the Late Blancan locality 3036 C, Froman Ferry, southwestern Idaho (Grand View Local Fauna), dorsal (occlusal) and lateral views (from J. C. Merriam, 1918). *C, I. johnstoni*, W. T. 1239 (Type), from the Cita Canyon, Texas, locality, UCMP-V 3721, dorsal (occlusal) and lateral views. *D, I. crusafonti*, UNSM 25493 (Type) (from Schultz and Martin, 1970, fig. 1B), Broadwater Local Fauna (appendix 2), with posterior restoration (after Martin, Schultz, and Schultz, 1988, fig. 10), lateral view.

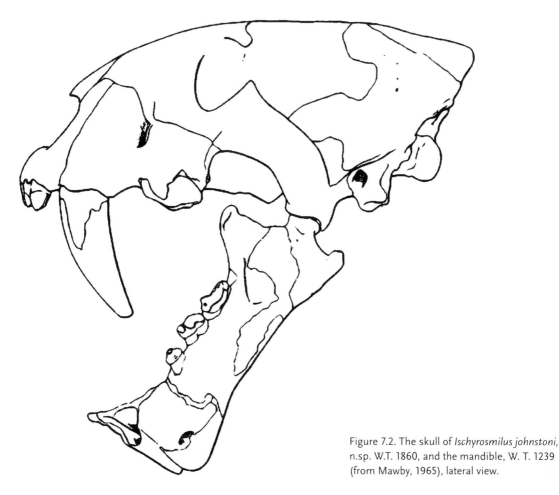

Figure 7.2. The skull of *Ischyrosmilus johnstoni,*
n.sp. W.T. 1860, and the mandible, W. T. 1239
(from Mawby, 1965), lateral view.

(1970) named an additional species of *I. crusafonti* (UNSM 25493, fig. 7.1D) from the Broadwater Local Fauna. The viewpoint of Mawby, that *Ischryosmilus* and *Homotherium* were congeneric, was shared by R. Martin and Harksen (1974) when they reported the nearly complete *Homotherium* skull (AMNH 95297) from the Pliocene of Delmont, South Dakota (fig. 7.3). Kurtén and Anderson (1980) only recognized one species, *I. ischyrus,* as did Churcher (1984), but unlike Churcher, they synonymized the genus with *Homotherium.* They also suggested that saber-tooth taxonomy has historically fluctuated between numerous species, often based on fragmentary material, and a smaller number of species grouped with anatomically more complete specimens.

How much taxonomic diversity is actually present in the several million years represented by the North American Pliocene? Can all the morphological diversity be easily encompassed by a single species? We think that at least the small form represented by the skull from South Dakota (Martin and Harksen, 1974) is morphologically distinct from the Idaho skull (IMNH 900-11862), and represents a separate species more closely related to *H. serum.* This skull was found only a few hundred miles from the Broadwater Local Fauna and appears close to the same age. The holotype mandible of *H. crusafonti* (UNSM 25493) from that fauna occludes well

with the Delmont skull, and comparison with the Birch Creek skull (fig. 6.3) shows that it is distinct from *H. ischyrus* (fig. 7.3). For the time being, we are content to include the remaining Blancan *Homotherium (H. idahoensis, H. johnstoni)* in *H. ischyrus.*

SYSTEMATIC PALEONTOLOGY

Family: Felidae.
   Subfamily: Machairodontinae.
   Included tribes: Machairodontini, Homotheriini, Smilodontini.
   Tribe: Homotheriini (Kurtén, 1962).
   Diagnosis: Felids with relatively short faces, short, coarsely serrated upper canines; procumbent incisors; serrated incisors and cheek teeth; protocone reduced or absent on P^4; short dependent flange on the mandible; short tail.
   Included genera: *Homotherium, Xenosmilus.*
   Homotherium ischyrus
   Diagnosis and Description: Provided in detail in chapter 6 of this book.
   Holotype: Partial ramus UCMP 8140 (fig. 7.1A).
   Diagnosis and Description: Based on the Delmont specimen, *H. crusafontis* is smaller than *H. crenatidens* or *H. ischy-*

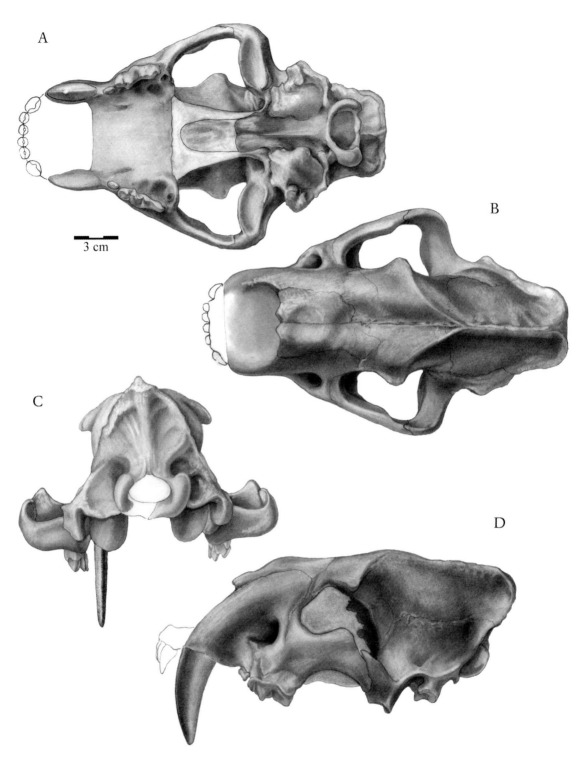

A

B

3 cm

C

D

Figure 7.3. The skull of *Homotherium crusafonti*, AMNH 95297, from the Pliocene Delmont Local Fauna: *A*, ventral view; *B*, dorsal view; *C*, posterior view; *D*, lateral view. The Delmont Local Fauna is roughly equivalent to the Broadwater Local Fauna (Martin, Schultz, and Schultz, 1988) (chapter 6 appendix, tables 6.A.1, 6.A.2). (drawing Mary Tanner)

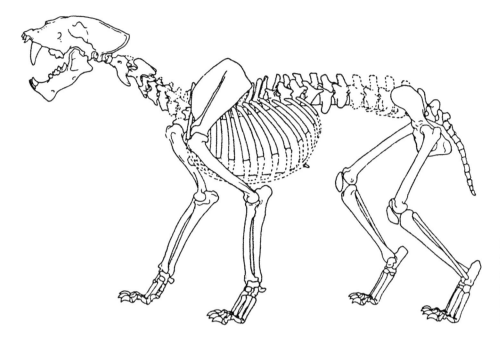

Figure 7.4. The restoration of the skeleton of *Homotherium serum* in lateral view (based on the figures from Meade, 1961), taken from Martin, Schultz, and Schultz, 1988.

rus, with the nasals more retracted and the skull more domed, having a lower skull roof. The enlargement of the external nares contrasts with *H. ischyrus* and resembles *H.serum.* The posterior palatal edge is more anterior than in *H. ischyrus.* The occiput is more perpendicular than in *H. ischyrus.* The auditory bullae extend below the mastoids in *H. crusafonti* and above them in *H. ischyrus.* The site of attachment of the M. sternomastoideus is reduced compared to *H. ischyrus.* The mastoid process is separated from the auditory bulla by a more distinct groove in the Delmont specimen, a feature less well developed in the Idaho homothere.

COMPARISONS OF THE DELMONT SKULL TO OTHER HOMOTHERES

The anterior-posterior distance of the base of the canine of the South Dakota specimen is about 30 mm, and the length of the exposed tooth crown is about 73 mm. The labiolingual diameter at the base is 12 mm. The basal thickness of the canine in the Idaho specimen is 14 mm labiolingually. It is 34 mm in anteroposterior distance along the base of the canine tooth. The canine in *H. crusafonti* is 30 mm in anteroposterior length. The canine associated with the holotype of *H. ischyrus* is a little more elongated than is that of the Delmont specimen. In the Delmont homothere skull, the anterior margin of the nasals is even with the anterior edge of the carnassial tooth, while in the Idaho homothere they are significantly farther forward.

In the Delmont *Homotherium,* the width of the skull just anterior to the infraorbital foramen measures 75.5 mm. The same measurement across the Idaho skull is 66.5 mm, reflecting the narrower muzzle in the Idaho specimen. This feature results from the enlarged external nares in the Delmont skull

and unites this specimen with *H. serum.* A measurement across the postorbital processes of the frontal bones in the Idaho homothere is 121 mm and only 87 mm in the Delmont skull. The processes in the Idaho skull are more blunt and thicker dorsoventrally than in the Delmont specimen. In the Idaho specimen, the shape of the dorsal surface of the head is more squared, while the sharply pointed and downwardly tapering postorbital processes in the Delmont skull give it a domed shape in anterior view. The Idaho skull shows a height of 111 mm, measured posterior to the carnassials, while the Delmont skull measures 96 mm at this point. This indicates that the Idaho specimen has a higher skull roof. The distance from the base of the occipital to the anterior portion of the sagittal crest in the Idaho specimen measures 110 mm, while the same measurement in the Delmont specimen is 91 mm. The edge of the occiput is about even with the back of the condyles in the Delmont specimen, but the occiput in the Idaho specimen overhangs the condyles by at least 15 mm, indicating that this animal had a longer temporal fossa. In the Idaho specimen the glenoid is significantly ventral to the mastoid process, and they are more nearly even in the Delmont specimen. The mastoid process terminates above the bulla in the Delmont specimen and below it in the Idaho specimen. The postglenoid process is more highly developed in the Idaho homothere than in the Delmont specimen. The mastoid process differs in shape between the two animals. In the Idaho specimen, it is more oval and tilted outward, while in the Delmont specimen, the anterior part of the mastoid process is flattened horizontally and the posterior part is tilted forward, with the two parts separated by a notch. The mastoid process is separated from the auditory bulla by a more distinct groove in the Delmont specimen, a feature less well developed in the Idaho homothere.

Anterior to the infraorbital foramen in the Idaho homothere, there is a distinct pocket for the origin of the muscle that elevates and retracts the upper lip (M. levator nasolabialis). This feature is lacking or reduced in the Delmont specimen. The anterior edge of the internal nares is about even with the posterior edge of the tooth row in the Delmont skull. It is more posterior to the tooth row in the Idaho specimen. In the Idaho homothere in lateral view, the attachment for the M. sternomastoid measures anteroposteriorly 35 mm. This feature is reduced to approximately 25 mm in the Delmont skull.

The holotype of *Homotherium crusafonti* is also from a small homothere, and fits almost exactly to the Delmont specimen. We suggest that the Delmont skull be referred to *H. crusafonti*, and that the species is distinct from either *H. ischyrus* or *H. crenatidens*. As was suggested in Martin, Schultz, and Schultz (1988), *H. crusafonti* is more similar to *H. serum* than is *H. ischyrus*. This may mean that *H. serum* evolved in North America from the Blancan form. The Broadwater Local Fauna in Nebraska also produced the holotype of the giant beaver, *Procastoroides sweeti*. The giant beaver from the Grandview Local Fauna in Idaho is more advanced and belongs to a separate species, *P. idahoensis*. The Grandview Local Fauna also produced the holotype of *H. idahoensis*. The Grand View Local Fauna is Late Blancan (Senecan) in age. The Delmont Local Fauna contains the same giant beaver, *P. sweeti*, as does the Broadwater Local Fauna, and is presumably similar in age.

The Broadwater Local Fauna also includes a small number of homothere postcrania that probably belong to *H. crusafonti*. These include a radius, partial ulna, distal humerus, metapodial, and an astragalus (Schultz and Martin, 1970). They are more slender and elongate in comparison with the similar elements of the Idaho specimen of *H. ischyrus*. The Lisco Local Fauna, while geographically close to the Broadwater Local Fauna, both coming from adjacent counties in northwestern Nebraska, produced a premaxillary of a homothere (Schultz and Martin, 1970) that is larger than *H. crusafonti*. This fauna also produced a partial forelimb of a probable homothere, consisting of the humerus, radius, and ulna. This forelimb is shorter and more robust than that of the Idaho homothere.

The Broadwater Local Fauna also includes a lower mandible of *Megantereon hesperus*, sometimes assigned to the European species, *M. cultridens* (Turner and Antón, 1997), so that we have cats belonging to both the scimitar- and dirktooth lineages in the Broadwater fauna. The skeleton (elongated, slender humerus; radius; ulna; and metatarsal) associated with the *H. crusafonti* holotype from the Broadwater Local Fauna contrasts with shorter more robust skeletal elements found associated with a scimitar-tooth premaxilla in the nearby Lisco Local Fauna. Certainly two kinds of homotheres were present in these two faunas.

SYSTEMATIC PALEONTOLOGY

Homotherium serum (Cope, 1893).

Type species: *Dinobastis serus*. Cope (later amended to *H. serum*, Churcher, 1966).

Holotype: Three metacarpals, three phalanges, the head of a femur, five incisors, two upper canines, and two upper carnassials, presumably from one individual, Academy of Natural Sciences Philadelphia (ANSP) collection.

Type Locality: Western Oklahoma.

Horizon: Presumably Late Pleistocene.

Diagnosis: 1^3 separated by a diastema from the upper canine; upper canine shorter than in *H. ischyrus*; P^3 reduced to a single posteriorly directed cusp; upper carnassial lacking the protocone, including the root (protoradix); ramus with an anterior masseteric pocket; short olecranon process on the ulna; shortened femur and tibia as compared with the elongate front limbs of *Homotherium serum*; reduced ability to retract the claws.

COMPARISONS OF *HOMOTHERIUM SERUM* TO OTHER HOMOTHERES

This species, *H. serum*, is known from a large sample from Friesenhahn Cave in Texas that is well described and illustrated by Meade (1961) and Rawn-Schatzinger (1992) (figs. 7.4–7.5, tables 6.A.1–6.A.4). The skull is similar to that in *H. ischyrus* and less so to that of *Xenosmilus* (fig. 4.6A–B, tables 6.A.1–6.A.2). This species has been commonly restored with plantigrade feet and a sloping, hyena-like back based on the original interpretation by Kurtén (1952). It should be noted that animals with plantigrade hind feet commonly are digitigrade on the front feet (for instance, bears), and Rawn-Schatzinger (1992) indicates that the front feet were digitigrade. The front limbs are elongate and slender, even when compared to the other homotheres, and the ulna has an unusually short olecranon process. Martin, Schultz, and Schultz (1988) thought that a plantigrade restoration of the hind feet would be incongruous with the species' relatively long legs and restored the skeleton based on Meade's (1961) figures (not on his measurements, as suggested in Rawn-Schatzinger, 1992) in a digitigrade posture. Because of the shortened hind legs with a robust femur, they thought that it was less adapted for running than was *H. crenatidens*. This was misconstrued by Rawn-Schatzinger (1992) as indicating that they doubted its cursorial abilities. However, a careful reading of both papers shows that they were in agreement that it was cursorially adapted, although probably not as much so as the Pliocene examples of *Homotherium*. Rawn-Schatzinger (1992) presented an argument that the claws were poorly retractile, and this would seem to be a difference between *H. serum* and the Blancan forms. Rawn-Schatzinger (1992) illustrated the metatarsals widely separated from each other, but this is

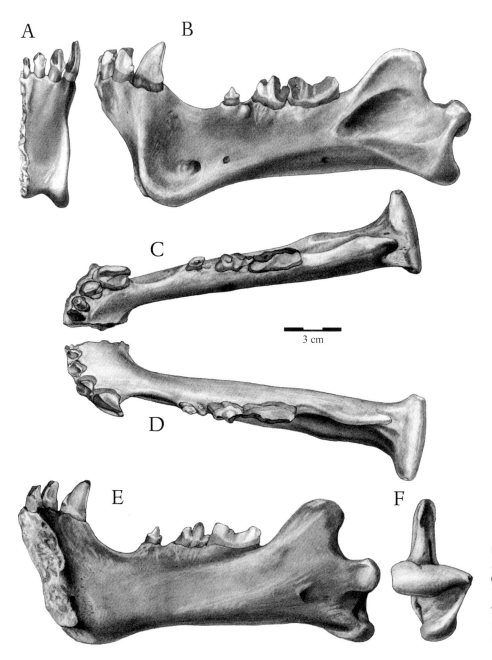

Figure 7.5. The mandible of *Homotherium cf. serum*, BIOPSI 0112, left ramus: *A*, anterior view; *B*, left lateral view; *C*, right dorsal view; *D*, left dorsal view. Right ramus: *E*, medial view; *F*, posterior view. (drawing Mary Tanner)

unlikely, and they were probably appressed to each other, as is normal for cats and the Pliocene homotheres. Both Radinsky (1975) and Rawn-Schatzinger (1992) mentioned that the optic center on the endocranial cast was unusually large. This feature is also seen in the extant cheetah and was thought to be correlated with fast running. However, when we examine the femur and the tibia we see features more closely related to strength than to running. We thus see a combination of highly cursorial traits, with other traits toward a less cursorial mode. How did this combination come about? We suggest that cursorial traits continued to improve until sometime in the Middle Pleistocene, when there was a reversal of trends from a more cursorial pursuit strategy to

one with a greater emphasis on ambush. We think that this occurred after the extinction of the large ambush homothere, *Xenosmilus,* and the appearance of a new pursuit predator, *Panthera,* in the North American fauna.

The type, and almost all of the figured and described material of *H. serum* is from the Late Pleistocene. Early and Middle Pleistocene examples are reported but have not been critically figured or described. Among the most important of these is a cranium from the "McPherson *Equus* Beds" in south central Kansas that was originally reported as a "small example of *Smilodon*" by Hibbard (1952). It was correctly identified as a homothere by Churcher (1966), who based the posterior portion of his *H. serum* skull restoration on it. When

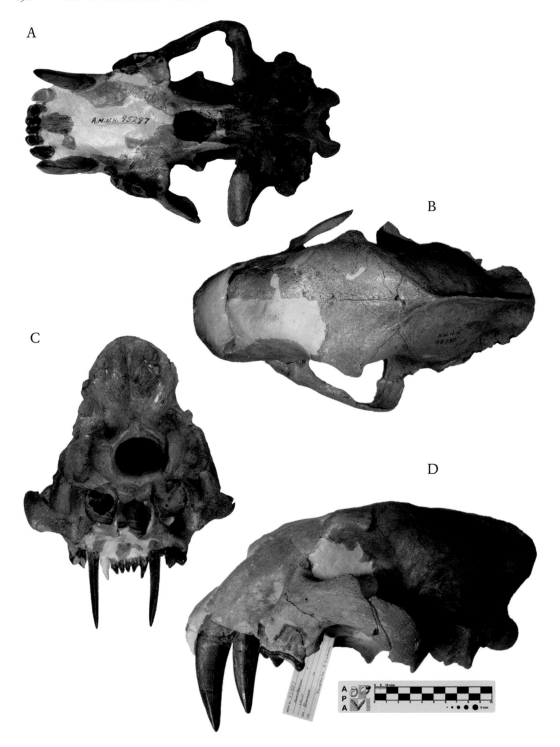

Figure 7.6. Photographs of the skull of *Homotherium crusafonti*,
AMNH 95297: *A,* ventral view; *B,* dorsal view; *C,* posterior view;
D, lateral view. (courtesy Ascanio D. Rincón)

compared to the *H. ischyrus* skull from Idaho, the pedestal
for the glenoid fossa is longer, the sagittal crest is higher, and
the occiput is more triangular. The posterior contact region
with the parietals shows a large and inflated frontal sinus as
was present in the earlier homothere. The smaller size of the

occipital condyles indicates a smaller body size than in the
examples of *H. ischyrus* and is closer to *H. crusafonti.*

Recently a mandible of a homothere was recovered during
an excavation project near the Arkansas River in southwestern Kansas (BIOPSI 0112, fig. 7.6, tables 4.A.3–4.A.4). This

specimen lacks a clear geological context, but morphologically it seems primitive when compared to the Late Pleistocene examples and advanced when compared to the Blancan (Pliocene) ones. It seems reasonable to suspect that it is an early Irvingtonian (Pleistocene) form. The mandible is from a large robust individual. The first incisor is elongated and peg-like, the second is broader and has a small labial cusp, and the third shows a distinct anterolingual wear-groove from the upper incisor. It is a roughly triangular tooth with a small labial cusp. The lower canine shows a pronounced anterolingual groove produced from wear with the third upper incisor. It is serrated. Although the jaw is larger than that of *Xenosmilus,* the incisors and canines are distinctly smaller. The canine depression on the flange is bordered anteriorly by a ridge and is so deeply indented that the anterior mental foramen opens anteriorly. The posterior mental foramen opens laterally, just under the third premolar. The dorsal margin of the diastema is rounded, as in other examples of *Homotherium* (more sharply edged in *Xenosmilus).* The lower third premolar is single-rooted and is separated from the lower fourth premolar by a short diastema. As with other *Homotherium,* the P$_4$ has a large vertical paracone (Martin, Schultz, and Schultz, 1988), and the whole tooth is tilted backward against the anterior margin of the carnassial. All the cheek teeth are tilted labially and slightly overhang the outer margin of the dentary.

There is a deep dorsal depression within the masseteric fossa, but the fossa is not pocketed anteriorly as is characteristic for *H. serum.* The lower margin of the masseteric fossa extends outward to form a distinctive labial shelf. The coronoid process is low (below the canine tip) and rounded. The mandibular condyle is wide, and the angular process is large and turned inward. The absence of the anterior pocket is significant, as the occurrence of an anterior pocket is a prominent feature of the Late Pleistocene specimens, including the ones from Alaska, Friesenhahn Cave in Texas, the Gassaway Fissure in Tennessee, and Madison County, Nebraska (Rawn-Schatzinger and Collins, 1981; Martin, Schultz, and Schultz, 1988). The pocket is also present in the 28,000-year-old jaw of *H. latidens* from the North Sea (fig. 9.1A). That ramus is remarkably small, but is more similar to the Late Pleistocene American examples than either is to the material from the Villafranchian and Blancan. The North American *H. serum* also differs from the Pliocene forms in showing a slight shortening of the tibia. Recently, *Homotherium* has been discovered in a Middle or Early Pleistocene tar pit in Venezuela (A. Rincon, personal communication, 2008).

Pinsof (1998) reported an additional cranium, along with some limb bones of *H. serum,* from the Late Pleistocene of the American Falls locality in Idaho, and there is a skull and some lower jaws of a small *Homotherium* from a locality near Fairbanks, Alaska, in the American Museum of Natural History collection. These show a deep anterior pocket in the masseteric fossa. It is likely that the Alaskan specimens actually represent an eastern extension of the range of *H. latidens* into North America. This could probably be established with additional study and publication of the American Museum materials.

SYSTEMATIC PALEONTOLOGY

Genus: *Xenosmilus* (Martin et al., 2000).

Type species: *Xenosmilus hodsonae* (Martin et al., 2000).

Diagnosis: An extinct felid with short, broad, coarsely serrated, saber-like upper canines; third lower premolar (P$_3$) absent; anterior margin of masseteric fossa on ramus not pocketed as in *Homotherium*; lower mandible with two large mental foramina below the diastema; relatively small head with large occipital condyles; cranium elongated with frontals much narrower than in *Homotherium*; temporal fossa elongated; apex of the occipital crest much posterior to the condyles (nearly even in *Homotherium*); legs greatly shortened; hind feet plantigrade.

Etymology: *xenos* (Greek) strange; *smilos* (Greek) knife.

Type Species: *Xenosmilus hodsonae* (Martin et al., 2000).

Etymology: *hodsonae,* for Debra L. Hodson, the wife of one of the paleontologists naming the species.

Holotype: BIOPSI (Babiarz Institute of Paleontological Studies, Inc.) 101; a nearly complete skeleton except for the lumbar vertebrae, sternum, ribs, and tail. The specimen is deposited in the Robert S. Dietz Museum of Geology, Arizona State University.

Paratype: University of Florida, UF 60,000; partial skeleton, including skull and mandibles.

Type locality: University of Florida collecting site (Commercial limestone quarry, Haile 21A) Alachua County, Florida.

Horizon: Undifferentiated sinkhole fill sediments in Eocene Ocala limestone.

Age: Early Irvingtonian (Early Pleistocene, 1.0–1.2 M.Y.A.).

Diagnosis: Only species in the genus.

Discussion: *Xenosmilus* is thoroughly described and illustrated in chapter four of this volume.

SUMMARY

Recently both *Homotherium* and *Xenosmilus* have been discovered in South America, where they remain rare members of that fauna. In North America, we recognize two Pliocene species of *Homotherium, H. ischyrus* (figs. 6.3, 6.4, 7.1A) and *H. crusafonti* (figs. 7.1D, 7.3). The Late Pleistocene *H. serum* may represent a later immigrant, as it is closely similar to the Late Pleistocene *H. latidens* in Europe. *Homotherium* is also known from several Late Pleistocene specimens from Alaska. We suggest that the Alaskan material might be more closely related to *H. latidens* than to *H. serum.*

Reconstruction of the Late Pliocene specimen of *Homotherium crenatidens* from Tajikistan, PIN 3120-610 (*top*). Reconstruction of the facial musculature (*middle*). The skull and mandible; the lower mandible is a composite of *H. cf. crenatidens* and is based on BIOPSI 0154 (*bottom*). (drawing Mark Hallett)

8

A Saber-tooth Cat Skull from Tajikistan, Central Asia, and the Relationships between Eurasian and North American Homotheres

PETER E. KONDRASHOV

LARRY D. MARTIN

A REFERRED SPECIMEN (Paleontological Institute of Russian Academy of Sciences, PIN 3120-610) is a nearly complete, but has a crushed and distorted skull, lacking the zygomatic arches, from Kuruksay, Tajikistan, Late Pliocene. The skull shows the broad forehead characteristic of *Homotherium* (figs. 8.1–8.3, tables 4.A.1–4.A.2) with the maximal width at the level of the glenoid fossa and the minimal width at the postorbital constriction. The snout is approximately the same length as the braincase.

The occiput is narrow and well defined by a strong occipital crest. There is a well-developed vertical crest that begins from the top of the occiput and runs down toward the foramen magnum, but does not reach the latter. This crest widens and disappears at about one-third the height of the occiput, forming a low external occipital protuberance. The condyles embrace the foramen magnum for 80% of its height. They are almost cylindrical, and widen slightly mediolaterally. The foramen magnum is relatively large and almost round. The condylar foramen is round, faces ventrally, and is situated in the occipitale laterale anterior to the medial side of the condyle.

The basioccipital bears a strong medial crest with a well-developed pharyngeal tubercle. There are two well-expressed rugosities along the basioccipital crest, medial to the tympanic bullae, for the insertion of the M. longus capitis. The well-developed basioccipital crest confirms the relatively strong development of that muscle (Antón and Galobart, 1999). The rugose area for the fleshy attachment dominates the scar area for the tendinous attachment. The development of the M. longus capitis correlates with the strength of vertical head movement, and suggests that *Homotherium* was capable of driving the sabers downward with great force.

The basisphenoid is flattened ventrally. The foramen lacerum anterior, including the opening of the auditory tube, lies lateral to the basisphenoid, anterior to the bulla, and medial to the glenoid fossa. The foramen lacerum posterior is situated anteriorly or anterolaterally to the condyloid foramen. It is slit-like and faces ventrally.

The alipshenoid bones are slightly reduced. They contact the maxillary and palatine bones anteriorly and the orbitosphenoid and squamosal bones ventrally and posteriorly. They reach the tympanic bulla and form part of the foramen lacerum anterior and the foramen ovale. Due to damage to this region, no trace of the pterygoid bones can be found. The foramen ovale is oval in shape and faces anteroventrally, with its long axis perpendicular to the long axis of the skull.

A

B

C

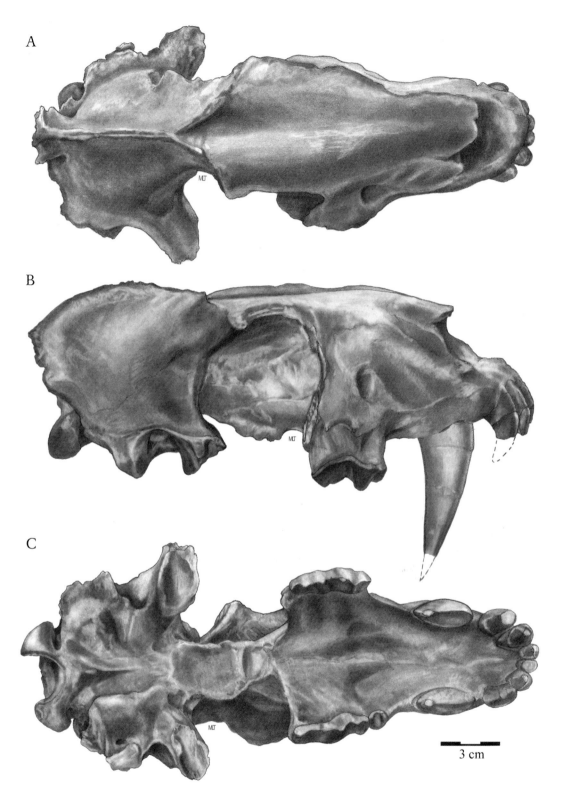

3 cm

Figure 8.1. The skull of the Late Pliocene *Homotherium crenatidens* from Tajikistan, PIN 3120-610, drawn from a cast of the skull: *A*, dorsal view; *B*, lateral view; *C*, ventral view. (drawing Mary Tanner)

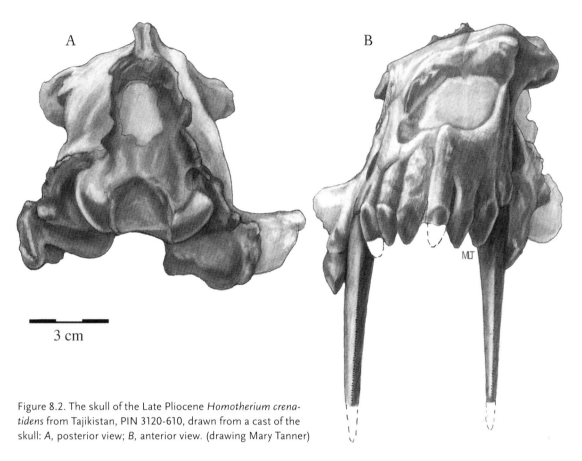

A

B

3 cm

Figure 8.2. The skull of the Late Pliocene *Homotherium crenatidens* from Tajikistan, PIN 3120-610, drawn from a cast of the skull: *A*, posterior view; *B*, anterior view. (drawing Mary Tanner)

The premaxillary bones are well developed for the support of large incisors. They wedge between the maxillary and nasal bones and have a short contact with the frontals posteriorly. This seems to be a derived character of this specimen. In *H. crenatidens,* the dorsal premaxillary process is considerably shorter (Ballesio, 1963, fig. 10; fig. 8.2).

The frontals are greatly inflated and wide, with strong postorbital processes. Each frontal is triangular, so two of them form a rhomboid-like structure on the roof of the skull. The wide postorbital process forms the roof of the orbit and has a hook-like jugal process. Two crests begin from the most lateral part of the postorbital process. One runs posteroventrally down to the anterior edge of the glenoid fossa, and the other runs posteromedially to join the same crest of the other side of the skull to form the sagittal crest. The nasal bones are relatively short and almost rectangular. The nasal opening is large and square, with rounded angles. The infraorbital foramen is relatively large, situated above the anterior root of the P[4]. It is oval and faces anteriorly. The infraorbital canal is short and straight. The anterior opening is larger than the posterior, so the canal broadens anteroposteriorly.

The muzzle is wide, and the maxillary bone is large, to house the huge canines; the maxillary bone forms the anterior part of the orbital ring.

The palate tapers anteriorly and is triangular and relatively wide between the premolars. There is a large palatine

foramen across from the upper P[3] that faces anteroventrally. The parietals form the roof of the braincase. The sagittal crest is relatively short and low. It begins at the postorbital constriction and fuses posteriorly with the occipital crest. The two parasagittal crests unite across from the external auditory meatus. The sagittal crest bears tuberosities on its lateral surface for the insertion of the temporalis muscle.

The squamosal forms the lateral walls of the braincase. They form strong zygomatic processes contributing to the zygomatic arch. The glenoid fossae are almost perpendicular to the long axis of the skull. The glenoid fossae are oval, concave, and face anteroventrally at about 45° to the base of the skull. The postglenoid process is well-developed, tall, and forms the posterior wall of the glenoid fossa.

The tympanic bullae are large and ellipsoid. The strong and wide mastoid processes project ventrally and lie lateral to the bulla. The mastoid processes are saddle-shaped and extend anteroposteriorly. The ventral surface of the saddle is the attachment area for the M. brachiocephalicus. The medial wall of the mastoid process bears a triangular-shaped facet for the insertion of M. sternocephalicus. A well-defined, sickle-shaped curved depression on the dorsolateral wall of the mastoid process forms the insertion of M. obliquus capitis cranialis. This muscle attaches posteriorly to the ventral surface of the wings of the atlas. The muscle was relatively strong and provided extra force to the vertical bite. The paroccipital

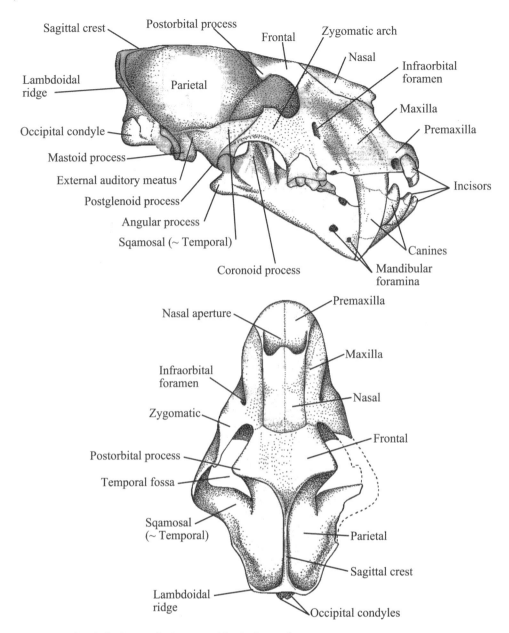

Figure 8.3. The skull of *Homotherium crenatiden* (redrawn after Ballesio, 1963, fig. 10). Skull features are shown in blue: *A*, lateral view; *B*, dorsal view. (courtesy Elizabeth Ebert)

process is partially broken on PIN 3120-610, although part of the facet for the M. digastricus insertion is visible. The external auditory meatus is situated anterodorsal to the mastoid process. The meatus is relatively wide, and its dorsal wall is formed by the sharp crest of the squamosal. The stylomastoid foramen is relatively large, oval, faces ventrally, and is situated between the tympanic bulla and the mastoid process. The tympanohyal foramen is situated in the same depression. It lies lateral to the latter foramen and faces posteroventrally.

The jugal bone is not preserved on PIN 3120-610. But judging from the strong development of the zygomatic processes of both maxillary and squamosal bones, the jugal was short and slightly reduced.

The incisor row is strong, much curved anteriorly, and forms an arcade. The incisors are relatively large and round in cross section, increasing in size from the first to the third. There are well-developed anterior and posterior cingula on all the incisors. The well-expressed lingual cingulum on I^1 connects the anterior and posterior cingula. The I^2 has a crown that is slightly compressed laterally. The lingual cingulum is well developed. The I³ is almost oval and has a large, laterally compressed crown. Only anterior and posterior cingula are expressed, the former being significantly larger, forming a vestigial cusp. There is slight diastema between I³ and the canine.

The canines are large, broad, and laterally compressed, with serrations developed on a sharp crest that forms the anterior and posterior edges. The crest turns inward proximally, reflecting the helical growth pattern present in this high-crowned tooth. The root of the canine extends to a point above the anterodorsal margin of the infraorbital foramen. The P³ is reduced, with a single broad root (missing from the right side). It has a large central cusp, bordered by smaller anterior and posterior cusps. The anterior cusp is more nearly vertical, while the other two tilt strongly posteriorly.

The P⁴ is large and double-rooted. The anterior root is much smaller than the posterior root, which is long and laterally compressed. The P⁴ consists of three lobes: parastyle, paracone, and metacone. The parastyle lobe is the shortest and the metacone lobe is the longest. The paracone lobe is higher than the other two. The paracone lobe is separated from the metacone lobe by a deep groove (carnassial notch), while the parastyle lobe is closely appressed to the paracone. The protocone and protoradix are completely absent, once again, demonstrating that this skull is from a relatively derived species.

CONCLUSIONS

This Late Pliocene specimen of *H. crenatidens* (PIN 3120-610) is of special interest because it comes from a relatively northern locality in Central Asia and gives insight into the northern population, connecting the Eurasian and African populations with those of North America. In that sense, it seems advanced when compared to the Late Pliocene Idaho homothere (IMNH 900-11862; Hearst et al., chapter 6 in this

volume). Some of these osteological advances are described as follows: the Asiatic specimen has a relatively and absolutely longer carnassial, and this is especially true for the paracone on that tooth. The protocone root (protoradix) is absent, while a trace of it remains on the Idaho *H. ischyrus* specimen. The P³ of the Asian *H. crenatidens* does not show separate roots, as in the Idaho specimen, and the cusps on that tooth are smaller and less distinct. The incisor arcade is as procumbant, but it is more rounded than in *H. ischyrus*. Overall, the skull is smaller than the Idaho specimen. The mastoid process is relatively smaller and is more widely separated from the glenoid pedestal, resembling the younger species, *H. latidens* and *H. serum* in this respect. The auditory bullae are shorter than in *H. ischyrus* and have straight anterior margins. These margins extend anteriorly past the mastoid processes, as in *H. ischyrus*. In the Idaho homothere, the stylohyoid pit is large and shallow, forming an embayment of the ventrolateral surface of the bulla that gives a "C" configuration, while in the Asiatic specimen, that surface is straight. The canines are relatively longer on *H. ischyrus* (tables 4.A.2, 6.A.2). There is little doubt that due to these osteological advances, we are dealing with two genetically separated populations. This is a remarkable confirmation of the speculation that the more northern populations were evolving characteristics at an earlier time that later became established farther to the south.

SUMMARY

New material of *H. crenatidens* from the Late Pliocene Kuruksay locality in Tajikistan, Central Asia, is described. The material includes an almost complete skull, lacking the zygomatic arches. It shows that a number of derived skull features that characterize the later species occurring farther to the south appeared earlier at this northern site than they did in more southern locations in North America and Europe.

SCIMITAR CAT
Homotherium serum

Reconstruction of the scimitar cat *Homotherium serum*. (drawing
Mark Hallett)

9

A Framework for the North American Homotheriini

LARRY D. MARTIN

JOHN P. BABIARZ

VIRGINIA L. NAPLES

ALTHOUGH SABER-TOOTH CATS are now fixed in the public mind by the notoriety of the dirk-tooth *Smilodon*, the original discovery was almost entirely centered on the scimitar-tooth cats, *Homotherium* and *Machairodus*. The first descriptions go back to Cuvier (1824) and the beginnings of vertebrate paleontology. Unfortunately Cuvier's discovery was based entirely on upper canines that he badly misinterpreted as those of an extinct bear *Ursus cultridens*. Still worse, he designated no holotype, and the syntypic series included almost all the genera to be recognized later (*Machairodus, Homotherium,* and *Megantereon*). Owen (1846) even indicated that many paleontologists had considered similar teeth in England as relating to the carnivorous dinosaur *Megalosaurus*. The result was confusion in the later literature, especially because isolated canines are *nomina vana,* as they are inadequate to define any of the species. This has not prevented later authors from trying to use stratigraphic context to assign them to species, thereby synonymizing better-established names that had been correctly used for long periods and in many publications. The sorting process began with an isolated canine in Cuvier's series from the late Miocene of Eppelsheim, Germany. Using material from the same locality, Kaup (1833) proposed the genus *Machairodus* and recognized it as a felid. Because Kaup's description was the first to recognize an actual sabercat, later species were commonly first described within that genus and separated when better material showed how they differed. This was the case when Owen (1846) described the British Pleistocene material as *Machairodus latidens*. Fabrini (1890) proposed the name *Homotherium,* but it did not get wide usage until Arambourg (1947) used it for an African species. As a result, new genera became necessary for species that were described later: *Dinobastis* (Cope, 1893), *Epimachairodus* (Teilhard de Chardin and Leroy, 1945). Kurtén (1962) synonymized these genera with *Homotherium,* while some later workers recognized that they were close to *Homotherium* but maintained their separation from that genus (De Bonis, 1975; 1976; Martin, Schultz, and Schultz, 1988).

The genus *Homotherium* is characterized by extreme breadth across the frontals, a nearly vertical occiput, and elongated distal extremities of the legs. Two species of *Homotherium* are generally recognized in Europe, *H. crenatidens* in the Villafranchian and *H. latidens* in the late Pleistocene. Turner and Antón (1997, p. 48) include all the named European taxa, including *H. crenatidens,* in the species *H. latidens,* giving that species a temporal range of around five million years and a geographic

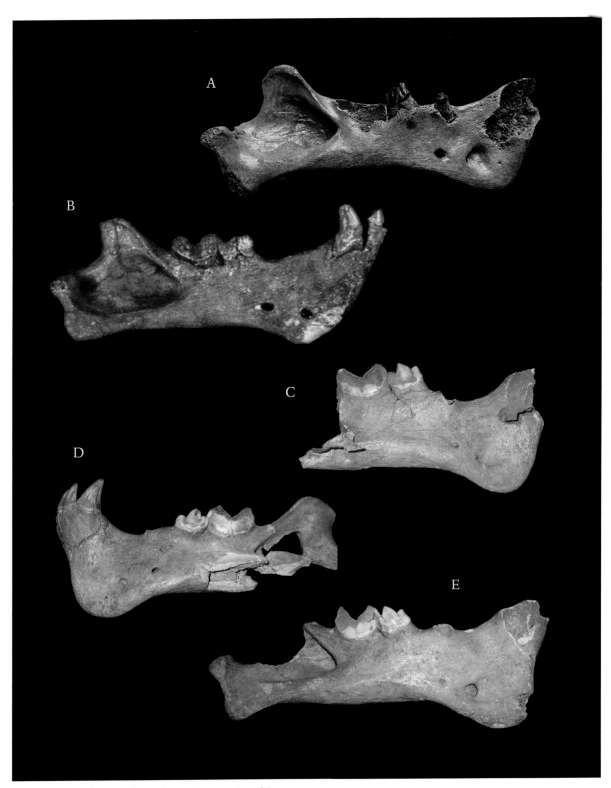

Figure 9.1. Lateral views of mandibles of two species of the genus *Homotherium*. *A*, Jaw from the North Sea, *H. latidens*, Rotterdam Natural History Museum, NMR 9991-01695. *B*, Specimen described by Backhouse (1886), which includes a lower right ramus with P_3–P_4. *C*, Partial right ramus with P_4–M_1 and the alveolus for P_3 of *Homotherium serum*, TMM 933-3533, from Friesenhahn Cave, Texas. *D*, Left ramus of the same specimen, with I_1, I_2, C_1, P_4–M_1, and the alveolus of P_3. *E*, Right ramus of *H. serum*, Friesenhahn Cave, TMM 933-1283, with P_4–M_1 with the alveolus for P_3. (illustrations W. M. S. van Logchem)

range that included the entire Holarctic. No other mammalian species shows such a combination of duration and geography.

Enormous temporal and geographic range of a single species usually results from fragmentary material that provides few diagnostic features. This can lead to the assignment of more complete but actually unrelated material to that species, and has resulted in many of the ambiguities and controversies that plague saber-tooth taxonomy. Later workers may ignore the holotype and only examine referred material to ascertain the morphology of the original species. For instance, the excellent life reconstruction of *H. latidens* material in Turner and Antón (1997) is actually based on the descriptions of *H. crenatidens* by Ballesio (1963). The referral of the excellent material described by Antón and Galobart (1999) to *H. latidens* from the Early Pleistocene Incarcal Fauna of Spain would also seem to extend the definition of the species beyond the normal bounds of variability. The nature of the problem becomes evident when we compare that material with the Late Pleistocene (28,100 ± 220 B.P.) jaw recovered from the North Sea (Reumer et al., 2003; fig. 9.1A) that is probably more nearly contemporaneous with the type material of *H. latidens* from a Late Pleistocene cave in England. The North Sea jaw is very different from the Incarcal material from Spain, and closely similar to a British ramus (Backhouse, 1886; fig. 9.1B), also from the Late Pleistocene, and jaws of *H. serum* from North America (fig. 9.1C–E).

The situation was even more complicated in North America. The first homothere to be described was *Dinobastis serus* (Cope, 1893). Cope's material was fragmentary and subsequently assigned to *Smilodon* by numerous authors (Adams, 1896; Matthew, 1910; Merriam and Stock, 1932; Simpson, 1945). Meade (1961) described most of the skeleton of *Dinobastis serus* from Friesenhahn Cave in Texas and demonstrated conclusively that it was distinct from *Smilodon*. Kurtén (1962) and Churcher (1966) affiliated *Dinobastis* and *Epimachairodus* with *Homotherium*, and these assignments are followed in subsequent works. The genus *Homotherium* itself was poorly known until Ballesio's classic monograph (Ballesio, 1963; fig. 9.2; tables 6.A.3–6.A.4). The other North American genus that has often been affiliated with *Homotherium* is *Ischyrosmilus* from the Blancan. *Ischyrosmilus* was erected by Merriam (1918) for *Machairodus ischyrus*. A historical summary of the taxonomic history surrounding *Ischyrosmilus* and its various species can be found in Churcher (1984), who came to the surprising conclusion that *Smilodon gracilis* actually belongs to *Ischyrosmilus* and that the latter genus is a Smilodontin. Schultz, Schultz, and Martin (1970) and Schultz and Martin (1970) had supported the distinctiveness of *Ischyrosmilus* from *Homotherium* but considered it a homothere on the following features: the large size of the lower incisors in *Ischyrosmilus* as compared to those in Smilodontins; the conical recurved shape of the incisors, with distinct serrations; the more rounded dorsal surface of the diastema;

the relatively shorter diastema; the greater reduction of the P$_3$; the higher paraconid on P$_4$ and the greater posterior inclination of that tooth; the relatively deep, rounded anterior margin of the masseteric fossa on the ramus; and the more elongate lower carnassial. All of these features can be found in the holotype rami of the various species of *Ischyrosmilus* and unite them with the Homotheriini rather than with the Smilodontini. Martin, Schultz, and Schultz (1988) also pointed out that *Smilodon gracilis* was definitely a Smilodontin but was distinct from other species in the genus and suggested that it might be better thought of as an advanced species of *Megantereon*. Martin (1980), after examining Ballesio's material, concluded that *Ischyrosmilus* was congeneric with *Homotherium*.

This conclusion was strongly criticized by Churcher (1984, p. 40), who seemed to indicate that Martin had been alone in such an assignment. Martin, Schultz, and Schultz (1988) showed that the lower jaws of all described species of *Ischyrosmilus* differed from the Smilodontins (*Megantereon and Smilodon*), and resembled homotheres. They accepted the assignment of these lower jaws to *Homotherium*, as had Beaumont (1978) and Savage and Russell (1983). Berta and Galiano (1983) assigned the type species of *Ischyrosmilus, I. ischyrus* to *Dinobastis,* and the remaining species (*I. idahoensis, I. johnstoni, I. crusafonti*) to *Homotherium*. Berta (1987) did the most thorough revision of *Smilodon gracilis* to date and rejected Churcher's (1984) assignment of that species to *Ischyrosmilus*.

Claiming sexual dimorphism and ordinary variability, many workers reduced the diversity of Blancan North American homotheres to a single species, *I. ischyrus,* just as the European workers had reduced the many species of Villafranchian *Homotherium* to *H. crenatidens.* Turner and Antón (1997) further reduced the European diversity to just *H. latidens.* Because lower jaws have limited taxonomic usefulness, this argument has a certain attraction in North America, as any fragmentary Blancan material might be assigned to the "only known Blancan species." However, the discovery of a nearly complete associated skeleton from the Late Blancan of Idaho (IMNH 900-11862) provided diagnostic characters beyond the lower jaw, and a comprehensive review of the various isolated remains reported in the literature (Hearst, chapter 6 in this volume). In particular, it allowed direct comparison with the Delmont Homothere skull (AMNH 95297) from South Dakota, European material, and the skeletons of *Homotherium serum* from the Late Pleistocene of Texas.

The genus *Homotherium* was originally based on material from the Villafranchian of Europe and is roughly contemporaneous with the North American Blancan, falling between 4.8 and 2.2 M.Y.A. Several decades ago, this time interval was considered Early Pleistocene (Ice Age), but now it is more usually considered to be Pliocene (latest Tertiary). Deposits formally included in the North American Pliocene

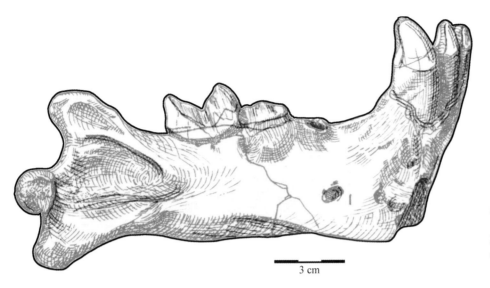

3 cm

Figure 9.2. Right mandibular ramus of *Homotherium crenatidens* (based on Ballesio, 1963, FSL 210991, fig. 13; from Martin, Schultz, and Schultz, 1988, fig. 9A).

Figure 9.3. Skull and mandible of *Megantereon cultridens,* BIOPSI 0185 (cast) CB 20, Bone Clones, Inc., showing the elongate canines and capability for achieving a large gape. (courtesy David Kronen, photo © Bone Clones, www.boneclones.com).

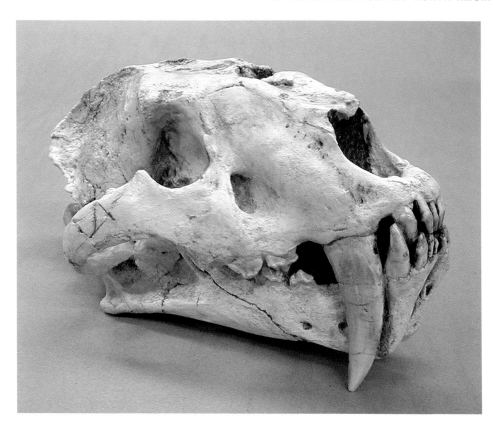

Figure 9.4. Skull and mandible of *Amphimachairodus giganteus*, BIOPSI 0180, at full occlusion, anterolateral view (cast) BC 102, Bone Clones, Inc. (courtesy David Kronen, photo © Bone Clones, www.boneclones.com)

(Hemphillian) are now generally assigned to the latest Miocene, although some would put the Late Hemphillian in the Pliocene, thereby placing the Miocene/Pliocene boundary in the middle of the Hemphillian Land Mammal Age. This doesn't seem reasonable, and supports the idea that the Hemphillian might be split into an earlier Kimballian "Early Hemphillian," based on the *Amebelodon fricki* Quarry (Barbour, 1929) in Nebraska, and a restricted Hemphillian, "Late Hemphillian," based on the Coffee Ranch locality in Texas, the type locality of the original Hemphillian. The earliest homotheres that are clearly identifiable are all younger (Blancan/Pliocene), although they, like the early dirk-tooth cat, *Megantereon* (fig. 9.3), probably have roots in the Late Miocene, conceivably from some common ancestor. We think that the derivation of *Homotherium* from the Miocene genera *Machairodus* (fig. 9.4) or *Amphimachairodus* (Kretzoi, 1927) is unlikely. It would require a shortening of the muzzle and the development of a short tail, while later forms of *Machairodus* or *Amphimachairodus* appear to be trending in the opposite direction. They also had long legs. It seems likely that the common ancestor of the short-legged *Xenosmilus* and the long-legged *Homotherium* should not share an ancestor that was already specialized toward long legs.

Xenosmilus is the sister taxon of *Homotherium* and has recently been shown to occur in South America (Mones and Rinderknecht, 2004), but in North America it is still restricted to Florida, with a possible radius from the Blancan of Arizona

(White, Morgan, and White, 2005). It may have found its greatest success in subtropical environments, including tallgrass savannah. It represents the intrusion of a homothere into the ambush predator niche at a time when that niche was only occupied by the relatively small *Smilodon gracilis*. Global cooling and drying during the Pleistocene may have ecologically stressed this genus, resulting in extinction. Several other large animals are known to have disappeared from the North American fauna at about this time (around 800,000 years ago) including the hunting dog *Xenocyon*, the giraffe-camel *Titanotylopus*, and the elephant-like gomphotheres. In the Late Pleistocene, the *Xenosmilus*-sized ambush niche was filled by greatly enlarged individuals belonging to *Smilodon*, while that previously held by *Smilodon gracilis* may have been taken over by a northern population of jaguars, *Panthera onca*.

The development of the Beringian Steppe Tundra in the Pleistocene must have inhibited the exchange of saber-tooth predators. The Late Pleistocene populations in unglaciated Beringia were isolated from the rest of North America by the continental ice sheet. The Alaskan lions, which are probably ecologically most like the homotheres, are cave lions *Panthera spelaea*, rather than the American lion *Panthera atrox* (Barnett et al., 2009; fig. 9.5). We might also expect the Alaska scimitar cat to reflect this biogeographical pattern. In that case, it is probably *Homotherium latidens* (fig. 9.6) rather than *Homotherium serum*.

Figure 9.5. Skull and mandible of *Panthera spelaea*, from the Late Pleistocene, Nome, Alaska, shown in anterolateral view at large gape (cast) BC 104, Bone Clones, Inc. (courtesy David Kronen, photo © Bone Clones, www.boneclones.com)

Pliocene *Homotherium* is known throughout the American West, and it continued to occupy this range during the Pleistocene. In the Late Pleistocene, we have a more detailed vegetational record, and Martin et al. (1985) related distinctive assemblages of animals to these vegetational distributions. They describe these distributions as faunal provinces that they named on the basis of characteristic animals occurring in those faunal provinces. The distribution of *Homotherium*, a long-legged pursuit predator, lies mostly within the *Camelops* Faunal Province, which was considered to be composed mostly of pine parkland and characterized by the most cursorially adapted large mammals. *Xenosmilus,* however, falls in the *Chlamythere* Faunal Province, characterized by subtropical animals, southern deciduous forest, and local areas of tall grass savannah (fig. 1.8B).

Because the oldest and most primitive *Homotherium* is from Mio-Pliocene deposits in Central Asia (Zdansky, 1924; Teilhard de Chardin and Piveteau, 1930; Teilhard de Chardin, 1939; Teilhard de Chardin and Leroy, 1945), we suspect that

Homotherium had a northern Asiatic origin, extending its range southward as climates cooled during the beginning of the Ice Ages. This is logical when we consider that open grasslands probably had an earlier origin in the northern part of their range, and it is to that habitat that long-legged running cats such as *Homotherium* would be especially adapted. We suppose that *Xenosmilus* also had a northern origin but became restricted to the Western Hemisphere, where it was especially successful in warmer latitudes. We doubt that either genus of homothere is derived from any of the advanced taxa of machairodontins, like *Amphimachairodus,* and we think that their similarity in jaw morphology is another case of ecomorphy. The Blancan and Villafranchian homotheres in North America and Europe are very similar, but not identical, suggesting the existence of a biogeographical filter that mitigated the faunal exchange between the two populations. In North America, there seems to have been a larger lineage *(Homotherium ischyrus)* and a smaller one *(Homotherium crusafonti).* The latter lineage seems more similar to *Homothe-*

Figure 9.6. Reconstruction of *Homotherium latidens*, lateral view. (drawing Mark Hallett)

rium serum. During the Pliocene, the Homotheriini and Smilodontini occur together, but during the Pleistocene this seems only to hold true in the Western Hemisphere.

Near the end of the Late Miocene, there was a major extinction that essentially destroyed the existing (Ogallala) Chronofauna in North America (Martin, 1985; Martin and Meehan, 2005). The later Hemphillian faunas that follow that extinction have a Plio-Pleistocene aspect, causing a few scholars to include parts of them in the Blancan. Although some characteristic Ogallala taxa may have made it into the earliest portions of this interval, the typical cat-like carnivores of the North American Ogallala *(Barbourofelis,* fig. 9.7; *Nimravides; Pseudaelurus)* seem to have become extinct. For a brief period, North America was visited by a few unusual Eurasian genera: the bear *Agriotherium;* the scimitar-tooth cat *Amphimachairodus;* and *Adelphailurus,* a close relative of the Asiatic cat *Metailurus.* This fauna intrusion was quickly replaced by the characteristic large felids of the Plio-Pleistocene Chronofauna *(Homotherium, Megantereon, Miracinonyx, Lynx, Puma,* and *Panthera).* Except for the mountain lions

(Miracinonyx, Puma) all of these genera had a Holarctic distribution, as did the feliform predators of the Ogallala Chronofauna, with the possible exception of *Nimravides.*

The Holarctic aspect of the Ogallala Chronofauna reflects an initial high latitude origin. As the Chronofauna extended farther south following the progressive cooling trend, the climatic filter at high latitude shifted to a cooler and drier ecology, becoming progressively less forested. This tended to isolate the earlier emigrants geographically, and they began to show independent evolutionary trends different from those of their Eurasian relatives. It was at this time that a new Plio-Pleistocene faunal complex probably developed in Beringia.

Traditionally, felid saber-tooth cats were assigned to a separate subfamily (Machairodontinae) and all extant cats to the subfamily Felinae. DNA evidence (Barnett et al., 2005; Barnett, 2006) suggests that the ancestor of *Smilodon* diverged from modern felids approximately 14.5 M.Y.A. This would give a basal date for the Machairodontinae at about the estimated date for the beginning of the Ogallala Chronofauna. The exact ancestry of *Smilodon* is not well defined, but this date fits well with a postulated relationship to *Paramachairodus* (Turner and Antón, 1997). It is not clear how useful the

Figure 9.7. Skull and mandible of *Barbouro-felis fricki*, University of Nebraska, U.N.S.M. 76,000 (Type), mandible partially open.

term Machairodontinae really is, and it might best be treated as a tribe containing the two genera, *Machairodus* and *Amphimachairodus*. Kurtén (1962) was one of the first to use a tribal classification for saber-tooth cats when he proposed that the saber-tooth cats of the Plio-Pleistocene be divided into a *Megantereon-Smilodon* lineage, Smilodontini "dirk-tooth cats," and the Homotheriini "scimitar-tooth cats."

A summary of our thoughts about felid saber-tooth phylogeny is represented in chapter 1 of this volume (fig. 1.8A). We think that felid cats first appear in North American in the Miocene with the genus *Pseudaelurus*. Early forms of this genus still have an elongated cranium, which they share with another earlier primitive form, *Proailurus*. These early cats were adapted for the ecotone between forest and grassland. Felids began to evolve larger body size as they moved into more open vegetation. In North America, *Pseudaelurus* appears to have split from a lineage that led directly to a saber-tooth cat, *Nimavides,* and a similar evolutionary event may have led from a large felid stock in Africa to *Dinofelis* (Werdelin and Lewis, 2001). Exactly how these forms relate to the other saber-tooth felids has yet to be resolved, and they are included in the Machairodontini, but with little confidence.

The Homotheriini are felids with relatively short, broad and coarsely serrated upper canines. They have serrations on all the dentition and a rounded arcade of large recurved incisors that almost contact the canines. They have relatively short muzzles, an elongated upper carnassial with enlarged paracone, and extreme reduction of the protocone and protocone root (protoradix). They have a small dependent flange on the ramus and a short tail. Kurtén's scheme of tribal designations was continued by Schultz, Schultz, and Martin (1970) and extended by Martin (1980). Martin (1980) also expanded the scope of the descriptive terms dirk-tooth and scimitar-tooth into an ecomorph designation, adding a new term, conical-tooth, for the condition found in modern felids. The differences in canine structure are correlated with proportional differences in the skeleton, and surely result from different methods of prey capture and killing, so that tooth construction is an important indicator of ecomorph status (Martin, 1980; Martin and Meehan, 2005). The discovery of *Xenosmilus* showed that even this indicator has limitations, and the scimitar-tooth cats were subdivided into another distinctive ecomorph type that we have termed "cookie-cutter cats" (Martin et al., 2000; Naples and Martin, 2008;

Naples et al., 2008). All four cat ecomorphs are represented by lion-sized predators in the North American Pleistocene and must have successfully partitioned the environment.

Contemporary with the homotheres, there is a largely African radiation of saber-tooth forms belonging to a more lion-like genus, *Dinofelis*, which has been recently restudied by Werdelin and Lewis (2001). Homotheres are the dominant scimitar cats in northern Eurasia and North America. *Dinofelis* is known from only a single specimen from the Blancan (Pliocene) of North America (Kurtén, 1972) and is rare in beds of equivalent age (Villafranchian) in Europe.

The oldest *Homotherium* is from Central Asia, and we think that the Homotheriini had a Northern Asiatic origin, extending its range southward as climates cooled during the beginning of the Ice Ages. We suppose that *Xenosmilus* also had a northern origin, but became restricted to the Western Hemisphere, where it was especially successful in warmer latitudes. We doubt that either genus of homothere is derived from *Amphimachairodus*, and we think that their similarity in jaw morphology is another case of ecomorphy. The Blancan and Villafranchian homotheres in North America and Europe are very similar, but not identical, suggesting that some biogeographical filter isolated the two regions during much of the evolutionary history of the two scimitar-tooth cat populations. In North America there seems to have been a larger lineage, *Homotherium ischyrus*, and a smaller one, *Homotherium crusafonti*. The latter lineage seems more similar to *Homotherium serum*. During the Pliocene, the Homotheriini and Smilodontini occur together, but during the Pleistocene this seems to hold true only in the Western Hemisphere.

SUMMARY

We propose that the Homotheriini began as a northern radiation of relatively unspecialized saber-tooth cats more closely related to the Smilodontini than to the Machairodontini. We think that this radiation extended across the Holarctic in the Pliocene and colonized North America. We suggest that there are two phases to this radiation, an earlier phase consisting of the fleet-footed species *Homotherium crenatidens* in Europe and *Homotherium ischyrus* in North America, and a later phase with shorter hind legs, *Homotherium serum* in North America and probably *Homotherium latidens* in Europe. In North America, a short-legged ambush genus, *Xenosmilus*, has a brief occurrence in the early Pleistocene, and it and *Homotherium* are now both known from South America.

APPENDIX A

The specimens studied in this book are housed in the following repositories:

American Museum of Natural History (AMNH)
Babiarz Institute of Paleontological Studies (BIOPSI)
Faculté des Sciences, Université Claude Bernard, Lyon (FSL)
Field Museum of Natural History (FMNH)
Idaho Museum of Natural History (IMNH)
John Day Fossil Beds National Monument (JODA)
Los Angeles County Museum (LACM)—present location
        of UC 22343 discussed in J. C. Merriam (1918)
Los Angeles County Museum of Natural History, George C.
        Page Museum, Hancock Collection (LACMHC)
Rotterdam Natural History Museum (NMR)
Russian Academy of Sciences at Moscow (PIN)
Texas Memorial Museum (TMM)
University of California Museum of Paleontology at Berkeley
        (UCMP), also previously identified as UC
University of Florida at Gainesville (UF)
University of Kansas Natural History Museum, Mammal
        Division (KU)
University of Kansas Vertebrate Paleontology (KUVP)
University of Nebraska State Museum (UNSM)
West Texas State College, Panhandle Plains Historical Museum,
        Canyon, Texas (WT), now known as PPHM

APPENDIX B

When measuring specimens, researchers often use different measurement points. Variation in measurement technique may result from a variety of factors that include the nature of the information being sought and comparisons to previous studies and publications. The measurement points shown in this appendix are intended to assist readers in understanding the basis of conclusions in this book and are those measurements that the chapter authors followed throughout. The measurement points and the corresponding numbers can be cross referenced to the specimen element measurement tables in the chapters.

Dorsal View

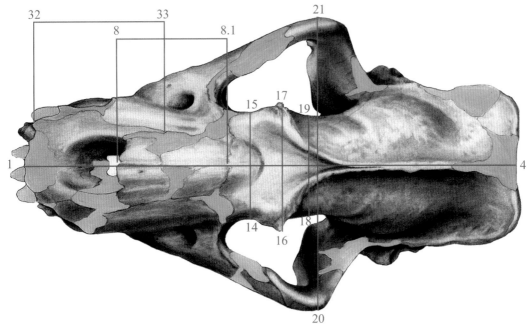

Ventral View

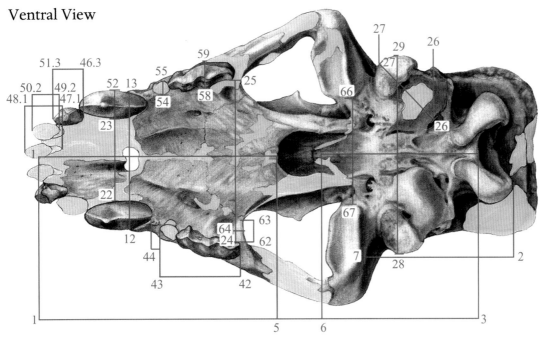

Figure A.B.1. (*above and overleaf*) Red-line measurement diagrams showing the points where skull measurements were taken for appendix tables, chapters 4 and 6. Indicated by brackets on five views of the skull using *Xenosmilus hodsonae*, BIOPSI 101, as a template.

Lateral View

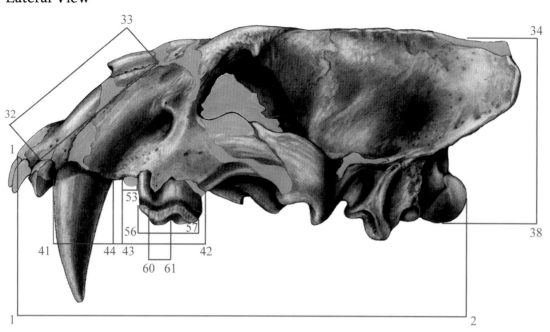

Anterior View

Posterior View

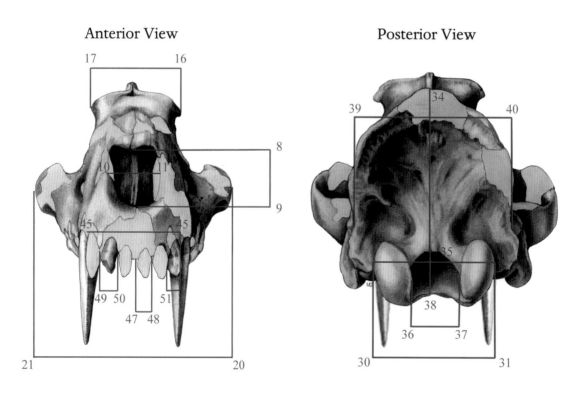

Anterior View

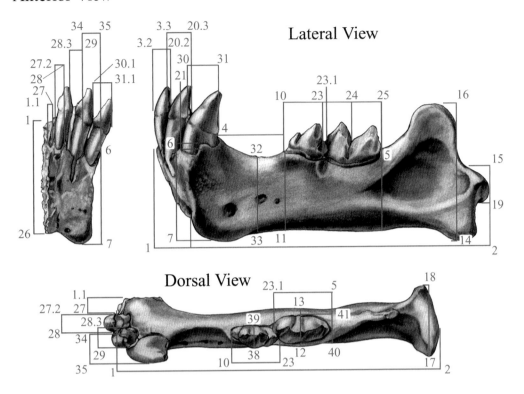

Lateral View

Dorsal View

Medial View

Posterior View

3 cm

Figure A.B.2. Red-line measurement diagrams showing the points where the mandible measurements were taken for appendix tables, chapters 4 and 6. Indicated by brackets on five views, using *Xenosmilus hodsonae*, UF 60,000, as a template. Measurements of features: anterior view *(top left)*; lateral view *(top right)*; dorsal view *(middle)*; medial view of the condyle *(bottom left)*; posterior view *(bottom right)*.

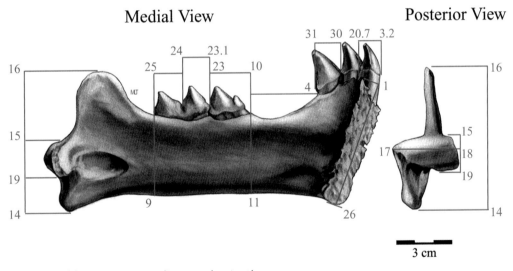

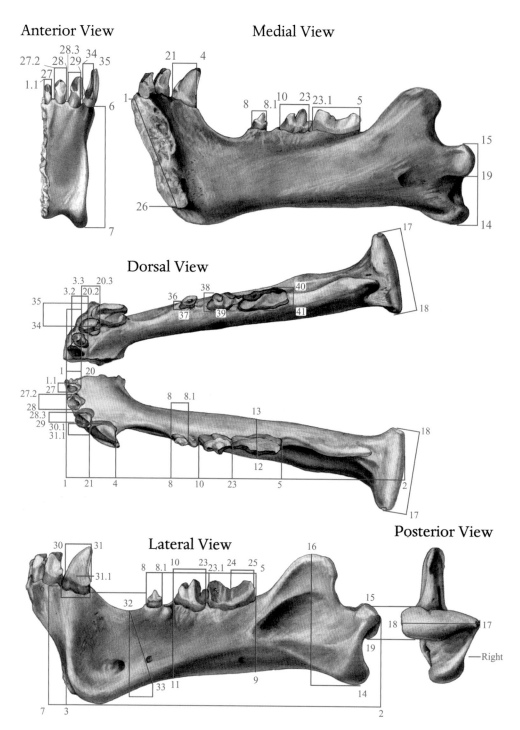

Figure A.B.3. Red-line measurement diagrams showing the points where mandible measurements were taken for the appendix tables, chapter 4. Indicated by brackets on five views of the mandible of *Homotherium cf. serum*, BIOPSI 112. Measurements of the features: anterior view *(top left)*; medial view *(upper right)*; dorsal views *(center)*; lateral view *(bottom left)*; posterior view of the right condyle *(bottom right)*.

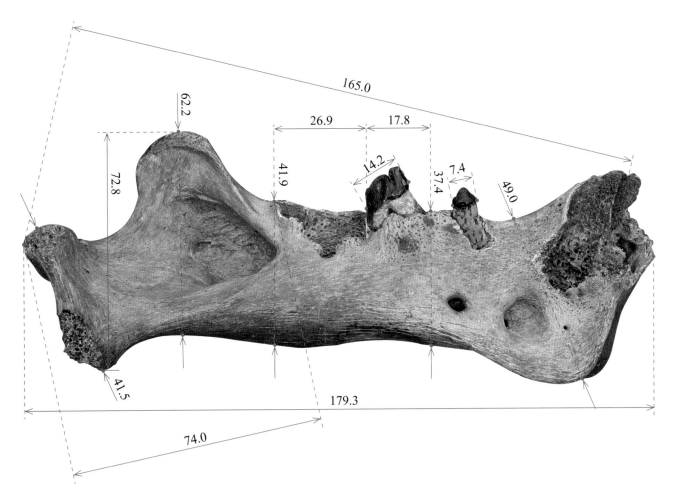

Figure A.B.4. This image of the North Sea Jaw, NMR 9991-01695 shows the actual points and measurements in brackets, as taken from the literature. These measurements represent *Homotherium latidens* from chapter 9.

GLOSSARY

Adductor. Any of the muscles that draw a part of the body toward its median line or toward the axis of an extremity.

Agenesis. Absence or failure of formation of any biological part during the embryonic stage.

Age of Mammals. The Cenozoic era. The most recent of three classic geological eras, covering a time period from 65.5 million years ago to the present, when mammals evolved to the dominant life form on earth, replacing dinosaurs from the previous Mesozoic era.

Alveolus. The cavity or sockets in tooth-bearing elements of bone (e.g., maxilla, dentaries) in which the roots and teeth are formed.

Anapophysis. Projection or protuberance located on the dorsal side of lumbar and thoracic vertebrae directed posteriorly.

Ankylosis. Fusion of bones to form a single unit.

Anomaly. Deviation from a rule or from what is regarded as normal.

APD. Anterior-posterior diameter. Distance measured from the closest end (proximal) to the farthest end (distal) of a bone or element; measurement from posterior to ventral side of an element. Mesiodistal diameter (dentition).

Apically. Toward the root.

Apomorphic. Derived character state.

Apophysis. Secondary center of ossification.

Arthritis deformans. Old name for rheumatoid arthritis, usually applied inaccurately.

Atlas. First vertebra of a tetrapod that articulates with the skull.

Auditory bulla. Bone or bones surrounding the inner ear chamber of a mammalian skull.

Axis. Second vertebra in a tetrapod following the axis.

Basicranium. Posterior and ventral portion of the skull. Includes the basioccipital, basisphenoid, petrosal, auditory bulla, and partial squamosal and pterygoid bones.

Beringia. An area of land periodically exposed during the Pleistocene (Ice Age), creating a land bridge connecting the Asian and North America continents.

Bering Strait. Shallow sea between Siberia and Alaska connecting the Chukchi Sea (Arctic Ocean) with the Bering Sea (Pacific Ocean).

Bison. Taxonomic group containing six species of large, even-toed ungulates within the subfamily Bovinae. Two extant species are *Bison bison* of North America and *Bison bonasus* in Europe.

Brachiocephalic. Skull short anteroposteriorly. Opposite of dolichocephalic.

Browser. An animal with a diet composed mostly of leaves and stems of shrubs, bushes, and trees.

Buccal. Labial or external/lateral side of a tooth.

Calcium Pyrophosphate Deposition Disease (CPPD). Form of arthritis related to the deposit of calcium crystals.

Carnassial. Specialized blade-like teeth of carnivorous mammals, best developed in the families Felidae and Nimravidae. They consist of a pair of teeth (upper P_4

and lower M$_1$) and are used to cut or slice through meat.

Carnivore. Animal with a diet mainly of vertebrate meat or flesh.

Carotid artery. The artery that supplies the head and neck with oxygenated blood.

Carpal. Any of the skeletal elements of the wrist.

Caudal. Of or relating to the tail.

CEJ (Cemento-Enamel Junction). The boundary where the enamel-covered crown of the tooth meets the cementum-covered root.

Cementum. Material overlying dentin on the root of a tooth to which the periodontal ligament or gingiva attaches.

Cenozoic. Most recent of three classic geological eras, covering a time period from 65.5 M.Y.A. to the present. The Age of Mammals.

Centrum. The spool or cylindrical-shaped body of a vertebra.

Cervical. Of or relating to the neck.

Cingulum. Ridge at the base of the crown on the outer margin of some mammal teeth.

Congeneric. Two or more plants or animals that are members of the same taxonomic genus.

Conical-tooth cat. One of three morphological groups of nimravid or felid-like cats with a dentition typical of all extant cat or cat-like carnivores. Upper and lower canines are oval, shorter, and non-serrated, with a blunt point on the anterior end. Bite and crush form of predators. Having a canine that is basically cone-shaped, as is usual in carnivorous mammals.

Conspecific. Of or relating to a single species. Two or more individuals belonging to the same species.

Cookie-cutter cat. Scimitar-tooth cat capable of cutting a bolus of meat or flesh from its prey by using a specialized bite that comprises the combined action by the procumbant serrated incisor arcades, in unison with the short, coarsely serrated sabers.

Costovertebral. Junction of rib and vertebral body.

Crown. Top or enamel-covered portion of a tooth above the root.

Cursorial. Well adapted for running at high speed or for long distances.

Deciduous incisor. First or primary incisor in the dental development of mammals (i.e., milk or temporary tooth).

Demi-facet. Posterior capitular facet of centrum body for capitulum (proximal head) of rib.

Derived characters. Characters of organisms more recently evolved, having the same function but changed from the original ancestral form.

Diaphysis. Main or midsection (shaft) of a limb bone in mammals, excluding the two epiphyseal ends or epiphyses.

Diastema. Space or gap between adjacent teeth.

Diatomite. Layer of sedimentary rock composed of the remains of diatoms, single-celled algae with cell walls made of silica (SiO_2).

Diffuse Idiopathic Skeletal Hyperostosis (DISH). Generalized ossification of ligament, tendon, and capsule insertions.

Digitigrade. Animal that walks or stands on its digits or toes.

Dire wolf. Extinct carnivorous mammal of the genus *Canis* that evolved in North America about 100,000 years ago during the Late Pleistocene (Ice Age).

Dirk-tooth cat. One of three morphological groups of saber-tooth nimravid or felid-like cats that have long and relatively slender, unserrated or finely serrated upper canines, associated with short, stout legs for stalking with short pursuit. Its bite is perpendicular to the tooth axis.

Dolichocephalic. Having a head that is disproportionately long. Opposite of brachycephalic.

Ecomorphy. Similar character traits developed on separate unrelated taxa, resulting from identical ecological pressures or circumstances.

Ecotone. Marginal habitat between two distinct environments.

Elastic modulus. Mathematical description of an object's tendency to be deformed elastically when a force is applied to it, defined as the slope of its stress-strain curve.

Embrasure. Area around the interproximal contact area of adjacent teeth.

Enamel. Hardest, most mineralized material in a vertebrate animal, composed of hydroxyapatite that forms the shiny outer covering of teeth.

Epiphysis. Rounded or expanded ends in mammalian limb bones that develop separately from the shaft or diaphysis in juveniles. They fuse as full (adult) growth is attained.

Epoch. Formal unit of geologic time and subdivisions of geologic periods.

Era. Formal unit of geologic time, consisting of two or more contiguous periods.

Erosion. Wear or reduction resulting from atmospheric conditions (wind/water abrasion). Indentation of bone (pressure atrophy).

Etiology. Cause or origination of why events occur. In medical theory, explaining the cause of a disease or pathology.

Eutherians. Group of placental mammals, extant or extinct, that are more closely related to living placentals than living marsupials.

Exostosis. Abnormal benign bone surface growth.

Extant. Living species or taxon.

Extinction. In biology and ecology, the death of every member of a particular species or group of taxa.

Extirpate. To wipe out, exterminate, or destroy completely.

Facies. Suite of distinctive characteristics that distinguish rock units or groups of fossils contained within the bed from other beds deposited at the same time and in lithologic continuity, as in mud-facies.

Family. Rank in taxonomic hierarchy between order and genus.

FEA. Finite element analysis.

Foramen. Any opening or hole in a bone for the passage of a nerve or blood vessel and, in some cases, muscle fibers (hystricomorphy).

Formation. Basic stratigraphic unit of rock in geology, consisting of one or more beds of the same lithology or distinctive, related lithologies (rock types).

Fossa. Depression, hollow area, or concavity in a bone.

Fusion. Union.

Generic. Pertaining or relating to a genus.

Genus. Lower level taxonomic rank immediately above the rank of species.

Glenoid cavity. Socket (fossa), of an articular depression, as in the scapulohumeral joint.

Glenoid process. Ventral area of the temporal bone/zygomatic arch where the condylar process of the mandible articulates with the glenoid cavity of the squamosal.

Graben. Elongate, depressed crustal unit or block bounded by normal faults on its sides. A rift, valley, or trough.

Grazer. Animal with a diet consisting mostly of grasses.

Group. In geology and stratigraphy, a unit of rock consisting of two or more formations.

Herbivore. Any animal with a diet consisting mostly of plant matter of any kind.

Holotype. Single specimen upon which the scientific name and description of a species is based.

Homeobox. DNA sequence. Related to regulation of patterns of development.

Hyena. Bone-crushing, dominant scavengers of the Old World. A mammalian family of the order Carnivora native both to the African and Asian continents, with an origin about 26 M.Y.A. in Eurasia.

Hyperapophysis. Lateral and posteriorly projecting process on the dorsal side of a vertebra.

Hystricomorphic. Condition where anterior masseter muscle fibers pass through the hypertrophied infraorbital foramen resulting in a shallow fossa, just anterior and lateral to the foramen, as seen in some extant rodents.

Ice Age. Denotes the geological period of long-term reduction in the Earth's surface and atmospheric temperatures. The most geologically recent Ice Age period (1.8 M.Y.A. to 10,000 years ago), known as the Pleistocene epoch.

Incisiform. Having the form of, or resembling, a typical incisor tooth.

Incisor. Anterior-most teeth in the mammalian dentition, with roots in the premaxillae.

Infectious spondylitis. Vertebral infection.

Inflammatory arthritis. Classification of arthritis characterized by inflammation and erosion.

Interspecific. Relating to two or more species.

Intraspecific. Relating to, or being within a single species.

Jugular vein. The paired veins, located laterally in the neck, that bring deoxygenated blood from the head back to the heart.

Labial. External, outside, or lateral side of a tooth, closest to the lips and cheeks.

Lingual. Internal side of a tooth facing the tongue. Opposite the labial or buccal side.

LLD (Labial-Lingual Diameter). Measurement of distance from the medial (inner) or tongue side to the opposite or labial-buccal (outside), as in a carnassial tooth. Inside to outside distance (buccolingual).

Mammoth. Any species of the extinct genus *Mammuthus*. A proboscidean, and close relative to the modern elephants that lived during the Pliocene and Pleistocene epochs, about 4.8 M.Y.A. to 4,500 years ago.

Manus. Forefoot or hand, composed of carpals, metacarpals, and digits (phalanges).

Marsupial. Group of mammals extending back approximately 125 M.Y.A. Pouched mammals in which the young are born in an immature state.

Mastodon. Members of the extinct genus *Mammut* of the order Proboscidea, similar to and contemporaneous with mammoths, but with teeth adapted to a browsing (chewing) woodland lifestyle.

Metapophysis. Tubercle projecting from the anterior articular process of some thoracic and lumbar vertebrae.

Milankovitch orbial variations. Cyclic changes in the Earth's axial tilt and precession of its orbit, theorized as the cause of approximately 100,000-year ice-age cycles during the past two million years of the Quaternary (Pleistocene-Recent) period.

Molar. Posterior or rear-most teeth in the mammalian dentition that are not replaced and less morphologically variable than the premolars.

Morphology. Form or features that make up and define the structure of an organism.

Myositis ossificans. Dystrophic ossification (calcification) of muscle.

NALMA (North American Land Mammal Age). Discrete geological units representing periods of time characterized by their unique fossil assemblages.

Nimravids. Eocene to Late Oligocene (Early Arikareean) saber-tooth cats with no ossified auditory bulla. In these cats, the entotympanic is not fused with ectotympanic.

Orbit. Skull concavity or depression that contains the eye.

Orbital fissure. Opening between the floor and lateral

wall of the orbit, serving as a conduit for nerves and blood vessels.

Osteoarthritis. Slowly progressive form of arthritis, characterized by cartilage deterioration and osteophytes.

Osteomyelitis. Bone infection.

Osteophyte. Bony overgrowth at joint or vertebral margins.

Overkill. Excessive harvesting or killing of a species to the extent that it interferes with reproduction of enough new individuals to maintain populations, resulting in decline and early extinction.

Paleontology. The scientific study of past life.

Paleopathology. The study of ancient disease.

Panamanian Isthmus. The narrow strip of land that lies between the Caribbean Sea and Pacific Ocean linking the North and South American continents. Formed about three million years ago during the Pliocene epoch.

Paratype. One or more referred specimens specifically designated in the original description of the holotype.

Parsimony. In a character-based phylogenetic tree, this is the least number of evolutionary steps required to explain observed morphological data.

Patella (kneecap). Sesamoid bone developed in the tendon of the quadricepts femoris muscle.

Pathological. Altered or abnormal condition caused by disease. Relating to pathology.

Pathology. Anatomical part of an individual with unusual morphology caused by illness, malnourishment, injury, or genetic defect.

Pectoral girdle. In mammals, typically consists of the clavicles and scapulae. The coracoid is usually lacking.

Period. Formal unit of geologic time between era and epoch (e.g., Cambrian Period).

Periosteal. Related to the dense fibrous membrane covering bone surfaces.

Pes. Rear or hind foot, consisting of metatarsals, tarsals, and digits (phalanges).

Phalanx/phalanges. Bone(s) of digit(s) of manus or pes composed of proximal (1st), middle (2nd), and distal (3rd) phalanx (claw).

Phylogeny. Genealogy of life.

Placental. Referring to mammals that bear their young live and at an advanced stage. Having an organ, the placenta that connects a developing fetus to the uterine wall and supplies nourishment and waste removal. Placental mammals. *See* Eutherians.

Plantigrade. In mammals, walking with podials and metatarsals flat on the ground.

Plesiomorphic. Primitive or ancestral character state, preceding an apomorphic or derived character state.

Postcranial. Relating to the portion of the skeleton other than the skull and mandible.

Premolar. In mammalian dentition, teeth posterior to the canines and anterior to the molars.

Pressure atrophy. Indentation of bone resulting from excessive applied pressure, resembling an erosion.

Primitive character. Individual shape or form having evolved earlier from another similar feature to which it is being compared. *See* Plesiomorphic.

Procumbent (procumbant). Leaning forward, as in procumbent lower incisors.

Prognathous. Having the lower jaw facing anteriorly and projecting beyond the face. Associated with procumbant incisor tooth arcades.

Propralinal. Jaw movement restricted to anterior-posterior direction and parallel to the main axis of the mandible.

Protoradix. Root of an undeveloped protocone on the upper carnassial P^4.

Pseudoarthrosis. Pathological entity characterized by a non-osseous union of bone fragments of a fractured bone.

Ramus. One side of a mammalian mandible. Hemimandible. The body or bone of the mandibles.

Regression. Sea level or lake level drop with respect to the surrounding land area.

Retractile claws. Mechanism whereby a cat's claws are withdrawn or retracted by pivoting the distal phalanx of the toe with the claw over the tip of the next more proximal phalanx when not in use. This same action extends or unsheathes the claws, spreading the toes at the same time.

RLB (Rancho La Brea). One of the best known fossil sites in the world, where the skeletal remains of animals trapped in asphalt have been recovered. It is in Los Angeles, California.

Root. Basal portion of a tooth that is not covered by enamel and attaches to the mandible by a periodontal ligament.

Saber-tooth cats. Scimitar- or dirk-tooth cats. Group of cats with enlarged (hypertrophied) bladed upper canines.

Sacral. Vertebra or vertebrae that articulate with the pelvis and may be fused to form the sacrum.

Sacralization. Developmental abnormality in which the last lumbar vertebra develops enlarged, elongated ala (wings or transverse processes), similar to the sacrum, but does not necessarily fuse to the sacrum or ilium.

Scimitar-tooth cat. Group of nimravid or felid-like cats that have relatively short, broad, coarsely serrated sabers with generally long legs typical of cursorial predators. Bite using inflected arc trajectory similar to conical-tooth cats.

Serrated. Having notched or jagged edges. A series of small denticles on the edge of the tooth.

Skull. Cranium and bones of the face.

Species. Group of morphologically and biologically similar individuals with one or more diagnostic characters that distinguish it from all other species. The basic rank in the taxonomic hierarchy.

Spondyloarthropathy. Form of inflammatory arthritis, characterized by subchondral erosions or fusion of peripheral or axial joint (including sacroiliac, costo-vertebral, and zygapophyseal) and vertebrae.

Spondylosis deformans. Osteophytes at vertebral margins.

Stress fracture. Tiny crack in bone caused by repeated application of a heavy load in excess of the bone's elastic modulus.

Subchondral. Area of bone underlying the cartilage of the joint.

Subluxation. Incomplete or partial dislocation.

Subspecies. Lowest taxonomic rank. Subdivision of species used to distinguish slight variations in populations within a species.

Supination. To turn or rotate a body segment such as the hand or forearm so that the palm faces up or forward. To turn or rotate the manus by adduction and inversion.

Suture. A frequently interdigitated, complex junction between the bones of the skull.

Sympatric. Occurring in the same range or general area without any identity reduction from breeding between species.

Symphysis. Area where two bones articulate, containing a fibrocartilage pad, as in the anterior connection of the right and left hemi-mandibles or ventral connection of the right and left pubic bones.

Synapomorphies. Shared derived characters used to derive relationships in cladistic analysis.

Synonymy. The suggestion that two or more described taxonomic units actually only represent one. In such a case the older name is retained and the later ones are suppressed.

Syntype/syntypic. In the case where the original describer of a species included several specimens as types, but did not designate one as the holotype. All examples have the same status as syntypes, until a later reviewer elects one to stand as the holotype. In cases where more than one species is represented in the syntypic series, that decision will fix the taxonomic position of the proposed species.

Systematics. The study of the diversification of life, both past and present, and the relationships among living organisms through time.

Tarsal. Any of the skeletal elements that compose the ankle (e.g., astragalus).

Tar seeps. Where liquid or gaseous hydrocarbons escape to the surface through fractures in the rock layers and form puddles, pits, or ponds of asphalt on the surface.

Taxon. Group of one or more phylogenetically related organisms belonging to a taxonomic unit, with common characters that differentiate them from other taxonomic units (e.g., population, genus, family, order).

Taxonomy. The study of classifying and naming biologic organisms.

Tertiary. Classic term for the geologic period 65 to 2.2 M.Y.A., beginning with the demise of the dinosaurs and the beginning of the Cenozoic era to the start of the most recent Ice Age or Pleistocene epoch.

Thoracic. Of or relating to the chest or thorax region.

TMJ (Temporomandibular Joint). The location where the condyle of the mandible articulates with the glenoid fossa of the temporal bone, allowing the mandible to be elevated, depressed, protruded, retracted, and moved mediolaterally.

Transgression. Sea level or lake level rise with respect to the surrounding land area.

Trenchant. Sharp cutting edge. Clearly or sharply defined. Distinct.

Tubercle. Small knob-like prominence on a bone. Smaller than a tuberosity.

Tuberosity. Knob or prominence on a bone. Larger than a tubercle.

Vertebrate. Common name for members of the subphylum Vertebrata. Animals with backbones or spinal columns.

Vestigial. Rudimentary remnant of a structure that functioned in a previous stage of species or individual development.

Volcaniclastic. Clasts or particles derived from volcanic rocks, deposited either from volcanic eruptions or reworked from earlier deposits of volcanic rock.

Zygapophyseal. Vertebral facet joint.

Zygomatic arch. Cheek bones. Serves as attachment site for muscles that help elevate the jaw.

LITERATURE CITED

Abler, W. L. 1992. The serrated teeth of tyrannosaurid dinosaurs, and biting structures in other animals. Paleobiology 18:161–183.

Adams, G. I. 1896. The extinct Felidae of North America. American Journal of Science 4:419–444.

Agenbroad, L. D. 1984. New World mammoth distribution; pp. 90–112 *in* P. Martin and R. G. Klein (eds.), Quaternary Extinctions: A Prehistoric Revolution, University of Arizona Press, Tucson.

Akersten, W. A. 1985. Canine Function in *Smilodon* (Mammalia, Felidae, Machairodontinae). Natural History Museum of Los Angeles County, Los Angeles.

———. 2005. The role of incisors and forelimbs in the shear bite and feeding of *Smilodon*. Journal of Vertebrate Paleontology 25:31A.

Alfaro, M. E., J. Janovetz, and M. W. Westneat. 2001. Motor control across trophic strategies: Muscle activity of biting and suction feeding fishes. American Zoologist 41:1266–1279.

Andersson, K. 2004. Elbow-joint morphology as a guide to forearm function and foraging behaviour in mammalian carnivores. Zoological Journal of the Linnean Society 142:91–104.

Andersson, K., and L. Werdelin. 2003. The evolution of cursorial carnivores in the Tertiary: Implications of elbow-joint morphology. Proceedings of the Royal Society, London 270, Suppl. 2:S163–S165.

Antón, M., and A. Galobart. 1999. Neck function and predatory behaviour in the scimitar toothed cat *Homotherium latidens* (Owen). Journal of Vertebrate Paleontology 19:771–784.

Antón, M., M. J. Salesa, J. F. Pastor, I. M. Sanchez, S. Fraile, and J. Morales. 2004. Implications of the mastoid anatomy of larger extant felids for the evolution and predatory behaviour of sabretoothed cats (Mammalia, Carnivora, Felidae). Zoological Journal of the Linnean Society 140:207–221.

Arambourg, C. 1947. Mission scientifique de l'Omo, 1932–1933. Vol. 1:438–443. Géologie-Anthropologie. P. Lechevalier, Muséum National d'Histoire Naturelle, Paris.

Backhouse, J. 1886. On a mandible of *Machaerodus* from the forest-bed. Quarterly Journal of the Geological Society of London 42:309–312.

Ballesio, R. 1963. Monographie d'un *Machairodus* du gisement villafranchien de Seneze: *Homotherium crenatidens* Fabrini. Traveaux du Laboratoire de Géologie de la Faculté de Sciences de Lyon 9:1–129.

Barbour, E. H. 1929. The mandible of *Amebelodon fricki*. Bulletin of the Nebraska State Museum 1, issue 15.

Barnett, R. 2006. Molecular evolution of extinct felids; *in* Department of Zoology, Oxford University.

Barnett, R., I. Barnes, M. J. Phillips, L. D. Martin, C. R. Harington, J. A. Leonard, and A. Cooper. 2005. Evolution of the extinct sabretooths and the American cheetah-like cat. Current Biology 15:R589–R590.

Barnett, R., B. Shapiro, I. Barnes, S. Y. W. Ho, J. Burger, N. Yamaguchi, T. F. G. Higham, H. T. Wheeler, W. Rosendahl, A. V. Sher, M. Sotnikova, T. Kuznetsova, G. F. Baryshnikov, L. D. Martin, C. R. Harington, J. A. Burns, and A. Cooper. 2009. Phylogeography of lions (*Panthera leo* ssp.) reveals three distinct taxa and a late Pleistocene reduction in genetic diversity. Molecular Ecology 18:1668–1677.

Beaumont, G. de. 1978. Notes complémentaires sur quelques félidés (carnivores). Archives des sciénces, Genéve 31:219–227.

Bell, C. J., E. L. Lundelius Jr., A. D. Barnosky, R. W. Graham, E. H. Lindsay, D. R. Ruez Jr., H. A. Semken Jr., S. D. Webb, and R. J. Zakrzewski. 2004. The Blancan, Irvingtonian, and Rancholabrean mammal ages; pp. 232–314 *in* M. O. Woodburne (ed.), Late Cretaceous and Cenozoic Mammals of North America: Biostratigraphy and Geochronology. Columbia University Press, New York.

Benton, M. J., and B. C. Emerson. 2007. How did life become so diverse? The dynamics of diversification according to the fossil record and molecular phylogenetics. Palaeontology 50, no. 1:23–40.

Berkovitz, B. K. 2000. Tooth replacement patterns in non-mammalian vertebrates; pp. 186–200 in M. F. Teaford, M. M. Smith, and M. W. J. Ferguson (eds.), Development, Function and Evolution of Teeth. Cambridge University Press, Cambridge.

Berta, A. 1987. The sabercat Smilodon gracilis from Florida and a discussion of its relationships (Mammalia, Felidae, Smilodontini). Florida State Museum, University of Florida, Gainesville.

Berta, A., and H. Galiano. 1983. Megantereon hesperus from the Late Hemphillian of Florida with remarks on the phylogenetic relationships of Machairodonts (Mammalia, Felidae, Machairodontinae). Journal of Paleontology 57:892–899.

Biknevicius, A. R., and B. Van Valkenburgh. 1996. Design for killing: Craniodental adaptations of predators; pp. 393–428 in J. L. Gittleman (ed.), Carnivore Behavior, Ecology, and Evolution. Vol. 2. Cornell University Press, Ithaca, NY.

Biknevicius, A. R., B. Van Valkenburgh, and J. Walker. 1996. Incisor size and shape: Implications for feeding behaviors in saber-toothed "cats." Journal of Vertebrate Paleontology 16:510–521.

Binder, W. J., and B. Van Valkenburgh. 2010. A comparison of tooth wear and breakage in Rancho la Brea sabertooth cats and dire wolves across time. Journal of Vertebrate Paleontology 30:255–261.

Bjorkengren, A. G., D. J. Sartoris, S. Shermis, and D. Resnick. 1987. Patterns of paravertebral ossification in the prehistoric saber-toothed cat. American Journal of Roentgenology 148:779–782.

Bohlin, B. 1940. Food habits of the Machaerodonts, with special regard to Smilodon. Bulletin of the Geological Institute of the University of Uppsala 28:156–174.

———. 1947. The saber-toothed tigers once more. Bulletin of the Geological Institute of the University of Uppsala 32:11–20.

Bryant, H. N. 1988. Delayed eruption of the deciduous upper canine in the sabertoothed carnivore Barbourofelis lovei (Carnivora, Nimravidae). Journal of Vertebrate Paleontology 8:295–306.

———. 1996. Force generation by the jaw adductor musculature at different gapes in the Pleistocene sabertoothed felid Smilodon; pp. 283–299 in K. M. Stewart and K. L. Seymour (eds.), Paleoecology and Paleoenvironments of Late Cenozoic Mammals. University of Toronto Press, Toronto.

Bryant, H. N., and T. J. Fremd. 1998. Revised biostratigraphy of the Nimravidae (Carnivora) from the John Day Basin of Oregon. Journal of Vertebrate Paleontology 18:30A.

Burton, R. F. 1884. The Book of The Sword. Chatto and Windus, London.

Caro, T. M. 1994. Cheetahs of the Serengeti Plains: Group Living in an Asocial Species. University of Chicago Press, Chicago.

Christiansen, P. 2006. Sabertooth characters in the clouded leopard (Neofelis nebulosa Griffiths 1821). Journal of Morphology 267:1186–1198.

———. 2008. Species distinction and evolutionary differences in the clouded leopard (Neofelis nebulosa) and Diard's clouded leopard (Neofelis diardi). Journal of Mammalogy 89:1435–1446.

Christiansen, P., and J. S. Adolfssen. 2007. Osteology and ecology of Megantereon cultridens SE311 (Mammalia, Felidae, Machairodontinae), a sabrecat from the Late Pliocene-Early Pleistocene of Seneze, France. Zoological Journal of the Linnean Society 151:833–884.

Churcher, C. S. 1966. The affinities of Dinobastis serus Cope 1893. Quaternaria 8:263–275.

———. 1984. The Status of Smilodontopsis (Brown, 1908) and Ischyrosmilus (Merriam, 1918): A Taxonomic Review of Two Genera of Sabretooth Cats (Felidae, Machairodontinae). Royal Ontario Museum, Toronto.

Coles, J. M. 1973. Archaeology by Experiment. Charles Scribner's Sons, New York.

Coltrain, J. B., J. M. Harris, T. E. Cerling, J. R. Ehleringer, M. -D. Dearing, J. Ward, and J. Allen. 2004. Rancho La Brea stable isotope biogeochemistry and its implications for the palaeoecology of late Pleistocene, coastal southern California. Palaeogeography, Palaeoclimatology, Palaeoecology 205:199–219.

Cook, R. D., D. S. Malkus, M. E. Plesha, and R. J. Witt, eds. 2002. Concepts and Applications of Finite Element Analysis. 4th ed. Wiley, Hoboken, NJ.

Cope, E. D. 1880. On the extinct cats of America. American Naturalist 14:833–858.

———. 1887. A saber-tooth tiger from the Loup Fork. American Naturalist 21:1019–1020.

———. 1893. A new Pleistocene saber-tooth. American Naturalist 27:876–877.

Cuvier, G. 1824. Recherches sur les Ossemens fossiles. Rev. ed. Dufour et d'Ocage, Paris. 4–5:517.

Davis, D. D. 1964. The giant panda: A morphological study of evolutionary mechanisms. Fieldiana, Zoology Memoirs, Chicago Natural History Museum 3:1–339.

De Bonis, L. 1975. Aperçu sur les Félinés Machairodontes; pp. 683–692 in Problèmes Actuels de Paléontologie-Évolution des Vertébrés. Actes du Colloque International du CNRS, Paris.

———. 1976. Un Félidé a longus canines de la colline de Perrier (Pui-de-Dome): Ses rapports avec les Félinés Machairodontes. Annales de Paléontologie (Vertébrés) 62:159–198.

Duckler, G. L. 1997. Adding insults to injuries: Evidence of excessive trauma in Late Pleistocene sabertooths at Rancho La Brea. Journal of Vertebrate Paleontology 17:44A.

Emerson, S. B., and L. Radinsky. 1980. Functional analysis of sabertooth cranial morphology. Paleobiology 6:295–312.

Emry, R. J., P. R. Bjork, and L. S. Russell. 1987. The Chadronian, Orellean and Whitneyan Land Mammals Ages; pp. 118–153 in M. O. Woodburne (ed.), Cenozoic Mammals of North America: Geochronology and Biostratigraphy. University of California Press, Los Angeles.

Erickson, G. M., and K. H. Olson. 1996. Bite marks attributable to Tyrannosaurus rex: preliminary description and implications. Journal of Vertebrate Paleontology 16:175–178.

Erickson, G. M., S. D. Van Kirk, J. Su, M. E. Levenston, W. E. Caler, and D. R. Carter. 1996. Bite-force estimation for Tyrannosaurus rex from tooth-marked bones. Nature 382:706–708.

Ewer, R. F. 1973. The Carnivores. Cornell University Press, Ithaca, NY.

Fabrini, E. 1890. Machairodus (Megantereon) del Val d'Arno superiore. Bollettino Comitato Geologico d'Italia, ser. 3, 1:121–144, 161–177, pls. 4–6.

Feranec, R. S. 2008. Growth differences in the saber-tooth of three felid species. Palaios 23:566.

Fink, W. L. 1993. Revision of the piranha genus *Pygocentrus* (Teleostei, Characiformes). Copeia 1993:665–687.

Frazzetta, T. H. 1988. The mechanics of cutting and the form of shark teeth (Chondrichthyes, Elasmobranchii). Zoomorphology 108:93–107.

Freeman, B., L. G. Nico, M. Osentoski, H. L. Jelks, and T. M. Collins. 2007. Molecular systematics of Serrasalmidae: Deciphering the identities of piranha species and unraveling their evolutionary histories. Zootaxa 1484:1–38.

Freeman, P. W., and C. A. Lemen. 2006. Puncturing ability of idealized canine teeth: Edged and non-edged shanks. Journal of Zoology 269:51–56.

———. 2007a. An experimental approach to modeling the strength of canine teeth. Journal of Zoology 271:162–169.

———. 2007b. The trade-off between tooth strength and tooth penetration: Predicting optimal shape of canine teeth. Journal of Zoology 273:273–280.

Gilroy, A. M., B. R. Macpherson, L. M. Ross, M. Schuenke, E. Schulte, U. Schumacher, M. Voll, and K. Wesker. 2008. Thieme Atlas of Human Anatomy. Thieme Publishers, New York.

Gonyea, W. J. 1976. Behavioral implications of saber-toothed felid morphology. Paleobiology 2:332–342.

Gorniak, C., and C. Gans. 1980. Quantitative assay of electromyograms during mastication in domestic cats (*Felis catus*). Journal of Morphology 163:253–281.

Graham, R. W. 1976. Pleistocene and Holocene mammals, taphonomy, and paleoecology of the Friesenhahn Cave local fauna, Bexar County, Texas. Ph.D. diss., University of Texas, Austin.

Heald, F. P. 1989. Injuries and diseases in *Smilodon californicus* Bovard, 1904 (Mammalia, Felidae) from Rancho La Brea, California. Journal of Vertebrate Paleontology 9:24A.

Heald, F. P., and C. A. Shaw. 1991. Sabertooth cats; pp. 24–27 in J. Seidensticker and S. Lumpkin (eds.), Great Cats: Majestic Creatures of the Wild. Weldon Owen, San Francisco.

Hearst, J. M. 1998. Depositional environments of the Birch Creek local fauna (Pliocene: Blancan), Owyhee County, Idaho; pp. 56–93 in W. A. Akersten, H. G. McDonald, D. J. Melrum, and M. E. Thompson-Flint (eds.), And whereas . . . Papers on the Vertebrate Paleontology of Idaho Honoring John A. White. Idaho Museum of Natural History, Pocatello.

Hemmer, H. 1978. Considerations on sociality in fossil carnivores. Carnivore 1:105–107.

Hennig, W. 1966. Phylogenetic Systematics. University of Illinois Press, Urbana, IL.

Hibbard, C. W. 1952. Vertebrate fossils from late Cenozoic deposits of central Kansas. University of Kansas Paleontological Contributions. Vertebrata 2:1–14.

Hill, F. W. G. 1977. A survey of bone fractures in the cat. Journal of Small Animal Practice 18:457–463.

Hoffman, P. 1993. Shooting saber-tooths. Discover 14, no. 4:50–59.

Hooijer, D. A. 1972. Varanus (Reptilia, Sauria) from the Pleistocene of Timor. Zoölogishe Mededelingen Rijksmuseum van Natuurlijke Histoirie te Leiden 47:445–448.

Hudson, R. J. 1985. Body size, energetics and adaptive radiation; pp. 1–24 in R. J. Hudson and R. G. White (eds.), Bioenergetics of Wild Herbivores. CRC Press, Boca Raton, FL.

Hulbert Jr., R. C. 2001. Florida fossil vertebrates; pp. 25–33 in R. C. Hulbert Jr. (ed.), The Fossil Vertebrates of Florida. University Press of Florida, Gainesville.

Janis, C. M. 1976. The evolutionary strategy of the equidae and the origins of rumen and cecal digestion. Evolution 30:757–774.

Jefferson, G. T., and J. L. Goldin. 1989. Seasonal migration of *Bison antiquus* from Rancho La Brea, California. Quaternary Research 31:107–112.

Jenkins Jr., F. A. 1973. The functional anatomy and evolution of the mammalian humero-ulnar articulation. American Journal of Anatomy 137:281–297.

Joubert, D., and B. Joubert. 1994. Lions of darkness. National Geographic Research 186:35–54.

———. 1997. Hunting With the Moon: The Lions of Savuti. National Geographic Society, Washington, DC.

Kaup, J. J. 1833. Description d'ossements fossiles de mammifères inconnus jusqu'à présent qui se trouvent au Museum Grand-Ducal de Darmstadt. 2 vols. 40, Darmstadt.

King, J. E., and J. J. Saunders. 1984. Environmental insularity and extinction of the American Mastodont; in P. S. Martin and R. G. Klein (eds.), Quaternary Extinctions: A Prehistoric Revolution, University of Arizona Press, Tucson.

Kitchener, A. 1991. Natural History of the Wild Cats. Comstock Publishing Associates, Ithaca, NY.

Kleiman, D. G., and J. F. Eisenberg. 1973. Comparisons of canid and felid social systems from an evolutionary perspective. Animal Behaviour 21:637–659.

Kretzoi, M. 1927. Materialen zur phylogenetischen Klassification der Aeluroideen. Paper presented at the Xe Congrès International de Zoologie, Budapest, 1927.

Kruuk, H. 1972. The Spotted Hyena. University of Chicago Press, Chicago.

Kurtén, B. 1952. The Chinese Hipparion fauna: A quantitative survey with comments on the ecology of the machairodonts and hyaenids and the taxonomy of the gazelles. Commentations of Biology (Society of Science, Fennica) 13:1–82.

———. 1962. The sabre-toothed cat *Megantereon* from the Pleistocene of Java. Zoologische Mededelingen 38:101–104.

———. 1963. Notes on some Pleistocene mammal migrations from the Palearctic to the Nearctic. Eiszeitalter Gegenwart 14:96–103.

———. 1972. The genus *Dinofelis* (Carnivora, Mammalia) in the Blancan of North America. University of Texas Bulletin, Pearce-Sellards Series 19:1–7.

Kurtén, B., and E. Anderson. 1980. Pleistocene Mammals of North America. Columbia University Press, New York.

Lange, K. I. 1960. The jaguar in Arizona. Transactions of the Kansas Academy of Sciences 63:96–101.

Leyhausen, P. 1965. Über die Funktion der Relativen Stimmungshierarchie (Dargestellt am Beispiel der phylogenetischen und ontogenetischen Entwicklung des Beutfangs von Raubtiernen). Z. Tierpsychol. 22:412–494.

Macdonald, J. R., and G. Sibley. 1969. Paleopathological ponderings, or how to tell a sick saber-tooth. Museum Alliance Quarterly 8:26–30.

Manning, P. L., D. Payne, J. Pennicott, P. M. Barrett, and R. A. Ennos. 2005. Dinosaur killer claws or climbing crampons? Royal Society Biology Letters 2:110–112.

Marean, C. W., and C. L. Erhardt. 1995. Paleoanthropological and paleoecological implications of the taphonomy of a sabertooths den. Journal of Human Evolution 29:515–547.

Martin, L. D. 1979. The biostratigraphy of arvicoline rodents in North America. Transactions of the Nebraska Academy of Sciences 7:91–100.

———. 1980. Functional morphology and the evolution of cats. Transactions of the Nebraska Academy of Sciences 7:141–154.

———. 1985. Tertiary extinction cycles and the Pliocene-Pleistocene boundary; pp. 33–40 in Institute for Tertiary-Quaternary Studies, TER-QUA Symposium Series 1, W. Dort Jr. (ed.), Nebraska Academy of Sciences, Lincoln, NE.

Martin, L. D., J. P. Babiarz, V. L. Naples, and J. M. Hearst. 2000. Three ways to be a saber-toothed cat. Naturwissenschaften 87:41–44.

Martin, L. D., and T. J. Meehan. 2005. Extinction may not be forever. Naturwissenschaften 92:1–19.

Martin, L. D., V. L. Naples, and H. T. Wheeler. 2001. Did mammoths have nonhuman predators? pp. 27–34 in D. West (ed.), Proceedings of the International Conference On Mammoth Site Studies, University of Kansas, Lawrence.

Martin, L. D., and A. M. Neuner. 1978. The end of the Pleistocene in North America. Transactions Nebraska Academy of Sciences 6:117–126.

Martin, L. D., A. R. Rogers, and A. M. Neuner. 1985. The effect of the end of the Pleistocene on man in North America; in J. I. Mead and D. J. Meltzer (eds.), Environments and Extinctions: Man in Late Glacial North America. Center for the Study of Early Man, University of Maine, Orono.

Martin, L. D., and C. B. Schultz. 1975. Scimitar-toothed cats, Machairodus and Nimravides, from the Pliocene of Nebraska and Kansas. Bulletin of the University of Nebraska State Museum 10:55–63.

Martin, L. D., C. B. Schultz, and M. R. Schultz. 1988. Saber-toothed cats from the PlioPleistocene of Nebraska. Transactions of the Nebraska Academy of Sciences 16:153–163.

Martin, P. S. 1967. Prehistoric overkill; pp. 75–120 in P. S. Martin and H. E. Wright Jr. (eds.), Pleistocene Extinctions: The Search for a Cause. Yale University Press, New Haven.

Martin, R. A. 1993. Patterns of variation and speciation in Quaternary rodents; pp. 226–280 in R. A. Martin and A. D. Barnosky (eds.), Morphological Change in Quaternary Mammals of North America. Cambridge University Press, Cambridge.

Martin, R. A., and J. C. Harksen. 1974. The Delmont local fauna, Blancan of South Dakota. Bulletin of the New Jersey Academy of Science 19:11–17.

Matthew, W. D. 1901. Fossil mammals of the Tertiary of northeastern Colorado: American Museum collection of 1898. Memoirs of the American Museum of Natural History 1:446–447.

———. 1910. The phylogeny of the Felidae. Bulletin of the American Museum of Natural History 28:289–316.

Mawby, J. E. 1965. Machairodonts from the late Cenozoic of the panhandle of Texas. Journal of Mammalogy 46:573–587.

McCall, S. A., V. L. Naples, and L. D. Martin. 1997. Social constraints on saber-toothed carnivores. Journal of Vertebrate Paleontology 17, no. 3:63.

McCall, S., V. L. Naples, and L. D. Martin. 2003. Assessing behavior in extinct animals: Was Smilodon social? Brain Behavior and Evolution 61:159–164.

McDougal, C. 1977. The Face of the Tiger. Rivington Books, London.

Meade, G. E. 1961. The saber-toothed cat, Dinobastis serus. Bulletin of the Texas Memorial Museum 2:23–60.

Meehan, T. J. 1998. Evolutionary Trends and Iterative Evolution of North American Cenozoic Mammalian Faunas. University of Kansas, Lawrence.

———. 2003. Extinction and re-evolution of similar adaptive types (ecomorphs) in Cenozoic North American ungulates and carnivores reflect van der Hammen's cycles. Naturwissenschaften 90:131–135.

Merriam, J. C. 1917. Relationships of Pliocene mammalian faunas from the Pacific Coast and Great Basin Provinces of North America. University of California Publications, Berkeley, Publications of the Dept. of Geology 10, no. 22:421–443.

———. 1918. New Mammalia from the Idaho Formation. University of California Publications, Berkeley, Bulletin of the Dept. of Geology 10, no. 26:523–530.

Merriam, J. C., and C. Stock. 1932. The Felidae of Rancho La Brea. Carnegie Institution of Washington, Washington, DC.

Miller, G. J. 1983. Some new evidence in support of the stabbing hypothesis for Smilodon californicus Bovard. Carnivore 3, no. 2:8–26.

Molloy, B. 2008. Martial arts and materiality: A combat archaeology perspective on Aegean swords of the fifteenth and fourteenth centuries BC. World Archaeology 40:116–134.

Mones, A., and A. Rinderknecht. 2004. The first South American Homotherini (Mammalia, Carnivora, Felidae). Communicaciones Paleontologicas, Museo Nacional de Historia Natural y Anthropologica 2, no. 35:201–211.

Moodie, R. L. 1923. Paleopathology: An Introduction to the Study of Ancient Evidences of Disease. University of Illinois Press, Urbana, IL.

———. 1926a. La Paléopathologie des mammifères du Pléistocene. Biologie Médic. 16:431–440.

———. 1926b. Studies in Paleopathology: XIX. Pleistocene examples of traumatic osteomyelitis. Annals of Medical History 8:413–418.

———. 1926c. Studies in Paleopathology: X. Vertebral lesions in the sabertooth, Pleistocene of California, resembling so-called myositis ossificans pathology compared with certain ossification in the dinosaurs. Annals of Medical History 8:91–102.

———. 1927. Studies in Paleopathology: XX. Vertebral lesions in the sabre-tooth, Pleistocene of California, resembling the so-called myositis ossificans progressiva, compared with certain ossification in the dinosaurs. Annals of Medical History 9:91–102.

———. 1930a. The phenomenon of sacralization in the Pleistocene saber-tooth. American Journal of Surgery 10:587–589.

———. 1930b. Studies in Paleopathology: XXV. Hypertrophy in the sacrum of the saber tooth, Pleistocene of southern California. American Journal of Surgery, New Series 8:1313–1315.

———. 1930c. Studies in Paleopathology: XXVI. Pleistocene luxations. American Journal of Surgery, New Series 9:348–362.

Naples, V. L., J. P. Babiarz, and L. D. Martin. 2002. Redesigning the saber-tooth paradigm; pp. 1–4, frontispiece in W. Dort (ed.), TER-QUA Symposium Series 3. Institute for Tertiary-Quaternary Studies, Kansas City, KA.

———. 2003. Why does Xenosmilus look like a panda? Journal of Vertebrate Paleontology 23:82A.

Naples, V. L., and L. D. Martin. 2000. Evolution of hystricomorphy in the Nimravidae (Carnivora, Barbourofelinae): Evidence for complex character convergence with rodents. Historical Biology 14:169–188.

———. 2008. Cookie-cutter cats: Another saber-tooth morphotype. Journal of Vertebrate Paleontology 23:127A.

Naples, V. L., L. D. Martin, and J. P. Babiarz. 2008. Muscle maps as a guide to the real saber-tooth. Paper presented at the Saber-tooth Symposium, University of Idaho, 2008.

Naples, V. L., H. T. Wheeler, and L. D. Martin. 2000. Non-human predation of mammoths: Mammoths as cat food? Journal of Vertebrate Paleontology 20:60A.

Nowak, R. M. 1999. Walker's Mammals of the World. 6th ed. Vol. 1, Johns Hopkins University Press, Baltimore and London.

Osborn, H. F. 1936. The Proboscidea: A Monograph of the Discovery, Evolution, Migration and Extinction of the Mastodonts and Elephants of the World. 2 vols. The American Museum Press, New York.

Owen, R. 1846. A History of British Fossil Mammals and Birds. John Van Voorst, London.

Pauly, D. 1994. Quantitative analysis of published data on the growth, metabolism, food consumption, and related features of the red-bellied piranha, *Serrasalmus nattereri* (Characidae). Environmental Biology of Fishes 41:423–437.

Pilgrim, G. E. 1931. Catalogue of the Pontian Carnivora of Europe. The British Museum, London.

Pinsof, J. D. 1998. The American Falls local fauna: Late Pleistocene (Sangamonian) vertebrates from southeastern Idaho; pp. 121–145 *in* W. A. Akersten, H. G. McDonald, D. J. Meldrum, and M. E. T. Flint (eds.), And Whereas . . . Papers on the Vertebrate Paleontology of Idaho Honoring John A. White 1. Idaho Museum of Natural History Occasional Paper.

Radinsky, L. B. 1975. Evolution of the felid brain. Brain Behavior and Evolution 11:214–254.

———. 1982. Evolution of skull shape in carnivores, 3: The origin and early radiation of the modern carnivore families. Paleobiology 8:177–195.

Rawn-Schatzinger, V. 1992. The scimitar cat *Homotherium serum* Cope: Osteology, functional morphology, and predatory behaviour. Illinois State Museum Reports of Investigations 47:1–118.

Rawn-Schatzinger, V., and R. L. Collins. 1981. Scimitar cats, *Homotherium serum* Cope from Gassaway Fissure, Cannon County, Tennessee, and the North American distribution of *Homotherium*. Journal of the Tennessee Academy of Science 56:15–19.

Reighardt, J., and H. S. Jennings. 1930. Anatomy of the Cat. Henry Holt and Co., New York.

Repenning, C. A. 1987. Biochronology of the microtine rodents of the United States; pp. 236–268 *in* M. O. Woodburne (ed.), Cenozoic Mammals of North America: Geochronology and Biostratigraphy. University of California Press, Berkeley.

Repenning, C. A., O. Fejfar, and W. D. Heinrich. 1990. Arvicolid rodent biochronology of the Northern Hemisphere; pp. 385–417 *in* O. Fejfar and W. D. Heinrich (eds.), International Symposium: Evolution, Phylogeny and Biostratigraphy of Arvicolids (Rodentia, Mammalia). Geological Survey, Prague.

Repenning, C. A., T. R. Weasma, and G. R. Scott. 1994. The Early Pleistocene (Latest Blancan-Earliest Irvingtonian) Froman Ferry Fauna and History of the Glenns Ferry Formation, Southwestern Idaho. U.S. Geological Survey, Denver, CO.

Resnick, D. 2002. Diagnosis of Bone and Joint Disorders. W. B. Saunders Company, Philadelphia.

Reumer, J. W. F., L. Rook, K. Vanderborg, K. Post, D. Mol, and J. De Vos. 2003. Late Pleistocene survival of the saber-toothed cat *Homotherium* in North Western Europe. Journal of Vertebrate Paleontology 23:260–262.

Riviere, H. L., and H. T. Wheeler. 2005. Cementum on *Smilodon* sabers. The Anatomical Record 285A:634–642.

Rothschild, B. M. 2008a. Diffuse idiopathic skeletal hyperostosis. eMedicine Obstetrics, Gynecology, Psychiatry and Surgery. www.emedicine.com/med/topic 2901.htm.

———. 2008b. Lumbar spondylosis (Spondylosis deformans). eMedicine Obstetrics, Gynecology, Psychiatry and Surgery. www.emedicine.com/med/topic 2901.htm.

Rothschild, B. M., and M. A. Bruno. 2008. Calcium pyrophosphate deposition disease. eMedicine Radiology. www.emedicine .com/radio/topic125.htm.

Rothschild, B. M., and R. Laub. 2006. Hyperdisease in the Late Pleistocene: Validation of an early 20th-century hypothesis. Naturwissenschaften 93:557–564.

Rothschild, B. M., and L. D. Martin. 2006. Skeletal Impact of Disease. New Mexico Museum of Natural History, Albuquerque.

Sardella, R. 1998. The Plio-Pleistocene Old World dirk-toothed cat *Megantereon* ex gr. *cultridens* (Mammalia, Felidae, Machairodontinae), with comments on taxonomy, origin and evolution. Neues Jahrbuch für Geologie und Palaontologie Abhandlungen 207:1–36.

Savage, D. E., and D. E. Russell. 1983. Mammalian Paleofaunas of the World. Addison-Wesley Publishing Co., Reading, MA.

Schaller, G. B. 1967. The Deer and the Tiger: A Study of Wildlife in India. University of Chicago Press, Chicago.

———. 1972. The Serengeti Lion: A Study of Predator-Prey Relations. University of Chicago Press, Chicago.

Schultz, C. B., and L. D. Martin. 1970. Machairodont cats from the early Pleistocene Broadwater and Lisco local faunas. Bulletin of the University of Nebraska State Museum 9:33–38.

Schultz, C. B., L. D. Martin, L. Tanner, and G. Corner. 1978. Provincial land mammals ages for the North American Quaternary. Transactions of the Nebraska Academy of Sciences and Affiliated Societies 5:59–64.

Schultz, C. B., M. R. Schultz, and L. D. Martin. 1970. A new tribe of saber-toothed cats (Barbourofelini) from the Pliocene of North America. Bulletin of the University of Nebraska State Museum 9:1–31.

Scott, W. B., and G. L. Jepsen. 1936. The mammalian fauna of the White River Oligocene, Part I: Insectivora and Carnivora. Transactions of the American Philosophical Society 28:1–153.

Seawright, A. A., and P. B. English. 1967. Hypervitaminosis A and deforming cervical spondylosis of the cat. Journal of Comparative Pathology 77:29–43.

Seidensticker, J., and C. McDougal. 1993. Tiger predatory behaviour, ecology and conservation; pp. 105–125 *in* Mammals as Predators. Symposium of the Zoological Society of London.

Shaw, C. A. 1992. Old wounds: The paleopathology of Rancho La Brea. Terra 31:17.

Shaw, C. A., and A. E. Tejada-Flores. 1985. Biomechanical implications of the variation in *Smilodon* ectocuneiforms from Rancho La Brea. Natural History Museum of Los Angeles County Contributions in Science 359:1–8.

Shermis, S. H. 1983. Healed massive pelvic fracture in a *Smilodon* from Rancho La Brea, California. PaleoBios 1:121–126.

Simpson, G. G. 1941. The function of saber-like canines in carnivorous mammals. American Museum Novitates 1130:1–12.

———. 1945. The principles of classification and a classification of the mammals. Bulletin of the American Museum of Natural History 86:1–350.

Slater, G. J., and B. Van Valkenburgh. 2008. Long in the tooth: Evolution of sabertooth cat cranial shape. Paleobiology 34:403–419.

Stout, T. M. 1978. The comparative method in stratigraphy: The beginning and end of an ice age. Transactions Nebraska Academy of Sciences 6:1–18.

Sunquist, M. E., and F. Sunquist. 2002. Wild Cats of the World. University of Chicago Press, Chicago.

Teilhard de Chardin, P. 1939. On two skulls of *Machairodus* from the Lower Pleistocene beds of Choukoutien. Part 1, The *Machairodus* of locality 13. Bulletin of the Geological Society of China 19:235–256.

Teilhard de Chardin, P., and P. Leroy. 1945. Les Felides de Chine. Part I, Les formes fossils. Publications de l'Institut de Geobiologie 11:1–58.

Teilhard de Chardin, P., and J. Piveteau. 1930. Les mammiferes fossiles de Nihowan (Chine). Annales de Paléontologie 19:1–154.

Therrien, F. 2005. Feeding behaviour and bite force of sabretoothed predators. Zoological Journal of the Linnean Society 145:393–426.

Thomason, J. J. 1991. Cranial strength in relation to estimated biting forces in some mammals. Canadian Journal of Zoology 69:2326–2333.

Thomason, J. J., L. E. Grovum, A. G. Deswysen, and W. W. Bignell. 2001. In vivo surface strain and stereology of the frontal and maxillary bones of sheep: Implications for the structural design of the mammalian skull. The Anatomical Record 264:325–338.

Thomason, J. J., A. P. Russell, and M. Morgelli. 1990. Forces of biting, body size, and masticatory muscle tension in the opossum *Didelphis virginiana*. Canadian Journal of Zoology 68:318–324.

Turner, A., and M. Antón. 1997. The Big Cats and their Fossil Relatives: An Illustrated Guide to their Evolution and Natural History. Columbia University Press, New York.

Van Dam, J. A., H. A. Aziz, M. Á. Álvarez Sierra, F. J. Hilgren, L. W. van den Hoek Ostende, L. J. Lourens, P. Mein, A. J. van der Meulen, and P. Pelaez-Campomanes. 2006. Long-period astronomical forcing of mammal turnover. Nature 443:687–691.

Van Lawick, H., and J. Goodall. 1971. Innocent Killers. Houghton Mifflin, Boston.

Van Valkenburgh, B. 1988. Incidence of tooth breakage among large, predatory mammals. American Naturalist 131:291–302.

———. 2001. Predation in Saber-Tooth Cats; pp. 420–424 *in* D. E. G. Briggs and P. R. Crowther (eds.), Paleobiology II. Blackwell Science, Oxford.

Van Valkenburgh, B., and F. Hertel. 1993. Tough times at La Brea: Tooth breakage in large carnivores of the Late Pleistocene. Science 261:456–459.

Van Valkenburgh, B., and C. B. Ruff. 1987. Canine tooth strength and killing behaviour in large carnivorans. Journal of Zoology, London 212:379–397.

Van Valkenburgh, B., and T. Sacco. 2002. Sexual dimorphism, social behavior, and intrasexual competition in large Pleistocene carnivorans. Journal of Vertebrate Paleontology 22:164–169.

Webb, S. D. 1977. A history of savanna vertebrates in the New World. Part 1: North America. Annual Review of Ecology and Systematics 8:355–380.

Weithofer, A. 1889. Ueber die tertiaren Landsaugethiere Italiens. Jh. K. K. Geol. Reichsanstalt., Wien 34:55–82.

Werdelin, L., and M. E. Lewis. 2001. A revision of the genus *Dinofelis* (Mammalia, Felidae). Zoological Journal of the Linnean Society 132:147–258.

Werdelin, L., and R. Sardella. 2006. The *"Homotherium"* from Langebaanweg, South Africa and the origin of *Homotherium*. Palaeontographia Abt. A277:123–130.

Wheeler, H. T. 2000. Confirmation of saber-tooth killing bite theories by re-enactment. Journal of Vertebrate Paleontology 20:79A.

———. 2004. Machairodont canine functional morphology is about more than saber shape: The varied role of the incisors in different killing bite models. Journal of Vertebrate Paleontology 24:129A.

Wheeler, H. T., H. L. Riviere, T. J. Fremd, and J. P. Babiarz. 2004. Convergent evolution in the enamel and gingiva of the Nimravid *Pogonodon* and the Felid *Smilodon* revealed in new material from John Day Fossil Beds National Monument. Geological Society of America Abstracts with Programs 36:53.

White, J. 1984. Late Cenozoic Leporidae (Mammalia, Lagomorpha) from the Anza-Borego Desert, southern California; pp. 41–57 *in* R. M. Mengel (ed.), Papers in Vertebrate Paleontology Honoring Robert Warren Wilson. Carnegie Museum of Natural History, Pittsburgh.

White, J., and B. L. Keller. 1984. Evolutionary stability and ecological relationships of morphology in North American Lagomorpha; pp. 58–66 *in* R. M. Mengel (ed.), Papers in Vertebrate Paleontology Honoring Robert Warren Wilson. Carnegie Museum of Natural History, Pittsburgh.

White, J. R. S., G. S. Morgan, and R. S. White Jr. 2005. Arizona Blancan vertebrate faunas in regional perspective; pp. 117–138 *in* R. D. McCord (ed.), Vertebrate Paleontology of Arizona. Mesa Southwest Museum Bulletin, Mesa, AZ.

Williams, P. L., R. Warwick, M. Dyson, L. H. Bannister. 1989. Gray's Anatomy. 37th ed. Edinburgh: Churchill Livingstone.

Woodburne, M. O., ed. 1987. Cenozoic Mammals of North America: Geochronology and Biostratigraphy. University of California Press, Berkeley.

Wroe, S., C. McHenry, and J. J. Thomason. 2005. Bite club: Comparative bite force in big biting mammals and the prediction of predatory behaviour in fossil taxa. Proceedings of The Royal Society B: Biological Sciences 272:619–625.

Zdansky, O. 1924. Jungtertiäre Carnivoren Chinas. Published by the Geological Survey of China, Peking. Palaeontologia Sinica, Ser. C, 2:1–153.

Zoback, M. L., and G. A. Thompson. 1978. Basin and range rifting in northern Nevada: Clues from a mid-Miocene rift and its subsequent offsets. Geology 6:111–116.

INDEX

Page numbers in *italics* indicate figures and tables.

abdomen, attacks on, 6, 8, 35–36

abductor pollicis longus muscle of *Xenosmilus hodsonae, 111,* 112–113

adductor longus, adductor brevis, and adductor magnus muscles of *Xenosmilus hodsonae, 115,* 117

African lion (*Panthera leo*), 32

Age of Mammals, 12

Ailuropoda melanoleuca (panda), 32, 45

Akersten, Bill, 126

ambush model of predatory behavior, 35–36, *37,* 90, 120

American bison: cross section of neck of, *27;* as prey species, 9–10, 43; Robocat killing bite tests and, 23–24, *24, 25;* tuberculosis in, 40

Amphimachairodus genus, 185, 208, 209

Amphimachairodus giganteus, 205, *205*

anconeus muscle of *Xenosmilus hodsonae, 107,* 110

ankle (tarsal bones): of *Homotherium ischyrus,* 168; of *Xenosmilus,* 83, *84,* 85, *85, 86*

anterior-posterior diameter (APD): of conical-tooths, 20; of dirk-tooths, 20; of scimitar-tooths, 28, *28,* 29

appendicular skeleton of *Homotherium ischyrus:* forelimb and pectoral girdle, 146, 148, *151,* 152, *152,* 154, 156–160, 162–163; hindlimb and pelvic girdle, 164, 168, 170–171, 173–175, 177

appendicular skeleton of *Xenosmilus:* forelimb and pectoral girdle, 66–67, 69–76; hindlimb and pelvic girdle, 76–83, 85–86, 88, 90

articularis genus muscle of *Xenosmilus hodsonae,* 114, 116

astragalus: of *Homotherium ischyrus,* 168, 170, *171;* of *Xenosmilus,* 83, *84,* 85, *85*

atlas vertebra of *Homotherium ischyrus,* 131, *132*

auditory bullae of *Homotherium ischyrus,* 129

axial skeleton of *Homotherium ischyrus:* cervical vertebrae, 131, *132,* 133–134, *133, 134, 135;* lumbar vertebrae, 144, *145, 146, 147–150;* thoracic vertebrae, 134–135, *137,* 138–139, *138, 139–143,* 144

axial skeleton of *Xenosmilus:* cervical vertebrae, 54–55, *57, 59;* thoracic vertebrae, 59–61, 63–64, 66

axis vertebra: of *Homotherium ischyrus,* 131, *132, 133;* of *Xenosmilus,* 54, *55, 56*

Barbourofelis: canines of, 20; extinction of, *11,* 185

Barbourofelis fricki, 208

Beringian Steppe Tundra, 205

biceps brachii muscle of *Xenosmilus hodsonae,* 111–112, *111*

biceps femoris muscle of *Xenosmilus hodsonae,* 117, *118*

biomechanics of cranial musculature, 104–5, *105*

bison. *See* American bison

bite force: head shape and, 120; studies of, 32; temporalis muscle and, 100; of *Xenosmilus,* 91–92. *See also* killing bites

Blancan, 203, 205

body size: of herbivores, 3–4; prey size and, 8–9; of saber-tooth carnivores, 4

brachialis muscle of *Xenosmilus hodsonae, 106, 107,* 109

brachioradialis muscle of *Xenosmilus hodsonae, 106, 107,* 110, 112

Broadwater Local Fauna, Nebraska, 190

brown bear (*Ursus arctos*), 32, 45

bulldogging model of predatory behavior, 35

calcaneum: of *Homotherium ischyrus,* 168, *170;* of *Xenosmilus,* 83, *84, 85*

calcium pyrophosphate deposition disease, 41

Camelops Faunal Province, 206

canine teeth: continuous cutting surface and, 91; of dirk-tooths, 4, 9, 20, *21;* of *Homotherium crenatidens,* 199; of *Homotherium ischyrus,* 130; morphologies of, 19; as *nomina vana,* 201; rate of growth of, 9; of scimitar-tooths, 4, 9, 28–29, *28;* size and shape of, 3, *4;* of *Xenosmilus,* 51

Canis dirus (dire wolves), 8, 9

Carbon 14 and radioactive dating systems, 12

Carcharodon carcharias, 29

carnivores, home ranges of, 16–17

cemento-enamel junction (CEJ): of dirk-tooths, 20, *21*; of *Homotherium ischyrus*, 130, *131*; of scimitar-tooths, 28, 29

cervical vertebrae of *Homotherium ischyrus*: atlas, 131, *132*; axis, 131, *132*, *133*; fifth, 133, *134*, *135*; sixth, 133–134, *136*; third, 131, 133, *134*, *135*

cervical vertebrae of *Xenosmilus*: axis, 54, *55*, *56*; fifth, 55, *55*, 57, *57*; sixth, 57, *58*, 59, *59*; third, 54–55, *55*, *56*

Chinese water deer (*Hydropotes inermis*), 21, *22*, *23*

Chlamythere Faunal Province, 206

Cita Canyon, Texas, 185

cladistic taxonomy, 13–14

climactic cycles, 10–11, *11*, 12

clouded leopard (*Neofelis nebulosa*), 20

Coffee Ranch, Texas, 205

common extensor tendon of *Xenosmilus hodsonae*, 106, *107*, 110

common flexor tendon of *Xenosmilus hodsonae*, 106, *107*, 109–110

comparative anatomy, 13

comparison measurement tables: mandible measurements, *96*, *182*; mandibular dentition measurements, *97*, *183*; maxillary dentition measurements, *95*, *181*; overview of, 92, 179; skull measurements, *93–94*, *180*

conical-tooth cats: bite forces of, 32; canines of, *3*, *4*; killing bites of, 20, *33*; phylogeny of, *15*; predation of, 4, 6; taxa of, 43

conspecific antisocial interactions, 36

conspecific social interactions, 36

cookie-cutter bites: description of, 91–92; living proxies using, 29, *31*; mandible and, 54

cookie-cutter cats, 8. *See also* cookie-cutter bites; *Xenosmilus hodsonae*

cooling, global, and extinction, 10–11, *11*

coracobrachialis muscle of *Xenosmilus hodsonae*, *107*, 109

coronoid processes of *Xenosmilus*, 52–54

Correlation, Principle of, 13

cranial flexion of *Xenosmilus*, 50

cranial region: biomechanics of musculature of, 104–105, *105*; musculoskeletal reconstruction of, 100

cranial region of *Xenosmilus hodsonae*: facial expression muscles, 100, 102–4, *103*, *104*; masticatory muscles, 100–102; muscles of, *101*, *102*

cuboid of *Homotherium ischyrus*, 173–174, *173*, *174*

cuneiform of *Homotherium ischyrus*, 157–158, *158*

cursoriality, 83, 190–191

Cuvier, G., 201

Delmont, South Dakota, 187

Delmont *Homotherium*, 189–190

deltoideus muscle of *Xenosmilus hodsonae*, 106

dentary-maxillary gapes: bite forces and, 32; of dirk-tooths, 20; of nimravids, 20

dentition: of *Homotherium ischyrus*, 129–130, *131*; mandibular measurements, *97*, *183*; maxillary measurements, *95*, *181*. *See also* canine teeth; incisors

diffuse idiopathic skeletal hyperostosis (DISH) in dirk-tooths, 39

digastricus muscle of *Xenosmilus hodsonae*, *101*, 102, *103*

digits, flexion and extension of, 120

Dinictis, 14

Dinobastis, 123

Dinobastis genus, 43–44

Dinobastis serus, 203

Dinofelis genus, 208, 209

dire wolves (*Canis dirus*), 8, 9

dirk-tooth cats: as ambush predators, 6; bite models of, 21–28; canines of, *4*, 9, 20, *21*; evolution of, 28; extinction of, 28; paleopathologic analysis of, 37, 38–39; potential prey of, 32–33; terminology for, 19–20. *See also* killing bites of dirk-tooths; *Smilodon*

dispersal and distribution: of Homotheriini tribe, 209; of *Homotherium* genus, 205–7; overview of, 16–17, *16*; of *Xenosmilus hodsonae*, 90

ecomorphs, 208–9

ectocuneiform of *Homotherium ischyrus*, 171, *172*, *173*

eighth thoracic vertebra: of *Homotherium ischyrus*, 137, *138*, *139*; of *Xenosmilus*, 62, *63*, *63*

elbow morphology, 36

eleventh thoracic vertebra: of *Homotherium ischyrus*, 138, *140*, *141*; of *Xenosmilus*, 63–64, *65*, 66, *66*

Equus (Plesippus) stenonis, 122, 178

extensor carpi ulnaris muscle of *Xenosmilus hodsonae*, *107*, 112

extensor digitorum longus muscle of *Xenosmilus hodsonae*, 117, *118*, 119

extensor hallucis longus muscle of *Xenosmilus hodsonae*, *118*, 119

extinction: paleopathology and, 35; processes of, 43; of saber-tooth cats, 9–11

facial expression muscles, 100, 102–104, *103*, *104*

faunal provinces, 206

FEA (finite element analysis), 26–27

felid cats, first appearance of, 208

femur: of *Homotherium ischyrus*, 164, *166*; muscles associated with, 113–117, *114*, *115*, *118*; of *Xenosmilus*, 76–80, *79*, *80*, 113, *115*

fibula: of *Homotherium ischyrus*, 167, *168*, *169*; muscles of, *118*, 119–120; of *Xenosmilus*, 82, *83*, *83*, 117, *118*

fibularis brevis muscle of *Xenosmilus hodsonae*, *118*, 119–120

fibularis longus muscle of *Xenosmilus hodsonae*, *118*, 119

fibularis tertius muscle of *Xenosmilus hodsonae*, *118*, 120

fifth cervical vertebra: of *Homotherium ischyrus*, 133, *134*, *135*; of *Xenosmilus*, 55, *55*, 57, *57*

fifth lumbar vertebra of *Homotherium ischyrus*, 144, *147*, *148*

fifth thoracic vertebra: of *Homotherium ischyrus*, 135, *137*, *138*; of *Xenosmilus*, 58, *60*, *60*

finite element analysis (FEA), 26–27

first lumbar vertebra of *Homotherium ischyrus*, 144, *144*, *145*

flexor digitorum longus muscle of *Xenosmilus hodsonae*, *118*, 119

flexor digitorum profundus muscle of *Xenosmilus hodsonae*, 111, 112

flexor hallucis longus muscle of *Xenosmilus hodsonae*, *118*, 119

flexor pollicis longus muscle of *Xenosmilus hodsonae*, 111, 112

Florida, 44, 47, 90. See also *Xenosmilus hodsonae*

forearm bones, 110, *111*. *See also* radius; ulna

forelimb and pectoral girdle: bones and musculature of, 105, *106*, *107*–113, *107*, *111*, 120; of *Homotherium ischyrus*, 146, 148, 152, 154, 156–160, 162–163; of *Xenosmilus*, 66–67, 69–76. *See also* radius; scapula; ulna

fossil record: dating of, 12; expression of disease in, 35

fourth lumbar vertebra of *Homotherium ischyrus*, 147, *148*

Friesenhahn Cave site, Texas, *2*, 8, *9*, 33, 43–44. See also *Homotherium serum*
Froman Ferry, Idaho, 185

gapes. *See* dentary-maxillary gapes
gastrocnemius muscle of *Xenosmilus hodsonae*, *114*, *115*, 117
gemellus superioris and gemellus inferioris muscles of *Xenosmilus hodsonae*, *114*, *115*, 116
Glenns Ferry Formation, Idaho, 126
glenoids: of *Homotherium ischyrus*, 129; of *Xenosmilus*, 48, 49, 50–51
gluteus maximus muscle of *Xenosmilus hodsonae*, 117, *118*
gluteus medius muscle of *Xenosmilus hodsonae*, 113, *114*, *115*
gluteus minimus muscle of *Xenosmilus hodsonae*, 113, *114*, *115*
gomphothere, 44
gracilis muscle of *Xenosmilus hodsonae*, 117, *118*
Grand View Local Fauna, Idaho, 190
grappling while standing on hind feet, 36, *37*, 40, 92
greater trochanter of *Xenosmilus*, 77–78

Haile 21A locality, Florida, 44, *47*
hair coat, reconstruction of, 99–100
handedness, 36, 38
Hearst, Jonena, 123, 126
height of predators, 8
Hemphillian, 205
herbivores: body size of, 3–4; global cooling and, 10–11
hindlimb and pelvic girdle: bones and musculature of, 113–117, *114*, *115*, *118*, 119–120, 120–121; of *Homotherium ischyrus*, 164, 168, 170–171, 173–175, 177; of *Xenosmilus*, 76–83, 85–86, 88, 90. *See also* astragalus; calcaneum; femur; fibula; inominate; navicular; patella; pes; tibia
home ranges, 11, 16–17
Homotheriini tribe: characteristics of, 44, 208–209; comparison of skulls of, 189–190; origins and distribution of, 209
Homotherium crenatidens: in Europe, 201, 203; grappling activity of, *36*; mandible of, *204*; reconstruction of, *194*; skull of, 195, *196*, 197–199, *197*, *198*
Homotherium crusafonti: Delmont *Homotherium* and, 190; mandible of, *187*; skull of, *188*, 192–193, *192*
Homotherium genus: basis of, 123; characteristics of, 201, 203; mandibles of, *202*; origins and distribution of, 205–7; timeframe for, 203, 205
Homotherium ischyrus: Delmont *Homotherium* compared to, 189–190; description of, 177–178; discovery and recovery of, 126; *Homotherium* included in, 187; location of specimen of, *124*; mandible of, 130, *131*; predatory behavior of, *122*; skeleton and reconstruction of, *125*; skull of, 123, 125–129, *127*, *128*; systematic paleontology, 187, 189
Homotherium latidens: Beringian Steppe Tundra and, 205; in Europe, 201, 203; grappling activity of, *36*; jaw of, 193, *202*, *217*; mandible of, *202*, 203; reconstruction of, *207*
Homotherium serum: ambush behavior in, 36; calcaneus of, *40*; comparison of to other homotheres, 189, 190–193, *191*; confusion about skeletons of, 43–44; distributions of, 16; in Friesenhahn Cave, *2*; mandible of, *202*, 203; manus of, *34*; predation behavior of, *5*, *7*, 33, *184*; reconstruction of, *200*; restoration of skeleton of, *189*; size of prey and, 10; skeleton of cub, *9*; systematic paleontology, 190; vertebra T10 of, *40*

Hoplophoneus, 14
humerus: of *Homotherium ischyrus*, 152, *153*, 154; of *Xenosmilus*, 66–67, 68, 69, 105, *106*, 107–8, *107*
Hydropotes inermis (Chinese water deer), 21, *22*, *23*
hypervitaminosis A, in dirk-tooth young, 39

ichnology, 35
Idaho: Birch Creek specimen from, *124*, 126; Froman Ferry, 185; Grand View Local Fauna, 190. See also *Homotherium ischyrus*
iliacus muscle of *Xenosmilus hodsonae*, *114*, *115*, 116
incisors: of dirk-tooths, 20; of homotheres, 90–91; of *Homotherium crenatidens*, 198–199; of *Homotherium ischyrus*, 129–130, *131*; of scimitar-tooths, 28–29; of *Xenosmilus*, 51, 54; of *Xenosmilus hodsonae*, 8, *8*
infraorbital foramen of *Xenosmilus*, 49
infraspinatus muscle of *Xenosmilus hodsonae*, *106*, *107*, 108
inominate: of *Homotherium ischyrus*, 164, *165*; of *Xenosmilus*, 76, 78
Irvingtonian period, 92
Ischyrosmilus genus: mandibles of, *186*; skull of, *187*; Smilodontins compared to, 203; species assigned to, 123, 130, 187; type species of, 185

jaguar (*P. onca augusta*), 43
jaw opening, 6

Kaup, J. J., 201
killing bites: of conical-tooths, 20, 33; of saber-tooths, 6, 19. *See also* cookie-cutter bites; killing bites of dirk-tooths; killing bites of scimitar-tooths
killing bites of dirk-tooths: APD and, 20, 28, 29; CEJ and, 20, 28, 29; continuous arc trajectory of, 19, 33; draw-cuts and, 23; FEA of, 26–27; jaw adduction force and, 26; models of, 21–28; required precision of, 27–28; sequence and trajectory of, *27*; studies of, 21–23; tactile input and, 26. *See also* Robocat killing bite test fixture
killing bites of scimitar-tooths: bite forces and, 32; default bites, 29; draw-cuts and, 29–30, 32; evolution of dirk-tooths and, 28; inflected arc trajectory of, 19, 33; predatory behavior and, 8, 9; tactile input and, 29

labial-lingual diameter (LLD): of conical-tooths, 20; of nimravids, 20
Late Miocene, 207
latissimus dorsi muscle of *Xenosmilus hodsonae*, *107*, 109
leg planting in ambush attacks, 36, *37*
lesser trochanter of femur of *Xenosmilus*, 79–80, 120–121
levator nasolabialis muscle of *Xenosmilus hodsonae*, 103–104, *104*
lions: African lions, 32; canines of, 9; killing of elephants by, 33; mountain lions, 43; *Panthera spelaea*, 205, *206*; predation compared to, 4
Lisco Local Fauna, Nebraska, 190
lower leg, bones and muscles of, 117, *118*, 119–120
lumbar vertebrae of *Homotherium ischyrus*, 144, *145*, 146, *147–150*

Machairodontinae, 207–208
Machairodus: CEJ of, 29; characteristics of, 205, *205*; diastema, CEJ, and presumed gingival relationship of saber and incisor arcade, *30*; lineage of, 185, 201, 208
magnum of *Homotherium ischyrus*, 158, 159, *159*
mammals, extinction of, *11*

mammoths (*Mammuthus columbi*): as prey species, 5, 7, 8, 33, 184; size of, 10

Mammut americanum. See mastodons

Mammuthus columbi. See mammoths

mandible: of *Amphimachairodus giganteus*, 205; of *Barbourofelis fricki*, 208; comparison measurements of, 96, 182; of homotheres, 186; of *Homotherium crenatidens*, 204; of *Homotherium* genus, 202, 203; of *Homotherium ischyrus*, 130, 131; of *Homotherium latidens*, 193, 202; of *Homotherium serum*, 191; measurement points for, 215–216; of *Megantereon cultridens*, 204; muscles elevating, 100–101; of *Panthera spelaea*, 206; of *Xenosmilus*, 51–54, 52, 53

mandibular dentition measurements, 97, 183

manus (forefoot): of *Homotherium ischyrus*, 160, 162–163; of *Xenosmilus*, 71–72

Martin, Larry H., 44, 126

masseteric fossa of *Xenosmilus*, 53, 54

masseter musculature, 100–101, 101, 102

masticatory muscles of *Xenosmilus hodsonae*, 100–102, 101, 102

mastodons (*Mammut americanum*): predators on, 8; size of, 10; tuberculosis in, 41

mastoid processes: of *Homotherium crenatidens*, 197; of *Homotherium ischyrus*, 129; of *Xenosmilus*, 50

Matthew, William Diller, 14

maxillary dentition measurements, 95, 181

measurement points and techniques, 212, 213–217

meat surplus and pack structure, 8–9

Megantereon cultridens, 37, 204, 205

Megantereon genus, 185

Megantereon hesperus, 190

mental foramen of *Xenosmilus*, 52

mesocuneiform of *Homotherium ischyrus*, 172, 173, 173

metacarpal I of *Homotherium ischyrus*, 160, 160, 161

metacarpal II: of *Homotherium ischyrus*, 160, 160, 161; of *Xenosmilus*, 72–73, 74, 75

metacarpal III: of *Homotherium ischyrus*, 160, 160, 161, 162; of *Xenosmilus*, 73, 74, 75

metacarpal IV: of *Homotherium ischyrus*, 162–163, 162; of *Xenosmilus*, 73–74, 74, 75

metacarpal V: of *Homotherium ischyrus*, 162, 163, 163; of *Xenosmilus*, 74–75, 74, 76

metatarsal I of *Xenosmilus*, 86

metatarsal II: of *Homotherium ischyrus*, 172, 174, 175; of *Xenosmilus*, 86, 87, 88

metatarsal III: of *Homotherium ischyrus*, 172, 174, 175, 176; of *Xenosmilus*, 86, 87, 88

metatarsal IV: of *Homotherium ischyrus*, 172, 174, 176, 177; of *Xenosmilus*, 86, 87, 88

metatarsal V: of *Homotherium ischyrus*, 172, 174, 176, 177; of *Xenosmilus*, 86, 87, 88, 89

morphological distance in taxonomies, 14

mountain lion (*Puma concolor*), 43

muscles of *Xenosmilus hodsonae*: with attachments to forelimb bones, 110–111; with attachments to humerus, 66–67, 106, 107, 108–110; elevating mandible, 53–54; on extensor surface of forearm, 107, 111, 112–113; of femur, 78, 79, 113–117, 114, 115, 118; on flexor surface of forearm, 111–112, 111; of lower leg, 117, 118, 119–120; of mastication, 100–102, 101, 102; of quadriceps group, 80; on radial surface of forearm, 107, 112; radius and, 71; reconstruction of, 99; of shoulder, 106, 107, 108, 120; temporalis, 91–92; ulna and, 70–71

musculoskeletal reconstruction of cranial region of *Xenosmilus*, 100

muzzle of *Xenosmilus*, 49

navicular: of *Homotherium ischyrus*, 170–171, 171; of *Xenosmilus*, 84, 85, 86

Nebraska, Lisco Local Fauna, 190

Neofelis nebulosa (clouded leopard), 20

Nimavides genus, 20, 28, 185, 208

ninth thoracic vertebra: of *Homotherium ischyrus*, 137, 138, 139; of *Xenosmilus*, 62, 63, 64

North American Land Mammal Ages, 12

nuchal crest of *Xenosmilus*, 50

numerical taxonomy, 13, 14

obturator internus and obturator externus muscles of *Xenosmilus hodsonae*, 114, 115, 116

occipital condyles: of *Homotherium ischyrus*, 129; of *Xenosmilus*, 50, 51

occipital crest of *Homotherium ischyrus*, 126–127, 129

occipital region muscles of *Xenosmilus hodsonae*, 104, 104

Ogallala Chronofauna, 207

oral region muscles of *Xenosmilus hodsonae*, 102–104, 104

orbicularis oris muscle of *Xenosmilus hodsonae*, 103, 104, 104

overkill model for Pleistocene megafaunal extinction, 3

Owen, R., 201

pack structure and meat surplus, 8–9

palate: of *Homotherium crenatidens*, 197; of *Homotherium ischyrus*, 128, 129; of *Xenosmilus*, 48, 51

paleopathology: of dirk-tooths, 37, 38–39; extinction and, 35; predatory behavior and, 35–36; of scimitar-tooths, 39–41, 40

panda (*Ailuropoda melanoleuca*), 32, 45

Panthera leo (African lion), 32

Panthera spelaea, 205, 206

paroccipal process, 129

patella: of *Homotherium ischyrus*, 164, 167; of *Xenosmilus*, 78, 80, 81, 117

patellar groove of *Xenosmilus*, 80

peccary (*Platygonus vetus*), 42, 44, 47

pectineus muscle of *Xenosmilus hodsonae*, 115, 116

pectoralis major muscle of *Xenosmilus hodsonae*, 106, 107, 108–109

pelvis of *Xenosmilus hodsonae*, 113, 114

pes (hind foot): of *Homotherium ischyrus*, 174–175, 177, 178; of *Xenosmilus*, 86, 87, 88, 88, 89, 90

phalanges: front, of *Homotherium ischyrus*, 163, 164; front, of *Xenosmilus*, 75–76, 76, 77; hind, of *Homotherium ischyrus*, 173, 174–175, 177, 177, 178, 179; hind, of *Xenosmilus*, 88, 89, 90, 90

phylogeny, 13, 14, 15

piranha and cookie-cutter bite, 29, 31

piriformis muscle of *Xenosmilus hodsonae*, 114–115, 114, 115

pisiform of *Homotherium ischyrus*, 158, 158, 159

plantaris muscle of *Xenosmilus hodsonae*, 115, 117

plantigrade stance, 90, 190

Platygonus vetus (peccary), 42, 44, 47

Pleistocene: in North America, 205–206, 209; vegetation during, 90, 91. See also Friesenhahn Cave site; *Homotherium serum*

Pliocene, 185, 187, 203, 205. See also *Homotherium crenatidens*; *Homotherium crusafonti*; *Homotherium ischyrus*

Pliocene-Pleistocene timescale, 47

Plio-Pleistocene Chronofauna, 207

Pogonodon Cope, 92

Pogonodon specimen, 28

P. onca augusta (jaguar), 43

popliteus muscle of *Xenosmilus hodsonae*, 114, 115, 117

postcranial skeleton. *See* appendicular skeleton; axial skeleton

postorbital processes: of *Homotherium*, 50; of *Xenosmilus*, 46, 49

predation: ambush model of, 35–36, 37, 90, 120; grappling while standing on hind feet, 36, 37, 40, 92, 120–121; of *Homotherium ischyrus*, 122; of *Homotherium serum*, 5, 7, 33, 184; overview of, 3–4, 6, 8–9; paleopathologic analysis of, 35–36

prey species: extinction of, 9–10, 43; of *Homotherium ischyrus*, 122, 178; mammoths as, 5, 7, 8, 33, 184; overview of, 3; tuberculosis in, 40

Proailurus genus, 208

proboscideans of Ice Age, 4

pronator quadratus muscle of *Xenosmilus hodsonae*, 111, 112

pronator teres muscle of *Xenosmilus hodsonae*, 106, 107, 109

Pseudaelurus genus, 185, 208

psoas major muscle of *Xenosmilus hodsonae*, 114, 115, 116

pterygoideus musculature, 101–102, 103

Puma concolor (mountain lion), 43

Pygocentrus nattereri, 29, 31

quadratus femoris muscle of *Xenosmilus hodsonae*, 115, 116

quadriceps femoris muscle of *Xenosmilus hodsonae*, 115

radioactive dating systems, 12

radius: of *Homotherium ischyrus*, 155, 156, 156; of *Xenosmilus*, 71, 72, 73, 110, 111

Rancho La Brea tar pits, 43

reconstruction: of *Homotherium crenatidens*, 194; of *Homotherium ischyrus*, 125; of *Homotherium latidens*, 207; of *Homotherium serum*, 200; of *Xenosmilus hodsonae*, 45, 99–100

rectus capitus lateralis muscle of *Homotherium ischyrus*, 129

rectus femoris muscle of *Xenosmilus hodsonae*, 113, 114, 116

restoration of skeleton of *Homotherium serum*, 189

ribs of *Homotherium ischyrus*, 146, 151

Robocat killing bite test fixture: bison hide and, 23–24, 24, 25; bull elk hide and, 24–25, 25, 26; choice of, 22–23; design of, 18, 21–22; draw-cuts and, 26; mounted on Bobcat X-331, 24; purpose of experiments with, 33; real-world application of, 26; shear-bite sequence, 24, 25, 25, 26; use of, 23; wounds produced by, 27

saber-tooth cats: body size, 4; diversity and population density of, 4; extinction of, 9–11; killing bite of, 6, 19

sacrum of *Homotherium ischyrus*, 146, 149, 150

sartorius muscle of *Xenosmilus hodsonae*, 117, 118

scapholunar of *Homotherium ischyrus*, 157, 157

scapula: of *Homotherium ischyrus*, 146, 148, 151, 152, 152; of *Xenosmilus*, 65, 66, 67, 105, 106

scavenging behavior of scimitar-tooths, 41

scimitar-tooth cats: canines of, 4, 9, 28–29, 28; legs of, 6; paleopathologic analysis of, 39–41, 40; phylogeny of, 15; terminology for, 19–20. *See also Homotherium*; killing bites of scimitar-tooths

second lumbar vertebra of *Homotherium ischyrus*, 144, 144, 145

second thoracic vertebra of *Homotherium ischyrus*, 134–135, 136, 137

semimembranosus muscle of *Xenosmilus hodsonae*, 118, 119

semispinalis capitus muscle of *Homotherium ischyrus*, 129

semitendinosus muscle of *Xenosmilus hodsonae*, 117, 118

sesamoids of *Homotherium ischyrus*, 159–160, 159, 173

seventh lumbar vertebra of *Homotherium ischyrus*, 144, 149, 150

seventh thoracic vertebra of *Xenosmilus*, 61, 61, 62, 63

sharks and cookie-cutter bite, 29

shoulder muscles of *Xenosmilus hodsonae*, 106, 107, 108, 120

sixth cervical vertebra: of *Homotherium ischyrus*, 133–134, 136; of *Xenosmilus*, 57, 58, 59, 59

sixth lumbar vertebra of *Homotherium ischyrus*, 144, 148, 149

sixth thoracic vertebra of *Xenosmilus*, 61, 61, 62

skeleton: of *Dinobastis*, 43–44; of *Homotherium ischyrus*, 125; of *Homotherium serum*, 43–44, 189; of *Homotherium serum* cub, 9; of *Xenosmilus hodsonae*, 44, 98. *See also* appendicular skeleton; axial skeleton

skull: of *Amphimachairodus giganteus*, 205; of *Barbourofelis fricki*, 208; comparison measurements of, 93–94, 180; of Delmont *Homotherium*, 189–190; of *Homotherium crenatidens*, 195, 196, 197, 198; of *Homotherium crusafonti*, 188, 192–193, 192; of *Homotherium ischyrus*, 123, 125–129, 127, 128; of *Ischyrosmilus johnstoni*, 187; measurement points for, 213–214; of *Megantereon cultridens*, 204; of *Panthera spelaea*, 206; of *Xenosmilus*, 44, 46, 49–51, 49, 98

Smilodon fatalis: bite impression of, 8; defending kill, *xii*; knee with osteoarthritic osteophyte, 37; Rancho La Brea tar pits and, 43; size of prey and, 10

Smilodon genus: cast of skull and mandible of, 18; extinction of, 11

Smilodon gracilis: elements of, 47; in Florida, 44; *Ischyrosmilus* and, 203; size of, 90; *Xenosmilus* and, 205

Smilodontinae tribe, 44, 203

Smilodontini, 209

soleus muscle of *Xenosmilus hodsonae*, 118, 119

South Dakota, 187, 189–190

splenius capitus muscle of *Homotherium ischyrus*, 127–128

spondyloarthropathy in dirk-tooths, 39

spondylosis deformans in dirk-tooths, 38–39

standing posture of attack, 36, 37, 40, 120–121

subscapularis muscle of *Xenosmilus hodsonae*, 106, 107

Superposition, Law of, 12

supinator muscle of *Xenosmilus hodsonae*, 111, 112

supraspinatus muscle of *Xenosmilus hodsonae*, 106, 107, 108

Tajikistan *Homotherium. See Homotherium crenatidens*

taxonomy, 13–14, 15, 203

temporalis muscle: of *Homotherium ischyrus*, 129; of *Xenosmilus*, 91–92, 100, 101

temporomandibular joint (TMJ) of scimitar-tooths, 29

tensor fascia lata muscle of *Xenosmilus hodsonae*, 117, 118

tenth thoracic vertebra: of *Homotherium ischyrus*, 138, 140, 141; of *Xenosmilus*, 63, 64, 65

teres major muscle of *Xenosmilus hodsonae*, 107, 109

teres minor muscle of *Xenosmilus hodsonae*, 107, 108

Texas: Cita Canyon, 185; Coffee Ranch, 205; Friesenhahn Cave site, 2, 8, 9, 33, 43–44. *See also Homotherium serum*

third cervical vertebra: of *Homotherium ischyrus*, 131, 133, 134, 135; of *Xenosmilus*, 54–55, 55, 56

third lumbar vertebra of *Homotherium ischyrus*, 144, 144, 146

thirteenth thoracic vertebra of *Homotherium ischyrus*, 139, 142, 143, 144

thoracic vertebrae of *Homotherium ischyrus*: eighth and ninth, 137, 138, 139; eleventh, 138, 140, 141; fifth, 135, 137, 138; second, 134–135, 136, 137; tenth, 138, 140, 141; thirteenth, 139, 142, 143, 144; twelfth, 139, 142, 143 thoracic vertebrae of *Xenosmilus*: eighth, *62, 63, 63*; eleventh, *63–64, 65, 66, 66*; fifth, *58, 60, 60*; first through thirteenth, *58, 59–60, 59*; ninth, *62, 63, 64*; seventh, *61, 61, 62, 63*; sixth, *61, 61, 62*; tenth, *63, 64, 65*

throat, attacks on, 6, 35–36

tibia: of *Homotherium ischyrus*, 168, *168, 169*; muscles of, 117, *118,* 119–120; of *Xenosmilus*, 81–83, *81, 82,* 117, *118*

tibialis anterior muscle of *Xenosmilus hodsonae*, *118*, 119

tibialis posterior muscle of *Xenosmilus hodsonae*, *118*, 119

tooth breakage in dirk-tooths, 38

trapezium of *Homotherium ischyrus*, *158*, 159, *159*

trapezius muscle: of *Smilodon*, 129; of *Xenosmilus*, 128–129

trapezoid of *Homotherium ischyrus*, *158*, 159, *159*

traumatic injuries: in dirk-tooths, *37*, 38; in scimitar-tooths, 39–41, *40*

triceps muscle of *Xenosmilus hodsonae*, *106, 107*, 109

tuberculosis in scimitar-tooths, 39, 40, *40*

twelfth thoracic vertebra of *Homotherium ischyrus*, 139, *142, 143*

ulna: of *Homotherium ischyrus*, 154–156, *154, 156*; of *Xenosmilus*, 69–71, *69, 70*, 110, *111*

Ursus arctos (brown bear), 32, 45

vastus intermedius muscle of *Xenosmilus hodsonae*, *114, 115*, 116

vastus lateralis muscle of *Xenosmilus hodsonae*, *114, 115*, 116

vastus medialis muscle of *Xenosmilus hodsonae*, *114, 115*, 116

vegetation, North American Late Pleistocene, 90, *91*

vertebral pathology in dirk-tooths, *37*, 38–39

Villafranchian of Europe, 123

wrist (carpal bones) of *Homotherium ischyrus*, 157–160, *157, 158, 159*

Xenosmilus genus, 205, 206, 209

Xenosmilus hodsonae: diastema, CEJ, and presumed gingival relationship of saber and incisor arcade, *30*; distributions of, *16*; general appearance of, *45*, 92; head shape of, 120; killing bites of, 29, 54, 91–92; life reconstruction of, *45*; mandible of, 51–54, *52, 53*; origins and distribution of, 90; at partial gape, *98*; peccary remains associated with, *42*; photograph of holotype, *46*; postcranial anatomy of, 120; reconstruction of appearance of, 99–100; skeleton of, *31, 44, 98*; skull and mandible of, 8, *8*; skull of, *46, 46, 48*, 49–51, *49*; systematic paleontology, 193; teeth of, 8; three views of, *x*; zygapophyseal facet joint of, *40*. *See also* cookie-cutter cats

zygomatic arch: of *Homotherium ischyrus*, 129; of *Xenosmilus*, 49–50, 100

zygomaticomandibularis muscle of *Xenosmilus hodsonae*, 101, *103*